Springer-Verlag

Geschäftsbibliothek - Heidelberg

Titel: **NATO ASI Series, Vol. F 19, Pictorial Information Systems in Medicin**

Aufl.-Aufst.: **1. Auflage**

Drucker: **Beltz, Hemsbach**

Buchbinder: **Schäffer, Grünstadt**

Auflage: **900**   Bindequote: **900**

Schutzkarton/Schuber:

Satzart: **Schreibsatz**

Filme vorhanden:

Reproabzüge vorhanden:

Preis: **DM 298,-- $ 125.00**

Fertiggestellt: **2.4.1986**

Sonderdrucke: **ja**

Bemerkungen:

Berichtigungszettel:

Hersteller: **C. Berger** Datum: **16.7.1986**

Pictorial Information Systems in Medicine

NATO ASI Series

Advanced Science Institutes Series

A series presenting the results of activities sponsored by the NATO Science Committee, which aims at the dissemination of advanced scientific and technological knowledge, with a view to strengthening links between scientific communities.

The Series is published by an international board of publishers in conjunction with the NATO Scientific Affairs Division

A Life Sciences **B Physics**	Plenum Publishing Corporation London and New York
C Mathematical and **Physical Sciences**	D. Reidel Publishing Company Dordrecht, Boston and Lancaster
D Behavioural and **Social Sciences** **E Applied Sciences**	Martinus Nijhoff Publishers Boston, The Hague, Dordrecht and Lancaster
F Computer and **Systems Sciences** **G Ecological Sciences**	Springer-Verlag Berlin Heidelberg New York Tokyo

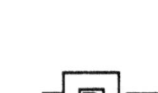

Series F: Computer and Systems Sciences Vol. 19

Pictorial Information Systems in Medicine

Edited by

Karl Heinz Höhne

Institute of Mathematics and Computer Science in Medicine
University of Hamburg, Martinistraße 52
2000 Hamburg 20, Federal Republic of Germany

Springer-Verlag Berlin Heidelberg New York Tokyo
Published in cooperation with NATO Scientific Affairs Division

Proceedings of the NATO Advanced Study Institute on Pictorial Information Systems in Medicine held in Braunlage/FRG, August 27–September 7, 1984

ISBN 978-3-642-82386-2   ISBN 978-3-642-82384-8 (eBook)
DOI 10.1007/978-3-642-82384-8

Library of Congress Cataloging in Publication Data. NATO Advanced Study Institute on Pictorial Information Systems in Medicine (1984 : Braunlage, Germany) Pictorial information systems in medicine. (NATO ASI series. Series F, Computer and systems sciences ; vol. 19) "Proceedings of the NATO Advanced Study Institute on Pictorial Information Systems in Medicine, held in Braunlage/FRG, August 27 – September 7, 1984"—T.p. verso. "Published in cooperation with NATO Scientific Affairs Division." Includes bibliographies. 1. Imaging systems in medicine—Data processing—Congresses. 2. Computer graphics—Congresses. I. Höhne, K.H. (Karl-Heinz), 1937-. II. North Atlantic Treaty Organization. Scientific Affairs Division. III. Title. IV. Series: NATO ASI series. Series F, Computer and system sciences ; no. 19. [DNLM: 1. Computers—congresses. 2. Information Systems—congresses. 3. Medicine—congresses. 4. Technology, Radiologic—instrumentation—congresses. W 26.5 N279p 1984] R857.O6N38 1984 610'.28'5 85-32111

© Springer-Verlag Berlin Heidelberg 1986
Softcover reprint of the hardcover 1st edition 1986

Bookbinding: J. Schäffer OHG, Grünstadt
2145/3140-543210

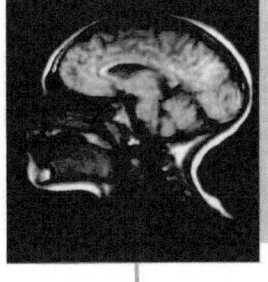

image data base

psychovisual issues

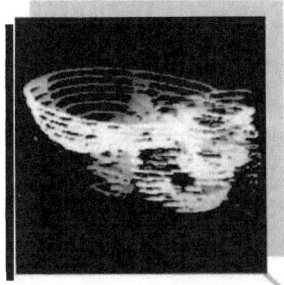

expert systems

computer graphics

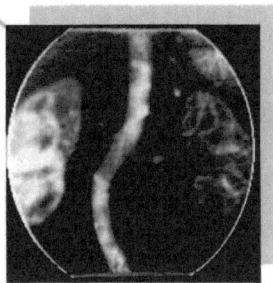

PICTORIAL INFORMATION SYSTEMS
IN MEDICINE

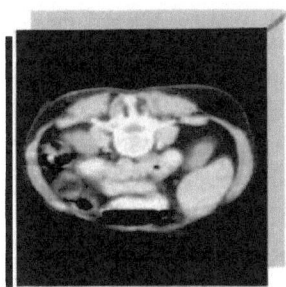

medical imaging

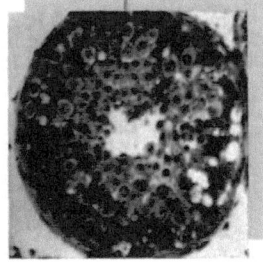

data compression

man-machine
interaction

Preface

This volume contains the proceedings of the NATO Advanced Study Institute on "Pictorial Information Systems in Medicine" held August 27–September 7, 1984 in Hotel Maritim, Braunlage/Harz, Federal Republic of Germany. The program committee of the institute consisted of K.H. Höhne (Director), G.T. Herman, G.S. Lodwick, and D. Meyer-Ebrecht. The organization was in the hands of Klaus Assmann and Fritz Böcker.

In the last decade medical imaging has undergone a rapid development. New imaging modalities such as Computer Tomography (CT), Digital Angiography (DSA) and Magnetic Resonance Imaging (MRI) were developed using the capabilities of modern computers. In a modern hospital these technologies produce already more then 25% of image data in digital form. This format lends itself to the design of computer assisted information systems integrating data acquisition, presentation, communication and archiving for all modalities and users within a department or even a hospital. Advantages such as rapid access to any archived image, synoptic presentation, computer assisted image analysis to name only a few, are expected.

The design of such pictorial information systems, however, often called PACS (Picture Archiving and Communication Systems) in the medical community is a non-trivial task involving know-how from many disciplines such as

– Medicine (especially Radiology),
– Data Base Technology,
– Computer Graphics,
– Man Machine Interaction,
– Hardware Technology and others.

Most of these disciplines are represented by disjunct scientific communities. It was the aim of the meeting to get experts in the different fields together to review the fundamentals, to identify the problems and to discuss the possible solutions.

It was an experiment, without any guarantee of success. Would the participants talk to each other? Would they understand their different terminologies? The organizers felt a lot of relief when they recognized a few days after the beginning of the meeting that the communication really succeeded.

The lectures and discussions of the institute fell into quite heterogenous categories. Starting from the fact, that the design of pictorial information systems certainly requires a careful analysis of the users needs on one hand and the special properties of the imaging modalities on the other, G.S. Lodwick looked at the task from the radiologists view, while S. Nudelman gave a detailed review of the specific properties of the different image aquisition devices.

The following paper of A.R. Bakker on hospital information systems was important for the following reason: When building PACS it is not only necessary to include know-how from hospital information system methodology; in fact PACS will not be of any value without a link to a patient information system. D. Meyer-Ebrecht followed in presenting his ideas on developing architectures for pictorial information systems in medicine.

Some of the most striking new possibilities opened up by digital imaging techniques are new kinds of image presentation such as 3D-displays. G.T. Herman reviewed this topic based on his long experience in this field, S.M. Pizer complemented his talk with an overview of 3D-imaging hardware. The large variety of available display techniques offered by image processing and computer graphics raises the new question of which is the optimal presentation. S.M. Pizer therefore also discussed the psychovisual issues in the display of medical images.

There is no doubt that a pictorial information system will not be successful unless it is as easy to use as a conventional film reading environment. The design of a suitable man machine interface is already non-trivial in general purpose computing. Man machine communication problems become even more severe when pictorial data are included. J. Nievergelt covered therefore the various issues in the design of man machine interfaces.

In addition complex systems such as pictorial information systems cannot be designed without appropriate software tools. J.W. Schmidt reported on novel concepts for supporting data intensive data base applications. K. Assmann and K.H. Höhne presented an implemented experimental programming environment for pictorial data management as a possible software solution.

Although the problem of designing systems, that can compete with the conventional image handling, is not yet solved, it is tempting to take a look into the future. While present PACS attempts deal with patient and pictorial data, future systems will also make use of knowledge about image interpretation, thus assisting the physician in the diagnostic process. I. Hoffmann gave a review of the state-of-art in the field of knowledge-based and expert systems.

The lecturer's papers were complemented by papers contributed by participants. Yet not all aspects could be discussed in depth. So the institute did not discuss very much the more technological aspects of storage and communication technology. Suitable hardware will certainly be developed by the engineers because it is also required urgently for other data intensive applications aside from medical imaging. The problem of image standardization was not covered. Although its solution is not trivial, it was not considered a scientific topic for this meeting.

One remarkable effect of the institute was the formation of a new interdisciplinary scientific community. As a visible result, EUROPACS, an association of European PACS researchers was founded during the meeting.

Many people contributed to the success of this institute. I am very grateful to the members of my department for their continuous help in the preparation of the meeting. The organizers would also like to express their gratitude to Dr. C. Sinclair, NATO Scientific Affairs Division, for his support and Dr. Tilo Kester, International Transfer of Science and Technology, for the help in the logistics. Last, but not least, the organizing committee would like to thank all the lecturers and participants, upon whose enthusiasm and technical expertise the institute was based.

Hamburg, September 1985   *Karl Heinz Höhne*

Table of Contents

Short Papers

PICTORIAL INFORMATION SYSTEMS AND RADIOLOGY

IMPROVING THE QUALITY OF COMMUNICATIONS

Gwilym S. Lodwick, M.D.
Department of Radiology
Massachusetts General Hospital
Boston, Massachusetts   02114

1. Introduction

Radiology is a diagnostic service which utilizes images to display disease. These images are interpreted by radiologists, who pass the diagnostic information on to the physician that takes care of the patient.   How information acquired in the practice of radiology is processed, stored, displayed, reported and transmitted to the referring physician is an application of medical informatics. The two management systems involved in this process are (1) The Radiology Infomations Management System (RIMS) which has a variety of essential management and control   functions,   and   (2) The Picture Archiving   and Communications System (PACS) which implies the management of digital images.   Each system is computer based, and when tightly   integrated   the   two comprise   the   communications structure of the totally digital department.   (1) The title of this conference, "Pictorial Information Systems" could imply the PACS concept, but in reality the presentations are broader, covering computer science and informatics in medical imaging. Pictorial   information systems had their origin with the discovery of x-rays by Wilhelm Konrad Roentgen in November 1895.   In the early years, the discipline of diagnosis from x-rays was called roentgenology, but as the goals of practice came to include treatment with x-rays and radium, radiology

NATO ASI Series, Vol. F19
Pictorial Information Systems in Medicine
Edited by K. H. Höhne
© Springer-Verlag Berlin Heidelberg 1986

seemed a better descriptive title. This has been especially true with the addition of imaging modalities using non ionizing radiation such as ultrasound, thermography, and nuclear magnetic resonance.

From the beginning, radiological images have been recorded on various media (glass plates, celluloid, plastic film) coated with light sensitive silver emulsion. Because of the microscopic dimension of the silver suspension, density variances appear continuous to the human eye, and films are therefore regarded as analog images. While in time many techniques have been invented for increasing the efficiency of film for recording the effect of x-ray photons, the fundamentally analog nature of film images has not changed. With the development of image amplification technology, televison cameras have been focussed on light emitting phosphors which display the ionizing effect of incident radiation. Film images are created by copying either the image on the phosphor or the one on the television display tube. For the first 75 years following the discovery of the x-ray, radiological imaging was exclusively analog, and radiology a film-handling discipline.

The combination of several inventions have conspired to introduce a dramatic change to medical imaging generally and to radiological imaging in particular. Most fundamental was the invention of the computer during the second world war. Next came the invention of the transistor, which made the computer both compact and reliable. Computers were programmed in languages which employed "binary digits" or "bits" of information that were "digital". A bit represents the choice between a "mark" or "space" (one or zero) condition. Bit rate is the speed at which bits are transmitted expressed in "bits per second". A byte is a small group of bits that is handled as a unit, an 8 bit byte being most usual.

Digital imaging technology was widely implemented in the exploration of space which began in the 1960s. Space probes carry television cameras which record previously unknown details of the universe. Image data are sent back to earth in a digital stream which is reformatted into images and processed by digital computers in order to enhance detail. By the late 60s film images were being experimentally modelled (2) and converted to digital format in order to investigate the value of this new imaging technology. Automated image analysis from x-ray images is a byproduct of this new technology. (3)

However, it was the invention of computed tomography by Haunsfield in the early 1970s which shook imaging technology to the core. Enthusiastically received by the radiological profession, CT made digital technology and digital imaging commonplace. However CT is a very complex and costly new technology, and certain trade-offs became necessary in the interest of getting better images faster. Two of these trade-offs have been provision of rapid reconstruction times and improved spatial resolution at the cost of remote displays and large capacity disc storage. As a result, reproduction of images from even a current case that has been stored on tape may require a 12 to 24 hour turn around time, which usually rules out the combination of an active file and a dynamic display as the source for routine CT interpretation. Another effect of the trade-off is that the radiologist's access to CT displays is usually of short duration and often interferes with the scheduled examinations.

The accepted solution to this problem is a compromise; CT images are copied on film by a multiformat camera which sacrifices the tremendously valuable contrast resolution available on the display for a "hard copy" image that can be carried away and studied at leisure. These multiformat images are a comfortable medium for the film oriented radiologist, and are the only satisfactory medium in the absence of a PACS system. However, they represent additional cost, they fail to reproduce the dynamic quality of the CT display and thus lose information, and by being a single set of images can become a source of conflict between the many consulting physicians who simultaneously need them for patient care. Clearly, CT has demonstrated the tremendous versatility of digital imaging technology, but by its very success CT has introduced a nagging unresolved problem into the radiologists' domain which will require an extensive application of medical informatics to eliminate. This unresolved problem is that we cannot afford to continue to downgrade the dynamic qualities of digital images by reducing them to fixed images on costly films, which themselves represent a major and largely unsolved management problem in radiology. As a result of adding all of the other digital imaging technologies including nuclear magnetic resonance imaging, digital images now represent around thirty percent of the total case load. This percentage is rapidly increasing. What were once analog pictorial information systems are now a combination of analog and digital, and we now have some new as well as the old problems of image management to resolve. The successful effort of Karl Heinz Hohne in organizing this Nato conference to provide indepth discussions of the problems and the future of such systems is indeed welcome and appreciated.

2. Radiology information management and costs

Radiology is a referral service, which means that whether the practice setting is in a hospital or in an office environnment, the departmental activities must be sensitive to the requirements of patients, referring physicians and consultants, to community standards and to a host of standards required either by the profession or by local, state and federal government. Radiology has experienced periods of unprecedented growth, demand for services, and expansion of technology requiring enormously expensive equipment and space. Radiological examinations are therefore costly, and the environment in which they are used increasingly cost sensitive. As a result of all of these forces and others, management in radiology has become an increasingly important issue. Inefficient management results in poor scheduling of patients, incomplete or unsatisfactory examinations, delayed reports or an inability to find the images which are important to medical decisions. Any or all of these problems conspire to affect the intrinsic economics of the department, through increasing the direct and indirect overhead costs for operations.

In a few words, efficient management of radiological images and information is a high level priority of the health care environment, which not only needs good radiology to ensure high quality patient care, but also needs cost control in this increasingly expensive technology. With the advent of diagnostic related groups (DRGs) as a basis for hospital reimbursement in the United States, operational efficiencies in radiology become even more pressing. (4)

Operational inefficiencies were already under intensive study in the late 60s. In July of 1968 a Task Force on X-ray Image Analysis and Systems Development reported on the methodology for improved production and utilization of image information. (5) With great foresight, the scientists comprising this task force modeled the complex interaction of patients, staff and images with realism and detail which have effectively withstood the test of time. (Figure 1) The task force also strongly supported the development of computer based information systems in radiology at a time when the focus of radiology was largely elsewhere. Many radiology information systems were designed and implemented. Only a few were successful. At least one such system was in full operational support of a clinical radiology department by 1970. (6) This system, which included report generation, was the subject of a comprehensive economic evaluation (7), and is now being marketed commercially. As of this writing, radiology information management systems are proliferating, although the state of the art is still evolving. (8) (9) (10) (11)

INFORMATION MANAGEMENT IN DIAGNOSTIC RADIOLOGY

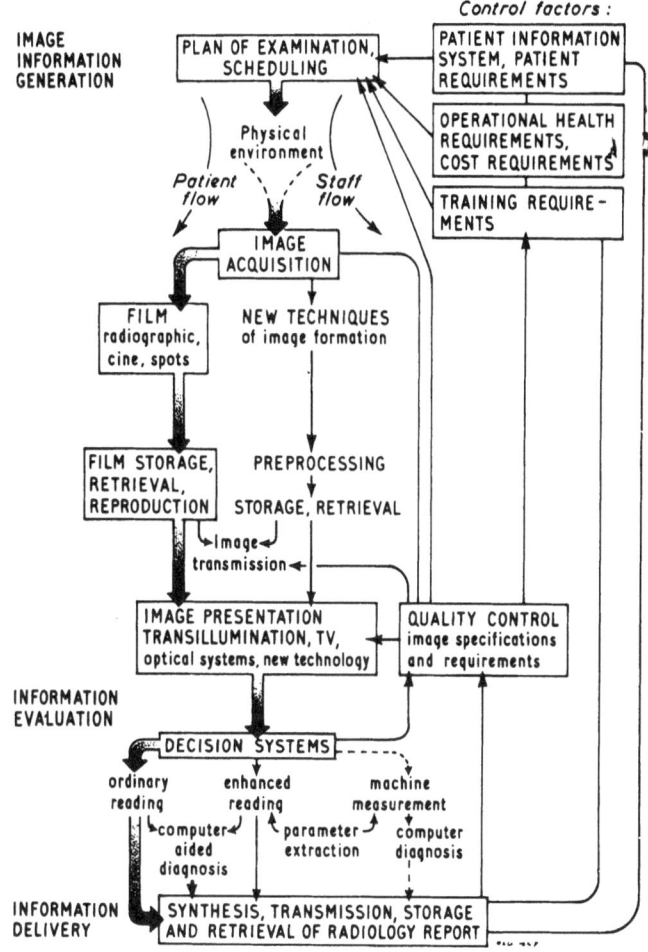

Figure 1.   From 1968 Report of Task Force, NCRH

3. File Management

In most radiology departments, film files are usually divided into at least two categories; (1) an active file of hospital inpatients, (2) an inactive file of the remainder. Films are shifted from active to inactive file status when patients are discharged from the hospital. Additionally, films on patients who have not been reexamined or readmitted for a predetermined length of time are regularly culled from the inactive file and placed in remote storage or disposed of. An information system with tracking capabilities makes culling a simple task.

As a rule, ten percent of the inactive files will have been reactivated by the end of the first year. For the remainder, the number drops to one percent or less by the end of the fourth year. Thus, files more than four years old stand little chance of being reactivated. This concept is important for future development of data base management systems to handle images stored on archival digital media.

The concept of the file room filled with shelves containing folders for 14 x 17 inch films evolved with Roentgen's discovery. In coping with multiformat images generated from digital displays, most radiology departments manually file the multiformat images in the individual film jackets which are retained in the central film library. A problem is that due to geographically separated sections of a department, very recent examinations of digital origin may not have been filed in the patients film jacket when they are most needed by various radiologists and clinical consultants. One at a time access to film jackets results in limitation of the diagnostic information to one user at a time. The film file room provides no opportunity to take advantage of the rapid access, multimodality imaging and multiple site displays which are a concept of the modern digital image management system.

The time is apparently past when single copy film images can provide an adequate communications medium for the mulidisciplinary and multispecialty practice environment. The pressures for rapid acquisition of diagnostic information, early evaluation and definitive therapy are just too great to be withstood by archaic image management technologies. Further, the pressures of peer review and government limitation of reimbursement call for a speeding up of the entire health care process; inefficiency and unexplained delay in communications will be poorly tolerated and correction expected. As the result of these pressures, there is an unremitting demand for control of films by non-radiologists. What cannot be obtained through legitimate channels is often taken by stealth. The file room staff is frustrated by demands which cannot be met. It is clear that film archives cannot be maintained without rigorously enforced policy. Until new technology can be put in place or some other solution found, here are some of the costs we experience with poor archiving:

1. Lost and stolen films

2. Incomplete medical records

3. Unsatisfactory "follow up" interpretations where previous film files are missing. This is an especially serious problem which can lead to medical malpractice suits.

4. Unhappy referring physicians

5. Wasted radiologist time

6. Costs of replacing lost films, including technical and equipment costs

7. Wasted patient time

8. Diminished quality of care

What are the solutions to file room anarchy?

1.   Proper department design with inherent internal security and direct information flow.

2.   Use of RIMS to monitor a vigorous film tracking policy.

3. Reasonable lending rules, which are enforced by staff and supported by the administration.

4. <u>Finally,</u> and <u>urgently,</u>

       <u>Phased conversion to digital image</u>

       <u>Management technology</u>

4. <u>Communications problems with referring physicians</u>

While our major focus has been on internal communications within the imaging specialty department, there are other important communications channels where at times the lines appear to be down.   These lapses relate principally to how we communicate with our professional colleagues, particularly those who send us patients.   Heilman, in a recent article in the New England Journal of Medicine asks, "Whats wrong with Radiology?"   and   observes   that   "Even   the   most   cursory evaluation of radiology as a system suggests that it has evolved with inherent flaws that virtually guarantee confusion, frustration, patient dissatisfaction and needless expense." (12) In this "golden age" of radiology, valuable new technologies are proliferating at a rate which can confuse and bewilder even the clinical radiologists who must use them, not to mention the clinicians who must ask for meaningful radiological consultations, often in vain. (13) The problem is compounded by the consultations received by the clinicians, when several conflicting radiological reports on the same patient can be received from different technique oriented sections of the department. We physicians within radiology seem at times not to be communicating with each other. We must clean up these problems: already for some time physicians generally have been under attack for their image of high professional ideals in conflict with selfish self interest (14).

5. How can we improve communications in Radiology?

So far, we have identified the major sites of systems failure. It is through the careful application of the principles of medical informatics that we can analyze these failures and provide new solutions through better communications.

1. Better departmental design. The early department of radiology often consisted of a few rooms in the basement of the hospital. As the discipline prospered, it was allowed to move into the lower corridors of the hospital, where it remains.

Unfortunately, the department consisting of examining rooms along either side of a busy hospital corridor is far from ideal for effective communications. First, corridor traffic consists of patients, doctors, technicians, cleaning personnel, visitors, couriers, administrators, all mixed together in a bewildering array. Second, patients often are waiting in wheel chairs and carts along the wall of the corridor, with utterly no privacy. Third, the product of the examination is often developed nearby, and whether x-ray films or multiformat images, they may never find their way to the file rooms. Fisher (15) believes that radiology departments should be designed by radiologists and other radiology personnel since they are the prime users of space. A modular design philosophy

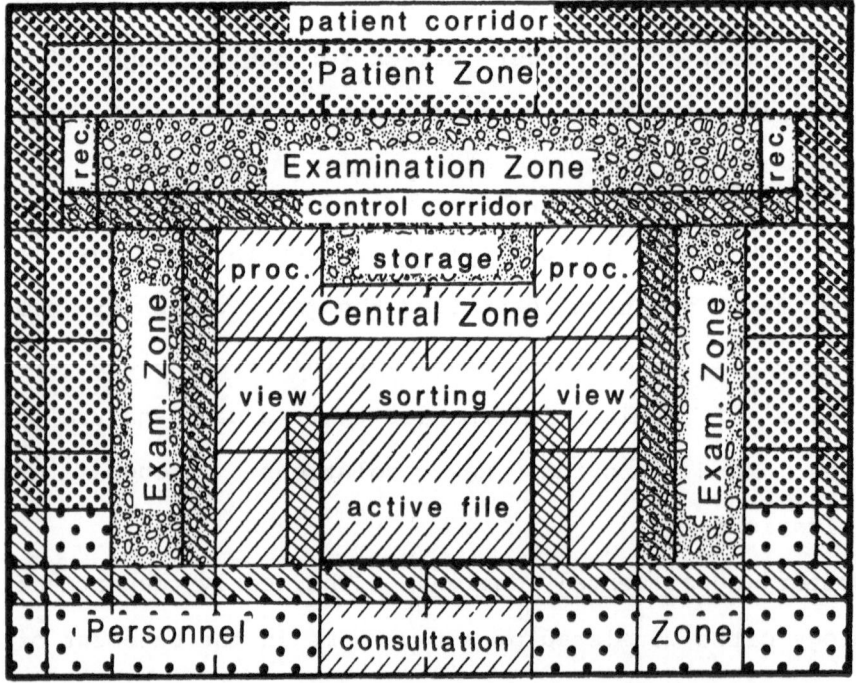

Figure 2. Functional design of radiology department. From Holm, ACR Report of Committee on Department Planning, 1977.

has evolved consisting of a peripheral zone for the patient, an intermediate zone for the examination, and a central zone for developing of films, viewing, interpretation and filing. (Fig., 2) This plan shortens lines of communication and provides internal security. Modular design has proved to be ideal for internal communications where x-ray films are used and filed. The large department may consist of several modules within a single large rectangular space, or in separate areas of the hospital. The design is also so effective for personnel traffic and patient flow that it may be expected to persist well into the future requiring only space reallocations in response to the changing needs of technology.

2. Establishment of imaging communications standards.

There is an urgent need to transfer images between various types and models of imaging equipment. Each manufacturer has its own design of electronic gear, and the output signals carrying image information are not compatable with imaging equipment of other manufacturers, and sometimes not with a manufacturers own equipment. The improvement of communication within the imaging department is dependent upon the development of networks, which would benefit greatly from standarization of the interface between the network interface unit and the imaging equipment. To meet the need for developing new communications standards in Radiology, the American College of Radiology (ACR) and the National Electrical Manufacturers Association (NEMA) have combined resources to form the ACR-NEMA Digitial Imaging and Communications Standards Committee. With the initial objective of designing an interface standard for imaging equipment, three working groups have been appointed by the parent committee. Work has progressed to the point where the proposed draft of the interface standard will be circulated for comment by November 1984. The Committee is considering future objectives and will continue to upgrade the interface standard as required by advances in technology.

3. Use of Radiology Information Management Systems. (RIMS) The major functions of the Information Management System are (1) scheduling, (2) registration of patients, (3) file room control, (4) case finding, (5) report generation and managemnent, (6) billing, and (7) the timely tracking and recording of the flow of certain information. Examples of the concept of tracking include following the location of the image beginning with its creation, and following of the reporting process, beginning with the time of delivery of the examination to the radiologist, and ending with the signing and delivery of the report. A good information system, can, on request, provide detailed information on any step of the examining process of any patient from time of scheduling and arrival to the delivery of the consultation report. Further, it can provide detailed information on overdue film loans, unreported examinations - all with speed and accuracy. A good system will have information available which the user may have never thought of requesting. Generation of charges can be tied to the consultation report. If the report is generated, the patient is billed and vice versa. However, slow response time

is unacceptable, otherwise radiologists will not use the system. Programs which degrade response time must be scheduled outside times of regular professional use.

Clearly, the functions of scheduling, patient registration, reporting and especially tracking provide detailed intelligence about radiological operations; the feedback necessary for corrective action and future planning. The role of the RIMS in department management becomes so pervasive that RIMS is recognized as an important instrument for exercising management policy. Its role in providing detailed management information is a valuable asset not only to the department but also to hospital administration. Therefore, while RIMS should be based within the department which it helps manage, the system needs to have a two way interface with the hospital information system (HIS). This communications interface can be even more important for transmission of consultation requests, clinical information about patients, and the results of consultative studies. Given the premise that RIMS, when properly used, is informed of most important details of the patient examination, particularly the numbers, location and dates of images, it follows that RIMS files will be essential to PACS operations. At this time it would seem ill advised to try to duplicate the RIMS system as a part of PACS design, particularly since the operational scope and technical requirements of the two systems are so dissimilar.

4. The Development of the Digital Image Management System (PACS)

Mission. The image management system is to be designed for

archiving, transmitting, and display of digital images in a manner best suited to their accurate interpretation by imaging consultants. Images will be simultaneously available to multiple users. They will be displayed rapidly, in proper sequence, with spatial resolution appropriate to the kind of image displayed, and with provision for utilizing the total range of contrast resolution inherent in the image data. Accuracy of interpretation will be enhanced by display of clinical history and findings, which means that the PACS and the RIMS functions must be tightly integrated. Ultimately, standards will be set for speed of transmission and other important variables, and it is expected that PACS will meet those design specifications.

Scope. The operational domain of PACS will range from (1) the non-networked interface between an imaging device and a display, quite possibly with a local relatively short-term archive. (2) A local area network consisting of several imaging devices quite possibly of the same technology, such as two CT scanners networked to a local archive and display. These may be known as Image Production and Display Centers (IPC). (3) A departmental network linking a number of IPCs to a central archive and possible central display and interpretation stations. (4) One or more large IPCs within a major community networked to a base center in a hospital or clinic, or possibly to a centralized billing and collection center. (5) Imaging departments networked to major remote health centers. (6) Combinations of any or none of the above, but designed to meet the special need of a health care plan.

Components of PACS. Clearly, (1) imaging equipment and devices (2) networks, (3) short and long term archiving devices, including magnetic tape and disc, optical disc, optical tape, erasable optical media, (4) computers for data base management, and (5) displays.

At least 3 levels of displays are seen. The largest and most elaborate digital display would include multiple CRTs, possibly of 2048 x 2048 pixel density, with rapid access and review, zoom, and various other kinds of image processing. This display would be human engineered with attention not only to the operational details of hand-eye coordination but also with attention to the principles of cognitive psychology, especially relating to what we know about how the mind perceives and handles images. A second level digital display may have one or perhaps two displays, less spatial resolution and more limited function, and cost about one-fifth to one-tenth as much as the larger display. The least expensive display will quite probably be analog supported by cable TV, but using a multi-functional digital display as its image source. The system will simultaneously transmit on multiple channels. Images one at a time will be zoomed, panned, or otherwise manipulated, possibly from the remote display.

Requirements of PACS. The development of PACS system is a high technology operation, each element of which is at or beyond the state of the art. For archiving, the early versions of optical digital disc technology are just entering the market place. Data base management systems for image archiving are envisioned but not perfected. Networking protocols and systems which work well for alphameric transmission are too slow for digital image transmission at the required data rates, but new systems are being tested. Interface standards are just being evolved. The pixel density requirements for displays will vary with applications; many are still unknown. The human engineering requirements of displays are in the state of conceptualization and testing. However, piece by piece, the components of PACS are being assembled and integrated, and we are seeing a dream rapidly becoming a reality. A major unanswered question is - "can we afford to pay for the reality?" (16) My hypothesis is that we cannot afford to not pay for the reality, yet this hypothesis must be tested in the medical marketplace. The future of PACS depends upon the following:

1. The speed with which digital images can be accessed as compared to film images.

2. The labor costs for digital/film image retrieval.

3. The production costs for digital images as compared with film images.

4. Differential costs of digital versus film imaging equipment.

5. Differential costs of storage required for digital compared to film.

6. How will radiologists accept digital displays as compared to film displays.

7. The relative accuracy of interpretation with digital/film displays.

8.   The costs of digital displays as compared to
conventional film displays.

9.   How cost effective and how practice effective will
the PACS system be as it is phased into the department?

Strategy for introduction. It would seem that the most logical
approach to implementing PACS would be through development of
Image   Production   Centers   (IPCs)   using   those   imaging
technologies which are now on-line. Archiving for the IPC could
be short to medium term, long term archiving could be provided
through concurrent or later acquisition of an archiving center.
Whenever   new   equipment   purchases   are   contemplated,
consideration should be given to specifying the ACR-NEMA
interface standard, and to the formation of a new IPC with
resources for production, archiving and display of digital
images, or alternatively, for inclusion in another center which
has similar resources in its possession. Obviously, the
feasibility of implementing this policy will depend upon the
state of development of digital imaging for that discipline.
It would seem logical to standardize to a network design of a
particular manaufacturer. A short and long term plan should be
evolved for PACS implementation, once that course has been
selected. In this way, the purchase of certain equipment items
can be optimized in terms of time of acquisition and in meeting
certain cost benefits.

6. Advantages of PACS

1. Multiple users can have access to the same examinations

simultaneously

2.   The penalty costs of poor archival control can be abolished

3.   The dynamic scale and contrast of the digitally formatted image will be fully available to all users

4.   Networking and multiple nodes will simplify the addition of displays as increased capacity is required.

5.   The role of multiformat images can be reduced to teaching and consultative applications

6.   The   production of multiformat images can be centralized and equipment costs reduced

5. Patterns of Care. Radiological practice is usually oriented along technological disciplines, such as nuclear medicine, CT, ultrasound, vascular radiology, etc.

A hypothetical example is used to illustrate a problem with this system.   Mrs. Jones, age 66, is a patient in the hospital suffering from abdominal pain of undetermined origin. Mrs. Jones' physician, Dr. Smith, is in the dark about the nature of Mrs. Jones' complaint since the symptoms do not totally make sense.   He is considering gall bladder disease and/or pancreatitis as possibilities.   Dr. Smith has had much help from CT but he knows this is a costly procedure, and that the schedule is probably fully booked for several days. He has heard that MRI has had great success in certain diagnostic situations, but has no experience. Who should he ask?

In   the   department of radiology   organized   along technological lines, everywhere he looks he finds a super specialist. Where is the radiological generalist? How can Dr. Smith decide which combination of examinations will lead most rapidly to a resolution of the diagnostic dilemma with the least expenditure of cost and time? Not only does Dr. Smith have a dilemma, but so does radiology. If Dr. Smith sends requests to several sections, he will get several answers which may not totally make sense. At the Massachusetts General Hospital, the chairman of radiology has circumvented this problem to a remarkable degree through organizing the staff along organ disciplines. (Figure 3) Dr. Smith can send his

PATTERNS OF CARE IN RADIOLOGY

	RAD	CT	US	VASC	MRI	NM
BONE	●	●		✻	✻	✻
CHEST	●	●		✻	✻	✻
GI	●	●	●	✻	✻	✻
ABDOMEN	●	●	●	✻	✻	✻
GU	●	●	●	✻	✻	✻
PEDS	●	●	●	✻	✻	✻
NEURO	●	●	●	●	✻	✻
CARDIAC	●	●			✻	✻

Figure 3. Section organization at Massachusetts General Hospital. Horizontal dots reflect organ orientation; vertical stars, technique orientation. Abdomen and GI are combined. MRI is in transition to organ orientation.

request to the Abdominal section, where the abdominal radiologist will advise Dr. Smith of which examinations will be most appropriate, and the abdominal radiologist will supervise the performance of all of the examinations including contrast studies, ultrasound, CT, MRI, even performing a biopsy if indicated. He will review and coordinate all reports that are sent to Dr. Smith, assuring that they make sense. And he will be able to provide Dr. Smith with a well planned, timely radiological workup.

Another approach to deciding which examinations for Mrs. Jones are most appropriate is through use of computer technology that depends upon statistical studies of outcome given certain sets of symptoms and findings. Artificial intelligence methodologies are being introduced into this process, so that Dr. Smith, through interacting with the computer, will be able to determine on his own the best choice of examinations based upon a computer program which uses prior knowledge, logic and probability.

SUMMARY: Radiology is in the "Golden Age" with features which make our profession very attractive.

1. Conventional and new non-interventional imaging technologies now provide detailed visualization of structure and physiology, localization of disease, better evaluation and understanding of function, and will in the near future, provide analysis of chemical abnormality in relation to site.

2. These precise imaging technologies are regularly combined

with percutaneous interventional techniques that allow rapid diagnosis, including biopsy in any organ system, and percutaneous correction of a number of abnormalities. Many of these procedures are performed on an outpatient basis, at a tremendous saving of cost and time.

3. To a great degree, progress in radiology has been related to a diversity of training and effort within the department which has brought a strongly interdisciplinary approach to the solution of problems. As a result radiology is in a position to be especially innovative, and is rapidly advancing through probing the boundaries of scientific knowledge.

4. The dynamic impact of radiology has not gone unnoticed within the disciplines of medicine. On the one hand, radiology is staffing its training programs with the best graduates of our medical schools, and also with young physicians who are shifting from other specialty fields. However, while radiology is moving out aggressively on the scientific front, we must take care that we do not lose our new technologies to other disciplines in medicine because of our deficiences.

5. While the greatest strength of radiology lies in remarkable imaging technologies and in our abilities to use these with great effectiveness, our greatest weakness lies in the deliberate pace imposed upon us by the systems we use for formalizing and delivering information. In the absence of reports, our referring physicians want to see the images and discuss them with us, but again, our systems place both the referring physician and the radiolgist in a catch-22 position if they do not allow us to produce the images. Regardless of

if they do not allow us to produce the images. Regardless of why this may be so, given the pressures of our new economic environment, we urgently need to find a substitute for images stored on film. The best candidate is the digital image which can be accessed immediately.

6. Continued failure to resolve this problem of control and communication will intensify the pressure from those who are uncompromising about needing and demanding both reports and images. Some are sufficiently unhappy with our system that they have set up competing operations. Our ultimate weapon is to combat this trend by excelling in _all_ aspects of our practice.

7. Conclusions

To deliver radiological services properly and to maintain a competitive position, we must:

1. Equip departments with state of the art information management systems which can track processes of scheduling, examining, and reporting, and very importantly, the location of films.

2. Decide, through mutual communication, what turn around time is acceptable for reporting consultations. Develop an on line reporting system which is capable of handling the speed you have accepted as a standard. Monitor speeds, periodically assess progress and review compliance.

3. Negotiate agreement on rules for lending films from the film library.

4. Monitor those examinations which the RIMS shows as

completed but not reported, in order to determine whose patients are involved and the probable cause. Evolve ways of solving problems, once they are detected.

5. Intiate the process of digital image management by designing and installing IPCs to transmit, store and display images which are digital de novo. Understand that short term improvements will evolve into long term systems.

6. Reverse the trend of converting digital images into film images. Use hard copies only when other alternatives do not solve a communications problem.

7. Communicate with referring physicians to determine how they can best use digital displays. Establish a long-range plan, with periodic assessment of progress.

8. Communicate with the hospital administration as to how the needs for the future can be financed and phased. Progress should be periodically reviewed.

9. Develop a "hot line" or some other mechanism for advising on referrals.

10. Keep in mind that while the ultimate outcome will probably be a totally digital department, there will be many years of a hybrid environment. Once a network is in place, we must develop methods of using it to transmit the information in film images, when the occasion demands.

References

1. Bauman RA, Lodwick GS, and Taveras JM: The digital computer in medical imaging: A critical review. Radiology

1984; 153:73-75.

2.   Lodwick GS,   Turner HH,   Lusted LB,   et al: J Chronic Dis 1966; 19:485-496.

3.   Harlow CA,   Dwyer SJ and Lodwick G: On radiographic image analysis. Topics in Applied Physics 1976; Vol 11, Digital Picture Analysis.

4.   Gempel PA, Girard R, McCann R, et al: Prospective payment, what it is/how to cope. International Health Services, Ltd 1983; 1-134.

5.   Lodwick GS and NCRH Task Force: Recommendations for obtaining the maximum benefit of radiation exposure in diagnostic   radiology   through   improved   production   and utilization of image information. Department of Radiology, University of Missouri, Columbia 1968, Vol. 1:1-98, Vol 2:1-517.

6.   Lodwick GS, Wickizer CR and Dickhaus E: Mars: Its Tenth Anniversary of operation and its future. Methods of Information in Medicine 1980:19, 125-132.

7.   Dickhaus EA, Economic Evaluation of Mars. Doctoral dissertation, University of Missouri 1974:1-297.

8.   Arenson RL, and London JW: Comprehensive analysis of a radiology operations managment system. Radiology 1979; 133:355-362.

9.   Bauman RA, Arenson RL, Barnett GO: Computer based master folder tracking and automated file room operations. Proceedings of the Fourth Conference on Computer Applications in Radiology. American College of Radiology, 3/75:469-480.

10. Barnhard HJ and Lane GB: The computerized radiology department: Update 1982. Radiology 1982;45:551-558.

11. Lehr JL: Installation of Mars II at the University of Chicago. Proceedings, 8th Conference on computer applications in Radiology. American College of Radiology. In Press.

12. Heilman RS: Sounding boards - What's wrong with radiology? The New England Journal of Medicine 1982;306:477-479.

13. Homer J: Commentary: A radiologist's point of view. JAMA 1981; 246:2581-2582.

14. Burham JC: American Medicine's Golden Age: What happened to it? Science 1982; 215:1474-1479.

15. Fisher HW: Radiology departments: Planning, operation and management. Edwards Brothers Inc. 1982:1-420.

16. Dwyer SJ, Templeton AW, Martin ML, et al: The cost of managing digital diagnostic images. Radiology 1982;144:313-318.

IMAGE ACQUISITION DEVICES AND THEIR APPLICATION

TO DIAGNOSTIC MEDICINE

S. NUDELMAN, PH.D.

DEPARTMENT OF RADIOLOGY - HEALTH CENTER
UNIVERSITY OF CONNECTICUT
FARMINGTON, CT   06032, U.S.A.

A. INTRODUCTION

This part of the NATO Institute Proceedings will be devoted to a broad
review of photoelectronic image acquisition devices used in diagnostic
medicine. To the extent that space permits, material covered will include
their identity and mechanisms of operation coupled to a discussion of
applications. It will not include for example, CT, MRI and ultrasonic
imagers which will be covered in other parts of the Proceedings.

The components of a pictorial information system comprise both analog
and digital devices (Ref. 1) as shown in Fig. 1. Photoelectronic image
acquisition devices are always analog. As used in diagnostic medicine,
they demonstrate the wide variety of devices which have found there way
into clinical practice, although originally conceived for application to
far ranging, diverse areas including entertainment, space and the military.
For example, the television tube was developed to implement the conception
of television which originated with A.A. Campbell Swinton in 1908; the
x-ray image intensifier has its origins in low light level intensifiers
developed for the military in World War II; and the components in gamma
cameras developed for nuclear medicine can be traced to the evolution of
nuclear physics.

Image acquisition devices used for diagnostic medicine operate over an
extraordinarily wide spectral range. Reference to Figure 2 reveals that
gamma cameras and x-ray imagers operate at the short wave length end of the
spectrum, endoscopes throughout the visible spectrum, thermal imagers in
the infrared and MRI with long wavelength radio waves. Accordingly, the
physical mechanisms used in the sensors vary widely as do the engineering
configurations in which they are installed to function as a camera. A
rather extensive list of imagers is presented in Table 1. They constitute
an extraordinary listing in terms of the diversity of the physical
mechanisms responsible for their operation; the wide range of
electromagnetic spectrum in which they function; their ability to image

NATO ASI Series, Vol. F19
Pictorial Information Systems in Medicine
Edited by K. H. Höhne
© Springer-Verlag Berlin Heidelberg 1986

photons, phonons and particles; and the procedure for viewing their images, i.e. with and without processing.

To date, imaging systems manufactured for diagnostic medicine are complete, that is they include the image acquisition device, image processor as needed, hard copy and display. They are made available on a stand alone basis and in general do not offer any opportunity for the user to extract the digital data in either raw or finished form. However, as the concept of PACS and its evolution matures into installed, operational systems, it will offer a local network to which users will want to interface all their "stand alone" systems. Furthermore, medical research centers wishing to improve on the diagnostic performance of their imagers, will desire access to data. Accordingly, there is good reason to anticipate that in the years ahead there will be emerging stand alone systems with communication ports designed to offer access to data to meet these needs. A broad knowledge of the manner in which image acquisition devices function will facilitate the ability of scientists and engineers to make appropriate use of data so derived.

During the past seven years, we have witnessed a surge in successful research and development for imaging in diagnostic medicine. There has resulted a rash of specialty meetings and associated proceeding which provide detailed coverage of the subject. References 2-14 are a selection of those proceedings which can serve well to present extensive reading material for a newcomer to the field; sufficient to acquire a real sense of clinical status and the state-of-the-art.

In the text that follows, descriptive material will be presented of the various types of photoelectronic imaging devices that have found application and some that have the potential for finding application to diagnostic medicine. In addition, pertinent examples of existent applications are included. The broad area of device specification merits separate consideration and was not included in the material covered by the Institute. The reader who needs to develop a working knowledge of this broad subject area is referred to the literature beginning with Ref. 15. In particular, he should understand the role of quantum efficiency, detective quantum efficiency, gain, signal, noise, dynamic range, spatial resolution, temporal resolution and gamma.

B. DEVICES AND APPLICATIONS

There are two basic types of photoelectronic imaging devices, which
are discussed in considerable depth in Ref. 15. The first is most often
referred to as an image intensifier and develops an image directly in a
manner similar to that of a simple lens; i.e. an image is projected on to
the input surface of the device and a subsequent image emerges from the
output surface. The device offers two very useful features: It functions
both as a spectrum converter and as a photon amplifier. The TV camera is
the second type of device. It uses a scanning electron beam to generate an
electrical video signal which reproduces the image projected optically on
the camera's input surface. The intensifiers output image can be recorded
on film or optically coupled to a TV camera. The output from the video
camera can be displayed directly, digitized and fed to a computer for image
processing, recorded and stored in digital or analog devices. Each of
these basic image acquisition devices has been used to acquire images for
diagnostic medicine over a wide spectral range, extending from high energy
gamma rays to the near infrared. Most recently has seen the emergence of
the laser scanner camera and its application to ophthalmology. This is a
new device and offers the potential for application to a range of
diagnostic procedures. The text that follows has been organized to
describe these devices and selected applications. It will begin with those
sensitive to light and end with x-rays.

1. Light and Photoelectronic Imaging Devices

a. Types

(1). Image Converters and Intensifiers

There are two basic functions served by image converters and image
intensifiers (II). The first is to convert input photons from one part of
the electromagnetic spectrum to output photons from another part of the
spectrum. An x-ray image intensifier (XII) is a prime example of that
function, in that incident x-ray photons are converted to visible photons
emerging from the XII. An II also provides intensification which can be
manifested as intrinsic photon gain when the device provides unity
magnification, or as an increased output photon flux resulting from a
combination of intrinsic photon gain and image demagnification. These
devices can take a variety of shapes and have been made in the form of

simple solid state layers as well as complicated vacuum tube devices. They
have been made responsive to particles and to electromagnetic radiation
extending from gamma rays to the near infrared, and to provide output
photons matched spectrally to the needs of vision, photography and video
sensors. The discussion presented below includes material that provides
background information as a matter of completeness. It is intended to give
the reader a better understanding of the field even though in some cases
there may not be a prevalent application to medical imaging and only faint
potential that one might arise.

(a) Solid State Devices
(a.1) Phosphor Screens

A screen comprises the simplest of structures. It consists of a layer
of phosphor particles embedded in a transparent bonding medium which has
been deposited in a uniform manner on an appropriate substrate. It has
long been used as a converter whereby, for example, exposure to ultraviolet
light causes the emission of light of some selected visible color.
Phosphors have been made to respond to high energy particles, gamma rays,
x-rays, ultraviolet, visible and infrared illumination. Applications
abound in areas such as entertainment, low level illumination, and
photography. The most practical of applications emerged when phosphors
were coated on the inside of cylindrical tubes and fabricated into the
important fluorescent lamp. The outstanding medical application is in
diagnostic radiology where the phosphor absorbs x-rays which causes the
emission of a proportionate amount of light. A detailed discussion of such
screens will be provided in Section B.2.a. below.

(a.2) Scintillators

Scintillators differ from phosphor particles in that they are used for
responding to individual high energy radiation events by generating
corresponding light pulses. The light pulses in turn are recorded
beginning with their detection by a matched photon detector. The
scintillators are available in small sizes to serve as elemental detectors
as well as large discs for direct imaging requirements. Their principal
application for diagnosis is in nuclear medicine. The scintillator is
commonly a uniform media which absorbs a photon of radiation (such as from

a gamma ray) and emits an output pulse of light. The pulse may provide information as a simple individual count, or it may in addition provide a brightness level whose magnitude is directly proportional to the energy of the absorbed photon. A detailed examination reveals that the pulse of light actually contains a number of individual scintillations of light whose number is dependent on the energy of the absorbed photon energy. A measure of pulse height permits discrimination of pulses generated by photons of different energies. Thus in nuclear medicine, it is common practice to count gamma ray photons which emerge directly from a radioactive isotope localized in a body organ, and discriminate against lower energy gamma rays which encountered scattering in passing through the body. Single crystals of sodium iodide both large and small are examples of inorganic scintillators while anthracene is an example of an organic scintillator which can function either as a solid or in a liquid solution.

(a.3) Photoconductive, Electroluminescent Structures.

These devices have evolved into different types of structures depending upon their application. A two layer structure comprising a layer of photoconductive material adjacent to a layer of electroluminescent material provided the first conceptual approach to a light amplifier. An electroluminescent phosphor is a substance which emits light (luminesces) when under the influence of an electric field. This type of device first appeared in the patent literature in 1952. By a strange coincidence, four independently conceived patents of essentially the same device appeared in three different countries. The earliest filing date was recorded by Stuermer in Germany, followed by Amalgamated Wireless in Australia, Westinghouse in the U.S.A., and General Electric in the U.S.A. (References 16-19).

The layers have been formed by evaporation, sputtering and painting technology. Their response has been demonstrated for broad regions of the spectrum and to particles. An exception is the work of Cusano who developed a doped evaporated single layer film of electoluminscent zinc sulfide which also served as an intensifier (Ref. 20).

The multiple layer approach is shown schematically in Figure 3. A potential difference is applied across the sandwich at the transparent electrodes. When the photoconductive layer is exposed to the incident radiation comprising the image, its resistance lowers locally to an extent dependent on the localized intensity of the image. This results in an increased potential difference across the electroluminescent layer whose light output increases with voltage. Thus, its light output depends upon the brightness of the incident image, on a pixel by pixel basis across the surface of the light amplifier. The result is a visible output image being observed through the output transparent electrode. With this approach, the sensor and display functions have been separated. It makes possible the selection of a photoconductor responsive to a desired spectral region and a phosphor with light emission from blue to red as desired. Devices have been developed responsive to x-rays and to visible radiation. They have not been placed in production, however, because of better performance derived from vacuum tube intensifier devices.

(b). Vacuum Tube Devices

The simplest construction is for a proximity focused II as shown in Figure 4. It comprises an input transparent disc on which is coated a transparent electrode and a film of some photoelectron emitter. Its output is another disc which also is coated with a transparent electrode on which is deposited a layer of phosphor. These two discs are closely spaced and their inner surfaces contain the transparent conducting layers, the photoemitter and the phosphor layer. A voltage difference is placed across these two surfaces so that an intense electric field exists in the gap between the discs. When incident photons are absorbed in the photoemitter, electrons are emitted into the vacuum. Then under the influence of the electric field, they accelerate across the gap and strike the phosphor with sufficient energy to cause it to luminesce. Furthermore, the intense field combined with the close spacing between the glass discs cause the electrons to move almost directly across the gap with little lateral displacement. This is the basis of proximity focusing since picture elements on the input side are duplicated on the output side with essentially identical geometry. Thus, when an image from some optical system is focused on the input face of the intensifier, there results an output image from the phosphor surface. This type of intensifier first appeared in low light level applications with diameters in the range of 15-35 mm. They achieved a spatial resolution in the range of 30-50 lp/mm. In recent years, it appeared as an x-ray image intensifier with a diameter of 225 mm. and a resolution in production of about 2.5 lp/mm: Special devices in the laboratory achieved an even better resolution of up to 4-5 lp/mm (Ref. 21).

A limitation on proximity focused II's is in the extent of gain as well as restricted opportunity to vary that gain. The gain is determined by the accelerating voltage applied across the gap and the ability of the phosphor to convert the energy of the incident photoelectrons into output photons of light. The higher the voltage, the greater the number of photons in the output light. Photon gain is the ratio of the number of photons out per photon absorbed by the input sensor and is limited by the voltage that can be applied across the gap. Furthermore, if there is a need to reduce gain, another limit is imposed by the voltage dependence of spatial resolution. As the voltage is reduced, the photoelectrons have more time to move across the gap and to increase their lateral displacement. The result is a loss of resolution.

The channel multiplier II overcomes the limitations of the proximity focused II by providing far greater gain as well as the ability to vary that gain. Its structure is shown in Figure 5 and is similar in geometry to the proximity tube. The major difference is the inclusion of the channel multiplier plate between the input and output surfaces. The plate is a glass disc with an array of minute holes extending from the front to the rear surface. The inner surfaces of these holes are treated to be good secondary electron emitters. Accordingly, when photoelectrons strike the input wall of a hole, they cause the emission of more electrons from the surface than are incident. The process is repetitive and many collisions can occur before the electrons emerge from the plate. Gains in the range of 100,000 to 1,000,000 are reported for miscellaneous applications. Spatial resolution is generally in the range of 25-36 lp/mm. This results from the spacing of the holes and double proximity focusing between the channel multiplier surfaces, the photoelectron emitter layer and the phosphor layer, respectively.

The most common structure used in a low light level intensifier (LLLII) is shown in Figure 6. It has the customary photocathode and phosphor surfaces, but they are now separated by sufficient space to incorporate electron optics. The device (without the channel multiplier) evolved during World War II. It was used for night vision applications and associated with the name "snooperscope". In later years it became known as a Generation 1 LLLII. The inclusion of the channel multiplier led to identification as a Generation 2 LLLII and when the device includes a photocathode made from gas or a variant, it is referred to as a Generation 3 LLLII. The latter is desirable for optimum response in the near infrared. A well designed Gen. 1 device has an advantage over proximity focusing in being able to provide better spatial resolution.

The very best resolution is not obtained with any of the intensifiers described above. They all incorporate some method of electrostatic focusing. However, it is possible to use magnetic field electron optics. Such devices are made in the form of cylinders and have been made available with diameters as large as 162 mm. One supplier specifies center resolution of 90 lp/mm. for a 90 mm. diameter tube. The disadvantages to these tubes are the bulk associated with the coils needed for the magnetic field which surround the LLLII and their unity magnification. They have not found there way into many applications.

Sensor response to spectra can be tailored to the application. In the range of 3000–900A, there are a host of photoelectron emitters available. Figure 7 provides a selection from one supplier of low light level intensifiers. There are other suppliers with their own listings of photosensors. Similarly, the spectral emission of the II light output can be tailored to the application. A selection of phosphors available from another vendor is shown in Figure 8. Thus, given the problem of diagnosing disease in some selective spectral region where the level of illumination is low, it is possible to request a supplier to provide a low light level intensifier having a photocathode with maximum response in that spectral region. Furthermore, if the image is to be acquired by an imaging system designed for acquisition, processing and display, the output phosphor can be selected to match the response of the video tube's sensor. This requirement on the phosphor can be quite different than that for vision.

(2) Video Imagers
(a). Electron Beam Scanners

These devices have been manufactured as vacuum tube and solid state imaging sensors. In general, vacuum tubes are best for applications requiring good to superior spatial resolution, charge integration for image development and snapshot operation. Solid state devices have the advantage of small size. This section will discuss only the vacuum tube devices since they are the most used in medicine today. However, their applications can serve as examples for potential use of solid state cameras as well.

There are two basic types of video tubes determined by whether they do or do not provide gain. Their basic features are shown in Figure 9 and 10. The "no gain" tube generally is referred to as a vidicon type tube having appeared with that name about 1950. It makes use of a photoconductive film deposited on the inside surface of the input window (Ref. 15,22).

The tube incorporates an electron beam which deposits electrons on the vacuum surface side of the film during the process of raster type scanning. The film is highly insulating and essentially holds the charge for the time required to scan out a raster. When the tube is illuminated by imaging radiation, the incident photons on being absorbed by the photoconductor cause a corresponding amount of electrons to disappear. The electron beam during the scanning procedure replaces the missing charge and in the process generates a video signal. The pattern of signal generation follows the image pattern on the photoconductor and is regenerated on a display whose raster scan matches that of the video sensor.

A tube with gain is also shown in Figure 10. It incorporates a Gen. 1 type of intensifier at its front end with the major difference being that the output phosphor is replaced by a highly insulating target. The target has been made in several structures and from a variety of materials. Its purpose is to provide gain and charge storage. The target functions by converting incident high energy photoelectrons into many more electrons being stored on the surface of the target facing the electron gun. Gains are available in different targets which are as low as 3-5 in the image isocon, as high as 2000 silicon intensifier target tube (SIT), and a middle level of 80-100 in the SEC tube. The gain at the target differs from that obtainable with a video amplifier. The former is achievable without introducing video electronic noise. It offers the opportunity to enhance the optical image to the extent that the photon noise associated with the optical image can surpass the electronic noise intrinsic to the video camera. This ensures that in low light level conditions subtle changes in contrast will not be lost because electronic noise is paramount. The image isocon provides additional gain by incorporating an electron multiplier at its rear. This multiplier has the attribute of being a low noise, high gain video preamplier. Nevertheless, its complexity, size and cost have limited its utility in recent years.

The spectral response of vidicon type tubes depends upon the characteristics of the photoconductor. Figure 11 shows typical performance expected from the tubes available today (Ref. 23).

Note, the very broad range of response available and the high values of quantum efficiency. The latter can approach 100% whereas the photoemitters used in intensifiers and high gain video tubes are in the main below 20%. One might expect that the higher quantum efficiencies would be better suited to the low light level requirement then the high gain tube. Even with this advantage, as the light level diminishes there comes a point where the preamplifier noise dominates the photon noise. The image must then be enhanced via target gain to continue to record images with minimal intrusion of electronic noise.

Although the information content in the signal might decrease by a factor of three to six from reduced quantum efficiency, the remaining signal benefits by gains that can amount to several thousand. There is a loss of basic information since few photons are being absorbed by the sensor. However, the gain ensures that those actually absorbed will generate a video signal well above electronic noise. Sometimes avoiding the need for gain can be managed by switching from a photoconductor with a limited range of spectral response to another with a wide range, such as from the saticon to the newvicon. Performance to be expected from vidicon tubes is summarized from Ref. 23 in Table 2.

The minimal signal that one can acquire is determined by the video electronic noise, whereas, the maximum signal is limited by the electron beam current and the characteristics of the target. An example of the dynamic range from a video camera is shown in Figure 12. Optimum performance is a function of the time it takes for the beam to scan across a picture element. This is reflected in the video bandwidth required for proper operation. Figure 13 illustrates how signal, noise and the signal to noise ratio are affected by the scan rate and associated bandwidth. Figure 14 shows the relationship that exist between video bandwidth, spatial resolution and TV line number.

(b). Laser Beam Scanner Cameras

The laser scanner camera (LSC) is new and has not yet played a significant role in diagnostic medicine. Nevertheless, its potential is very bright. Laser scanners have been well established in printing and display technology. They have recently begun to make an appearance as imaging sensors (cameras). Several years ago, a high powered LSC microscope came under development at the University of Arizona (Ref. 24).

More recently has seen the appearance of an instrument designed for imaging the retina (Ref. 25).

They have in common with printers and displays, operation dependent upon a laser scanning a surface usually in a raster mode of operation. The reflected photons scanning the surface can be recorded, however, by including an optical system with a photodetector. The latter on absorbing the reflected photons generates a video signal with a raster identical to the scanner which can then be recorded and/or displayed. The device can also be designed to operate in transmission as in a microscope.

The scanner camera's proponents point out that it offers the advantage of providing scatter free images, as compared to images acquired through conventional optics and video systems. Furthermore, far less illumination is required than for a video or photographic camera. This can have major impact where tissue is sensitive to heat or excess light. In the case of the microscope, a resolution of 0.5 microns has been achieved.

A schematic diagram showing the important components of an LSC is shown in Figure 15. It illustrates a system making use of a rotating polygon mirror to provide the horizontal scan and a vibrating mirror to provide the vertical scan of an xy raster. Systems have been developed that use vibrating mirrors for both scans as well as the exclusive use of rotating polygons. In general, the choice of the component depends upon the application and the extent to which jitter can be tolerated. The most difficult requirement is best managed by the polygon scanner with its favorable rotational moment of inertia. Accordingly, this design incorporates the polygon scanner in the horizontal direction to meet a high speed scan, jitter free requirement. This design is the basis of an instrument used in ophthalmology to image the retina and will be discussed in detail in section B1.b.(2). It is supported by a digital image acquisition, processing, storage and display system.

b. Applications

Photoelectronic imaging devices are not new to medicine in applications involving qualitative examination. Examples can be found in surgery, ophthalmology, radiology and internal medicine. About ten years ago, it became clear that a variety of superb video image tubes had been developed for science, space and military applications. They offered the precise performance required for diagnostic medicine and went well beyond the performance needs of broadcast television. A program was initiated at the University of Arizona to fully explore their potential. It led to Arizona's approach to intravenous angiography and to the concept of a photoelectronic digital radiology department (Ref. 26, 27). This offered the potential for the eventual emergence of an imaging department suited to meet the needs of diagnostic medicine in general. Imagers requiring light rather than x-rays, for example, present similar technological demands for digitization, processing, storage and display. Thus, there was reason to be optimistic that as the field matured, the economy of size as well as improved diagnostic performance could lead to an improved departmental structure. We are beginning to see that as the emerging digital imaging technology becomes apparent to the medical community, there is an appreciation for new avenues of research and application. Described below are examples of applications to internal medicine, ophthalmology, mammography and dentistry as evidence of this new experience.

(1). Internal Medicine - Endoscopy

Endoscopes are devices used in medicine to see below the surface of the skin. They can be made either to be rigid or flexible, and range in sizes from very small to very large. The smallest to our knowledge fits in the end of a hypodermic needle. Amongst the largest is a flexible endoscope used in gastrointestinal examinations. Their basic structure is illustrated in Figure 16. An endoscope includes an optical channel to transmit light for illumination of the interior, an optically coherent channel to transmit an image of the illuminated interior back through the endoscope lens optics at the interior (distal) end of the endoscope for proper illumination and image formation, and lens optics at the exit (proximal) end for coupling the image properly to the user. The latter can be selected to accomodate the human viewer, a photographic camera, the raster size of a television camera and/or some desired combination. Their arrangement as viewed from the distal end is shown in Figure 17. Notice that in addition to the optical channels, there is shown a channel serving to provide air and water and another channel to insert a forceps tool or to provide suction. These channels serve essential purposes particularly in keeping the surfaces of the lenses clear and a means for acquiring samples in vivo needed for pathology.

A light source and a light guide are required to be connected to the endoscope for internal illumination. A simple attachment can also be provided for instructing, called a teaching aide. It comprises a flexible imaging channel connected to the endoscope through a beam splitter. Figures 18 show the basic layout and mode of operation, respectively. As recently as 1975, the teaching aide was used as an easy way to connect a low light level color television camera to the endoscope (Ref. 28).

The eyepiece of the teaching aide was replaced by optics to interface to the television camera and resulted in excellent real time studies following the passage of a bronchoscope through the airways of the lung. Nowadays this approach is managed by a special beamsplitter attachment that connects a one tube newvicon color camera to the endoscope as shown in Figures 19. Notice that the camera tube fits into a tight cylindrical container at the attachment and the control electronics is in a distant separated package. This design ensures minimum space and weight being required for the camera in the immediate vicinity of the ongoing work. Its usefulness is also demonstrated for surgery in Figure 20.

Endoscopes for internal medicine are designed for specific applications. Their thickness, length, auxilliary channels and degree of maneuverability literally have to fit the requirement. Specific examples include the bronchoscope (lungs), coloscope (large bowel), gastro-duodenoscope (gastric and duodenal - first and second portion) and Lo Presti Panendoscope (entire upper GI tract). Flexible endoscopes always use fiber optics to transmit an image from its distal to proximal end. However, rigid endoscopes can use a solid optical chain to transmit an image. In general, the quality of an image is superior in the latter case since the procedure does not require sharing the image amongst many fibers. There are two ways in which fiber optics degrade an image: First, the spatial resolution is limited by the diameter of an individual fiber and the spacing between the centers of nearest neighbors. Second, the appearance of a chicken wire effect, which can seem to overlay an image. This is traceable to the manufacture of the fiber optic bundle. Over the years, these two problems have been reduced and the images obtained are diagnostic. Another problem that occurs with use is the breaking of individual fibers. The images suffer with the loss of corresponding picture elements. The endoscope must be replaced when to many fibers are lost. Examples of images obtained through an endoscope are shown in

Figures 21 a, b and c.

The development of a hypodermic fiberscope for the visualization of internal organs and tissue was first described in 1974 (Ref. 29). Its manufacture was based on the development of microlenses and high resolution imaging multifibers. In the smallest case, the lens diameter was 0.5 mm. to match a 0.5 mm. diameter bundle of multifibers. There were 11,000 individual fibers each having a diameter of 5 microns; two plano-convex distal lenses; over two hundred illumination fibers of 50 microns each; and an ancillary channel for a biopsy tool, aspiration or insufflation. It was encompassed by an 18 gauge hypodermic needle. Its design and construction are shown in Figures 22 a and b.

(2). Ophthalmology
(a). The Laser Scanner Ophthalmoscope

The newest addition to the array of instruments used for diagnosis is the laser scanner ophthalmoscope. According to its developers, it offers a substantial advantage over operation with a conventional ophthalmoscope. In the latter case, the whole retina is illuminated through the outer part of the enlarged pupil, while each picture element of its surface is observed through a small central part comprising only one tenth of the total exit pupil area. This reduces the intensity of the detectable light by a factor of ten and limits resolution. The new instrument uses a collimated narrow laser beam of 1 mm. diameter focused by the eye to twenty micrometers for illumination of a single part of the retina. The reflected light normally 3-5% of the incident light, is collected through 95% of the outer pupil. The basic arrangement was discussed in Section B.1.a.(3) and illustrated by Figure 15. The system also features an adaptive optical feedback system for laser beam control. The system essentially provides elimination of optical eye aberrations which diminish the fundus quality. The combination of the optical system and image processing makes possible new methods of diagnosis including measurement of eye aberrations, early determination of glaucoma and high resolution nerve fiber layers. Examples of images obtained with this instrument are shown in Figures 23a and b.

An approach developed for microscopy to obtain three dimensional imagery from a two dimensional laser raster scan also has potential for ophthalmology. Three dimensional images were recorded as a series of two dimensional images by varying the focal plane of the microscope.

Reconstruction is done by inverse filtering with the three dimensional optical transfer function of the microscope.

(b). The Funduscope

The conventional approach to studying the retina is with the use of the funduscope. It permits illumination of the patient's retina and its simultaneous visualization by the physician. It provides for illumination by white light for conventional imaging or by blue light for fluorescein angiography. In the latter procedure, liquid fluorescein is injected into a vein and is carried with blood flow through the heart and lungs. It passes in time through the blood vessels of the eye. It normally is not visible when the vessels are illuminated by white light. However, blue light penetrates through the vessel walls and causes the fluorescein to fluoresce a bright green. By using a proper combination of optical filters, it is possible to obtain sharp images of the arteries and veins embedded in the retina from the green fluorescence. Since the flow of fluorescein follows the flow of blood through the vessels, it is possible also to acquire a dynamic study. A sequence of images can be obtained to reveal the initial appearance of the fluorescein in the arteries followed by its spread through the capillaries and finally passing out of the retina through the veins.

The ability to obtain dynamic studies depended upon the evolution of low light level imaging sensors. Pioneering work was reported early in 1972 using the concept of the magnetically focused low light level image intensifier (Ref. 30). It is shown in Figures 24 a and b with an output window made of thin mica which replaced the usual phosphor disc. The mica was sufficiently thin to permit high speed photoelectrons to pass through the window and to expose the film. This device produced the earliest dynamic, real time retinal studies on film of which this writer is aware. Since that time more sophisticated approaches have evolved as video tubes were developed with improved sensitivity. A modern funduscope is shown in Figures 25a and b. The arrangement of the optical components and the light sources permits positioning of the patient, physician and cameras in a compact manner. The imaging sensor has changed from the original LLLII and cine camera into a higher quantum efficiency tube such as the newvicon, or the silicon intensifier target tube (SIT) which offers gains as high as 2000. In general, archival storage quality retinal images are still obtained using photographic film. Examples of the kinds of images that can be obtained are shown in Figures 26 and 27.

The development of video – digital systems has led to new, ongoing research to take full advantage of their features in striving for improved diagnostic performance. Images obtained with a SIT camera look very much like Figure 27. The spatial resolution is fine since 512 raster lines need only scan an inch of surface area, which corresponds to about 20 raster lines per mm., or about 50 microns width per raster line of image plane. This resolution can be improved by a factor of two by going to a 1,024 line raster and even further by using optical magnification. Processed images at the University of Arizona are reported to show retinal features that had not been observed otherwise. The digital system used at Arizona was first developed for intravenous angiography which has similar requirements in acquiring images from a video camera. The original system permitted acquiring images in a snap-shot mode, or at a repetition rate of up to one frame per second. Systems available today permit operation at speeds up to 30 f/s with a raster of 1024 x 1024 pixels. Further discussion of such systems is deferred to Section B.2.c.

(3) Diaphanography

This procedure is used as a means for detecting breast cancer (Refs. 31, 32, 33 and 34). It requires illumination of the breast on one side and examination of a shadowgraph type image viewed from an opposite side. Thus, it depends upon sufficient light being transmitted through the breast in the same manner as x-rays must pass through the body to provide a diagnostic image. A clinical system is shown in Figure 28. It points out the location of a video camera relative to the patient and the physician which is used to acquire an image for processing. Positioning the light probe against the breast and exerting pressure to shape the breast for examination is shown in Figure 29. The result is an ability to see vascularity and soft tissue structure that lends to detection of cancer. This system provides images obtained in two different spectral regions (in the red and near infrared), based on experimental evidence that optical transmission for malignant tissue in selected spectral regions differs from benign tissue. The light from the probe is filtered to provide two images acquired by the video digital system. These are digitized and processed.

The processing requires registration of the two images and comparison of their signal strength on a pixel by pixel basis. The differential information is displayed as a function of intensity and pseudocolor. Examples of images displayed are shown in Figure 30.

Diaphanography demonstrates a newly evolving procedure that takes advantage of processing to acquire diagnostically useful information. The procedure is relatively new and is experiencing controversy as to its accuracy in diagnosing cancer. Studies have been made which demonstrate that combined with x-ray examination of the breast, better detection statistics have been obtained than with either procedure alone. On the other hand other studies have reported poor performance. Unfortunately, it is not clear whether these mixed findings are the result of different degrees of diagnostic expertise from the participants, the use of different products with different specifications or perhaps the spectral differences reported for malignancy may not be an accurate, reliable indicator for diagnosis. Thus, this is an area for continuing research in image acquisition technology and the processing of images. It is our expectation that better quality images, to provide better contrast and depth of view, will eliminate the variable of examiner expertise and demonstrate an important, new contribution to diagnostic medicine.

(4). Dentistry

The time is opportune for new research in imaging and processing to strive for breakthroughs in diagnostic medicine. This statement also applies but more emphatically to dentistry. Common practice today is to use 35 mm. film for obtaining x-ray radiographic images. However, instrumentation has evolved over the years in new microfocal spot x-ray sources and video cameras that provide suitable spatial resolution and contrast. They couple easily to digital systems such as those developed for intravenous angiography and accordingly lend their images to processing, storage and display. There remains the research and development task of developing modifications of these image acquisition devices to meet the needs of dentistry.

Note, the fact that all the applications described above can be managed in a similar photoelectronic-digital manner and that they derive from procedures already established in radiology. They provide reason to believe that the concept of digital imaging centers for medicine and dentistry may find a place in man's future. Dentists, for example, send their patients to specialists for treatment of periodontitis and root canals. This new concept could have the dental patient go to the imaging center for his photoelectronic - digital images, which could then be returned to the dentist's interactive console via an electronic communications link. He would then carry out his usual diagnosis using a magnified image on a display supported by an image processing capability. Furthermore, as new imaging modalities other than x-ray are developed to serve dentistry, these can be readily implemented without any financial burden to the dentist. There appear to be possibilities that images obtained in spectral regions ranging from the ultraviolet through the visible to the infrared, coupled to spectral analysis might have impact in areas of dentistry such as periodontry. The imaging center approach is intended to bring down costs for delivering diagnostic medicine and dentistry to the patient while providing improved performance. One can hope that such lofty intentions are realized.

2. X-Ray Imaging Devices and Systems for Radiology

a. Shadowgraphs

X-rays are absorbed in passing through a body. The extent of the absorption varies according to the substance of a part. Bone for example, absorbs x-rays to a greater degree than any other substance found in the body. Tumors, fatty tissue and muscle also absorb differently from each other and to bone. As a result, the emergent x-rays intensity varies from point to point across the body surface in conformity with the magnitude of the corresponding degree of absorption.

The radiologist learns to read the shadows cast by the different organs and bones recognizing them by their shape and contrast. Sometimes, parts of the body that must be examined are essentially transparent to x-rays and do not cast a useful shadow. Examples include the blood vessels and the GI tract. To overcome this difficulty, the radiologist uses contrast media which are substances that are more opaque than the body part, and may add image processing for selected procedures. Examples of contrast media which provide useful absorption and are tolerated by the body include concentrations of liquid iodine, barium and gaseous xenon. Blood vessels require the injection of a concentration of iodine into an artery of a vein: It serves as the basis for angiography and intravenous angiography. A barium solution is swallowed to permit imaging of the stomach and intestines. Gaseous xenon has been tried with limited success to image the airways of the throat and lungs.

Fluoroscopy is a simple technique used in radiology to provide the means by which the radiologist can see the pattern of x-rays emerging from the patient. It consisted in the early years of placing the patient between an x-ray source and a phosphor screen. The x-rays passing through the body were absorbed in the screen which in turn emitted light in a pattern that corresponded to the x-ray pattern incident on the screen. The physician positioned on the side of the screen opposite the patient then can view the visible image from the screen as shown in Figure 31a. He could move the x-ray source and screen in tandem relative to the position of the patient so that there was opportunity to scan the body anatomy for best viewing of a region of interest. Unfortunately, the amount of light emerging from the screen was of low intensity and caused the physician to work in a darkened room. In addition, he was exposed to the x-rays that passed through the screen and was required to wear garments that contained lead shielding and leaded glasses. In subsequent years, overcoming the low light level condition was managed by a progression of steps. The first involved using a demagnifying x-ray intensifier with the disadvantage of having to look at the small phosphor disc. The second appeared within the past five years and consists of a proximity focused x-ray intensifier. It is manufactured in a 9" diameter so that there is a vast improvement in viewing as shown in Figure 31b.

Substantial improvement was managed further by optically coupling the output from the screen to an intensified video camera, as shown in Figure 31c. This permitted viewing of a TV display in a room removed from the place of the exposure so that the physician could avoid the problems of exposure and a darkened room. An example of a system that found wide application is shown in Figure 32. It used folded Bouwer mirror optics that permitted imaging a 12" screen on to the input surface of a LLLII whose output was then optically coupled to a video tube. The evolution of the large diameter x-ray image intensifier led to the replacement of this system. Coupling the output from the large diameter, demagnifying intensifier as shown in Figures 31d and 33, proved to be far better in efficient use of the x-rays. It resulted in reduced dose to the patient and improved resolution. However, each step in technological improvement came with increased cost in equipment. Thus, there is considerable incentive to find a breakthrough in new concepts for improving the system of Figure 33, which would be far less costly to manufacture and maintain.

Fluoroscopy in general is used for a preliminary viewing of the body. In selected procedures, the same kind of image acquisition components can be used also for detailed diagnosis. An example is intravenous angiography which will be described in the angiography section below. Its principal limitations are in the diameter of the intensifier and its resolution. Production intensifiers generally offer less than 6 lp/mm limiting resolution, and that for a diameter of about 6". They have been made in recent years in diameters up to 22". This increase in size is accompanied with a decrease in resolution, to not more than 3.5 lp/mm. This is far less than the intrinsic resolution of the x-ray sensor in the intensifier which is lmiting at about 10 lp/mm, or the limiting performance of high resolution phosphor screens which can approach 15 lp/mm. A primary goal in the research community is to find a new photoelectronic approach that can eliminate the resolution bottleneck. Meanwhile, any large dimension, high resolution imaging requirement is met by exposing a screen and recording on film.

b. Device Structures

(1) Screen-Film

Recording of an x-ray image on film provides the traditional radiograph. The radiograph of the chest is an example of this technologically simple procedure. Essentially, it requires placing film adjacent to a phosphor screen in a cassette. The x-rays passing through the body are absorbed in the screen. Each picture element of the phosphor in the screen emits a pulse of light whose magnitude varies proportionate to the amount of x-rays absorbed. Over the surface of the phosphor screen there appears the x-ray image converted to a visual image. The light from the screen exposes the film immediately adjacent which when developed provides an accurate reproduction of the x-ray image. Probably 85-90% of all diagnostic radiology is still managed with images acquired using the film screen cassette.

The screen with its high quantum efficiency and gain is the principal reason for the modern day success of the procedure described above. It functions with quantum efficiencies approaching 100% and photon gain can be in excess of 1,000. Film on the other hand, has a quantum efficiency less than 1% being a poor x-ray absorber and offers no significant gain. Accordingly as noted above, the phosphor screen is used to absorb the x-rays, convert them into a corresponding amount of light and expose the adjacent film. There is a trade off in the manufacture of screens between spatial resolution, quantum efficiency and gain. Increasing spatial resolution is accomplished by decreasing the phosphor particle size and the thickness of the layer. The effect is to reduce light scatter in the phosphor layer but at the cost of reduced x-ray absorption. Furthermore, there is also a point at which reducing phosphor particle size can adversely affect quantum efficiency and gain. As a result, given the same exposure, high resolution screen as used in mammography emit less light than lesser resolution screens as used in routine examinations of the chest. Since image detail depends upon the number of x-ray photons absorbed in the screen, the mammography screen requires a larger exposure than does the chest film. This causes the unfortunate requirement of a corresponding increased dose to the patient.

A goal of researchers in photoelectronic-digital radiology is to replace the film-screen cassette with a low cost photoelectronic device that can match if not improve on the image quality obtained from the cassette. Such a device would permit the generation of a direct video signal without loss of radiological information for digital image acquisition, processing, storage and display. Its success will lead to improvement in diagnostic performance and be cost effective. Examples of candidate devices being explored for their potential to replace film screen are described below. However, it should be noted in advance that none have approached ideal performance. That goal is proving difficult to achieve.

(2). Flat Panels

(a). The Selenium Plate

The simplest of the flat panels uses the xerographic type of selenium plate. It was described in 1977 as illustrated in Figure 34 and involves a two part operation (Ref. 35). The device is charged in the same manner as for a conventional xerox plate. It is transported to a station where the image from a patient exposure can be acquired. As a result of the exposure, the charge stored on the plate is depleted in a manner that establishes an electronic image across its surface. The plate is then transported back to a read-out station. There it is scanned by a laser beam in a raster, and in the process generates a video signal that can be recorded, displayed, processed and stored, as desired. A marketable system has not yet appeared indicating that this approach has not yet proven suitable for clinical operation.

(b). The Metal Haloid Plate

A recent innovation makes use of the property of certain materials to absorb energy on exposure by incident electromagnetic radiation, store that energy for substantial periods of time, and emit that energy on command in the form of electomagnetic radiation. Fuji makes use of these materials in demonstrating a system that incorporates metal haloids capable of absorbing diagnostic x-ray radiation which emit radiation in the range of 4,500 to 8,000 Angstroms when stimulated by laser radiation of shorter wavelengths (Ref. 36). The latter falls between 3,900 and 4,000A, and over a relatively narrow bandwidth (3,500 - 4,500 A). Thus, it is possible to select materials which can be simultaneously exposed with laser excitation emission in the ultraviolet to blue region of the spectra and observe the luminescent emission from the material in the longer wavelength region of the spectra (5,000 - 8,000A) without significant overlap between spectral regions. These spectral characteristics served as the basis for a new approach to imaging for diagnostic medicine.

Reference to Figure 35 demonstrates the photoelectronic mechanisms involved in the operation of the device. The material is prepared in the form of a phosphor powder which is deposited in an organic binder on a 1 mm. thick flat plate. The process begins with a freshly erased plate, then exposing to x-rays for recording an image, followed by read-out with illumination from a pulse of laser radiation and finally erasing any residual radiation with another exposure to light. Read-out is managed by scanning the surface of the plate with a laser in a manner similar to that described for the selenium plate. The significant difference is that the output signal from the selenium is electrical, whereas, in the metal haloid case it is optical. During the short dwell time of the laser beam on a pixel of the plate, it instantaneously causes the emission of a pulse of light whose magnitude is proportional to the x-ray exposure of that pixel. The arrangement for read-out is shown in Figure 36, where the laser beam is made to scan across the surface of the plate. The light emitted from the plate is picked up by a fiber optic broom, which transmits the light to a photomultiplier. There the optical signal is converted to an electrical signal, digitized and fed to a processor. The processed image is then converted back to an optical signal which can be used to record on film with another laser scanner as shown in Figure 37.

There are substantial advantages to using a flat panel imager. Geometric distortions inherent in the x-ray intensifier do not exist here. The panel itself can be made as large as desired and the laser scanner added without approaching the high cost of the largest diameter intensifiers. However, it involves both read and write from the same side of the panel which leads to the requirement to transport the panel from the place of exposure to a readout station. This can be done manually or automatically, but in either case is expensive in personnel, time or equipment. Equally important is a limitation on the speed with which one can obtain repetitive exposures. The ideal arrangement in which read and write can be managed inexpensively without moving the sensor panel has not yet been managed. It is the goal of researchers in the field. Meanwhile, the Fuji system offers a good opportunity to acquire and process images, and to carry out clinical research for improvement in diagnostic performance. It is reported able to absorb 50% of incident 80 keV x-rays, and with sufficient gain so that each x-ray photon is converted to a recorded pulse of electrons by the photomultiplier. Dynamic range is in excess of 10,000 which is also in excess of that found in most radiological images. A sampling raster is specified for the reader having an upper rating of 10 pixels per mm. which suggests a resolution of 5 lp/mm. This does not appear to be as high a resolution as obtainable from CsI or the rare earth screens, but this might be alleviated by a more favorable shape to the modulation transfer function.

(c). The Solid State Amplifier

This device was first conceived as a low light level amplifier and converter by Stuermer (Refs. 16-19). A brief description is given in Section B.1.a. and illustrated in Figure 3. He also developed this sandwich structure for radiology by using CdS and CdSe for the photoconductive layer and demonstrated images obtained with x-rays for non-destructive testing. Unfortunately, the device is slow and lacks the absorption of the more recently developed sensors. Thus, dose in medical applications would be unacceptable. However, it did demonstrate a gain of 400, with reason to believe that considerable improvement is possible. The device offers a controllable gamma, but suffers from relatively poor resolution of about 2.5 lp/mm.

An important reason for reviewing the history and performance of this device is because its successful development as a high gain, high resolution device, would open up the possibility of using the system shown in Figure 33. That system using a mirror or a lens is limited in large part because of limitations in optical efficiency. It causes excessive loss of optical photons emerging from the screen and imaged on the sensor. Accordingly, the dose has to be increased for good imagery which is an unsatisfactory solution. However, if the device can be made with x-ray imaging components that provide a photon gain probably of 3,000 or more and a DQE of 50% or more, this system would warrent re-examination for medicine. Its substantially lower cost than the large diameter x-ray II make it an attractive alternative.

(d). The Proximity Focused Tube

This device was described in Section B.1.a. and noted above for its application to fluoroscopy (Ref. 21). However, it too in a large diameter could serve to resurrect the optical coupled system represented by Figure 33.

(3). The Demagnifying X-Ray Intensifier (XII)

The basic structure of the XII is shown in Figure 38. It comprises an x-ray sensor adjacent to a photoelectron emitter. Modern tubes use an evaporated layer of CsI as the sensor and a layer of CsSb as the photoemitter. CsI emits light on exposure to x-rays which are absorbed in the CsSb layer and causes the emission of photoelectrons into the vacuum. The center section of the tube contains the electrodes which form the electron optics and the inner surface of the output glass disc supports a layer of phosphor. Photoelectrons emerging from the photoemitter are accelerated and reimaged to bombard the phosphor. The resultant output light image is a demagnified version of the incident x-ray image. Large diameter tubes can operate with voltage controls on their electron optics that permit switching the extent of demagnification. Thus, for example, a 12 inch input diameter intensifier can be set up to image only six or nine inches. The largest diameter tube available at this time is 22 inches, with popular sizes being 14, 12, 9 and 6 inches. Values of demagnification range from 12:1 down to 6:1. Two aspects of demagnification should be noted. First, the more the demagnification the brighter is the output light image. This results from the photoelectrons being concentrated in a smaller area and, thereby, causing more of them to strike the phosphor per picture element. This provides an apparent additional photon gain per picture element but is misleading since it can be accompanied by some loss of image information. Second, increased demagnification is accompanied by loss of some spatial resolution. Accordingly, a radiological procedure might well use the full intensifier diameter for fluoroscopy but will reduce the diameter to encompass the object being diagnosed and, thereby, ensure best picture resolution.

The paths that photoelectrons can take are illustrated in Figure 39. Note, that a lead disc is placed over the center of the intensifier sufficiently thick to absorb all of the incident x-rays. The x-rays that irradiate the remaining exposed face of the intensifier cause a corresponding disc of light to emerge from the phosphor output. However, the shadow cast by the lead disc is not black but a shade of gray representing some form of illumination. In fact, it is associated with scattering processes that occur in the input glass, the sensor, the electrodes and the back scatter from the output glass. They all combine to produce a veiling glare which results in the image being of lesser contrast than ideal. The ratio of the light emerging from the illuminated disc to that emerging from the shadow is called the contrast factor. The larger the number, the better is the intensifier. Factors reported for many intensifiers are in the range of 15 to 20, while the very highest is in the low thirties.

The sensor in modern intensifiers is a layer of CsI (Ref. 37). Its absorption and resolution characteristics are excellent for diagnostic radiology. The absorption of CsI is shown in Figure 40 and can be compared with the prominent screen phosphors gadolinium oxysulfide (which is representative of a family of rare earth phosphors) and calcium tungstate. The step like vertical transitions are representative of electronic transitions from the k energy levels in atomic structure and differs for different atoms. Absorption of x-rays from the k levels can only occur when the energy of the x-ray photons is sufficiently large to make the electronic transition occur. In CsI, two of the transitions are shown since the k levels in cesium and iodine are fairly close to one another. They appear at 33 and 36 keV respectively. Gadollinium at 50 keV and tungsten at about 69 keV characterize their respective atomic structure. The actual absorption properties of the CsI layer is superior to the others in that it is representative of bulk properties. Thus, it is more dense than the phosphor screen depositions. Accordingly, it's layer thickness can be thinner for the same degree of absorption even allowing for the differences in their mass absorption characteristics. Furthermore, CsI grows in a minute island, almost needlelike structure. They behave like a fiberoptic, in that light generated within an island tends to be reflected internally until absorbed by the photoemitter deposited at the vacuum end. As a result, CsI's layer structure is of optimum density and has a growth pattern that prevents the spread of light from the place of x-ray absorption. It is the choice sensor used in intensifiers today. Efforts to use it also for screens have not been successful. The material is hygroscopic. Efforts to embed it in a medium for insulation from the atmosphere have had mixed success: Perhaps, because any scratching of these surfaces breaking the seal leads to film deterioration, and because there are good phosphors available from which excellent screens can be made.

Notice that the k edges of gadolinium oxysulfide and of calcium tungstate not only exist at higher energy values, but from their respective edges toward increasing energy, they each have regions of dominant absorption. The substances of the body are essentially air, lung, fat, water, muscle and bone. Absorption of x-rays by the human body diminishes with increased energy, and the extent to which the individual substances absorb differs. In effect, in the lower regions of diagnostic energies, fat and soft tissue have sufficient absorption to present a diagnostic image. At the higher energies, their absorption is much reduced so that they become difficult to see on a radiograph. On the other hand, the contrast for bone relative to the others improves because its absorption has decreased at a much lesser rate. Thus, calcium tungstate is more attractive for procedures that are best performed at high energies relative to CsI, and the rare earths tend to be best in between.

Image processing introduces another aspect to selecting the best energy for imaging. An advantage to working at higher energies is reduced dose. Many more x-ray photons emerge from the body for the same exposure with increasing energy. Reference to Figure 41, reveals that increasing the number of countable photons makes possible the visualizing of lesser contrast for the same resolution. This might well be possible with image processing, for the principal is quite clear. Thus, research activities are being explored to work at higher energies with image processing as a means of maintaining diagnostic performance while reducing dose.

The spatial resolution for an image intensifier as characterized by the modulation transfer function (MTF) is shown in Figure 42. The intensifier has a 14" diameter and can function also with the electron optics adjusted for a diameter of 6.5". The MTF for the CSI layer is also shown. Quite clearly, the MTF of CSI is far superior to that from the intensifier. This loss is due to its other imaging components, namely, the electron optics and the output phosphor layer. Each has its own MTF. Notice also that the smaller diameter provides a better MTF than the larger diameter. This too, is a reflection of degradation in the performance of imaging components for the larger diameter. The output phosphor surface, for example, must be able to image more than twice as many pixels in any one dimension for the large diameter compared to the small and results in operation at a lower value of its MTF.

Most of radiological diagnostic information is contained within two line pairs per mm. This corresponds roughly to 0.25 mm. of object size. In general, radiologists reading from a radiograph do not look for a structure less than 1 mm. in size. They primarily detect its existence and also any pattern that might exist with more than one such object. However, having an imaging system that performs well at 2 lp/mm. ensures that the edges of a 1 mm. object are also clearly defined since this corresponds to the system operating at the second harmonic of the fundamental 0.5 lp/mm. The intensifier's MTF at 2 lp/mm., unfortunately, is substantially down from that of CsI, being about 30% and 10% for the 6.5" and 14" modes respectively. Thus, signal recorded at 2 lp/mm. is reduced by a factor of 3 to 9 from the optimum possible. Since the goal is to image the adult chest and abdomen, the factor of 9 is a serious penalty that really inhibits current systems from reaching ideal performance. The problem is further worsened when the MTF's of the coupling optics and the video tube are considered. The specifications of a 14" intensifier operating with three modalities are shown in Table 3. Notice that it defines a limiting resolution, which is based on the value of the MTF at the 5% level. Notice also that the contrast factor goes from a low of 18:1 to a high of 36:1 as the image diameter decreases.

Considerable clinical success has been achieved with systems manufactured to date. However, there is substantial room for improvement, sufficient to warrant research in new and creative directions on x-ray imaging sensors.

c. Subtraction Imaging

The technique of subtracting one image from another to obtain an image which would make visible subtle differences between them, has been practiced in early years with film and recent years with photoelectronic-digital systems. Using film imposed practical limitations on the speed with which images could be acquired and subtracted. Furthermore, precise digitization with a microdensitometer was a time consuming and expensive affair. The acquisition of images by a photoelectronic imaging sensor, however, offered the opportunity to acquire images rapidly and to transmit them to a computer for processing in a most efficient manner. Modern image tubes can be obtained with the dynamic range and spatial resolution meeting the requirements of most if not all those met in diagnostic radiology. There remain several negative factors that need to be overcome particularly in dealing with large dimension images such as for the full adult chest or abdomen. These have to do in particular with the limiting spatial resolution of large diameter x-ray intensifiers and the large number of raster lines required for a raster. Nevertheless, for a procedure such as angiography subtraction imagery has been demonstrated as a practical procedure capable of adding a significant step forward in the practice of diagnostic radiology.

Angiography is designed to permit radiological imaging of the body's blood vessels. Since the blood in the vessels and the vessel walls are essentially transparent to x-ray radiation, one cannot use conventional shadowgraphic techniques as described above. Contrast media must be injected into the blood if the fluid is to be sufficiently opaque to x-rays to make the blood radiographically visible. A conventional well established procedure uses the injection of iodine through a catheter into an artery selected for imaging. Iodine has attractive features as a media to enhance the contrast of blood. Its k edge was noted earlier with regard to CsI in Figure 40 and is at a sufficiently high energy to be attractive for radiology, is miscible with blood and safe in the quantities used for an examination.

(1). Temporal Subtraction

Intravenous injection of iodine combined with temporal subtraction was a new procedure which made its appearance in 1976 when a research program provided the first in vivo pictures of a dog's carotid arteries (Refs. 38-43). The purpose of the research program which began in 1975 was to develop a procedure that would permit non-invasive imaging of the coronary arteries. The scheme which evolved was straightforward, consisting of acquiring an image without contrast media followed by one with contrast media, subtracting the first from the second and enhancing the contrast of the difference image electronically. The original plan was to use a scan converter image tube that was popular at that time. Unfortunately, it could not manage the exact registration between read and write and between successive images to function properly in this application. The solution was to design and construct a digital memory. Prior to that time, they had been developed for and used in military and space applications. Digital memories were not available commercially. H.D. Fisher and M.M. Frost, two engineers at the University of Arizona designed and built a unit specifically for this application. It worked and permitted the acquisition of snapshot images from a video tube being read out at video rates, and the transfer of those images to the slower rate permitted by the computer. The system assembled is shown in Figure 43.

The first pictures taken with the system was of a hand phantom. It was overlaid with a sponge impregnated with iodine and plastic tubing. An exposure was taken with the tubing containing iodine as shown and the resultant image is shown in Figure 44. Another exposure was taken without iodine in the tubing and the image subtracted from the first image. The difference image was enhanced digitally and shown in Figure 45.

In-vivo imaging of a dog's carotid arteries was the next task. The neck area of an anesthetized dog was centered in the viewing field. Iodine based contrast media was injected into the dog's foreleg and an exposure was made before the arrival of the contrast media. The contrast media flowed through its natural course and in the time available before the next exposure, the first image is transferred to the computer. The flow of contrast media proceeded through the heart and lungs and then into the arterial system. In about eight seconds, a second image was acquired which contained the dilute iodine in the carotid arteries. It was transferred to the computer where the first image was subtracted from the second. The difference image was then displayed and then enhanced visually from an interactive display by simple image processing steps of windowing and stretching. Again, emphasis is made that the concentration of iodine in the arteries is dilute compared to that used in imaging with a direct arterial injection, and normally would be insufficient for direct imaging purposes. However, with the above system using only temporal subtraction of images it was possible to obtain clinically useful images.

The image shown in Figure 46 was taken with that first dog procedure. The digital memory had a 512 x 512 x 8 bit capacity and could acquire an image at the rate of 1 every two seconds. Present systems are available which can operate in real time. Expansion of digital memory with 1024 x 1024 pixel rasters has also been accomplished, and available with operation in real time. This fantastic progress has been accomplished with significant improvement in the performance from x-ray intensifiers and video tubes. The first human study reported was a study of a man's carotid arteries in 1979 and is shown in Figure 47 (Refs. 44-46).

An early 1977 effort to image the heart and coronary arteries of a dog is shown in Figure 48. It generated considerable enthusiasm and optimism that imaging the coronary arteries using an intravenous injection (considered a non-invasive technique) would be achieved within the three years. Unfortunately, the human coronary arteries proved to be a much more difficult target to achieve than any of the other body's arteries. The expansion of the intravenously injected bolus of iodine as it passes through the heart and the coronary arteries, the proximity of the heart with chambers containing iodine, and the presence of the lung field combine to thwart research efforts to image the coronary arteries. Concentrated research is continuing to accomplish this most important task.

It is ironic that substantial funding from the National Heart, Lung and Blood Institute has been expended for research devoted specifically to achieving non-invasive imaging of the human coronary arteries which has resulted in the successful non-invasive imaging of the other arteries of the body. Examples of image quality achieved to date are shown in Figures 49-58. The sequence shown in 49-52 provides an image of the neck beginning without contrast media followed by contrast media. The two images are essentially the same by direct visual examination. The low contrast difference image is shown in Figure 50 and the image enhanced by windowing and stretching is shown in Figure 51. Another sequence demonstrating the procedure for the lungs is shown in Figure 53. The kidneys and renal arteries are shown in Figures 54; a section of the lungs in 55; the head in arterial phase in 56; the head in venous phase in 57; and the iliacs extending into the femorals in 58.

In recent years, systems designed for intravenous angiography have improved to the point where they are also being used in conventional arterial injection angiography. Imaging for the coronary arteries and the heart, for example, can be managed for many cases with a six inch diameter image field where intensifier resolution can be quite good. Image rates are now up to 30 frames per second with a matrix of 512 x 512 pixels. This corresponds to 512 raster lines spanning a distance of 150 mm. or almost 3.5 raster lines per mm. It is also reasonable to expect that available real time 1024 systems will be installed in the near future, now that high speed digital discs are being manufactured. Resolution requirements imposed earlier were described as having to see a 1 mm. constriction in a 2 mm. vessel. This specification can be met with current systems and one can expect to achieve 0.5 mm. resolution within a few years. At that time, there will be good reason to expect that film based systems will begin to disappear in angiography.

Examination rooms for angiography tend to be very expensive, in the order of 0.75 to 1.5 million dollars depending on the features desired. An example is shown in Figure 59. It contains a bi-plane capability. The radiologist can image any two projections that he desires by manipulating the C-arms as desired. During a procedure, the arms would be turned on consecutively to avoid scattered x-ray radiation from one plane irradiating the x-ray intensifier in the other. Accordingly, they are fired consecutively and in a repetitious manner in the course of a dynamic study. Notice for the arm in the plane of the picture, the location of the x-ray source in its housing at the bottom and the intensifier in its housing at the top of the arm. The upper housing can include also a cine camera, a video camera and a 70 or 100 mm. photographic camera for individual exposures. There are a variety of designs in the arms and supports for angiography suites which offer advantages and some disadvantages. Thus, one should not be surprised to find that manufacturers have not settled on any one design and the radiologist does have a choice in striving for a room that suits his needs the best.

Although, intravenous angiography was described above using the experience of the University of Arizona, its history can be traced to research being carried out also at the Universities of Kiel and Wisconsin.

The earliest work reported using photoelectronic image acquisition, digital processing and temporal subtraction dealt with conventional angiographic studies of a pig's heart and aorta. It was carried out at the University of Kiel (Refs. 47-49). Their images were first stored on video tape and processed subsequently. They also reported a technique for real time imaging. It was managed by placing in digital memory an image obtained before the appearance of contrast media, then subtracting this image from all those that followed which contained contrast media. The image acquisition and subtraction rates were both in real time.

Wisconsin carried out work on energy subtraction and, as early as 1973, demonstrated the advantages of subtraction as a means of achieving contrast enhancement (Refs. 50-52). Their work was the first to use a scan converter type of photoelectronic storage tube as distinct from film based subtraction systems used, heretofore. By the end of 1977 and early 1978, their research led to reporting of temporal subtraction IA using multiple frame integration procedures and with analog storage similar to that demonstrated earlier by Kiel for angiography (Refs. 53-56). They also reported the video sensor-target integration approach similar to that used at Arizona but felt that multiple frame integration provided superior images. Over the following years, the latter approach became the Wisconsin "standard bearer".

(2). Energy Subtraction

Energy subtraction has been mentioned above without prior explanation. It is a procedure which takes advantage of the fact that the absorption of components of the body differ and that there relative values depend upon the energy of the irradiating x-ray photons. One can write a set of simultaneous equations which include the absorption characteristics of bone, soft tissue and muscle for different energies. These can be solved for any one of the body's components and, thereby, eliminating the others. Thus, with the use of the digital computer and the facility to handle the large pixel numbers involved in dealing with an image it is readily possible, for example, to derive an image of the body's skeletal structure by eliminating soft tissue and muscle. Alternatively, the skeletal structure can be eliminated and an image of soft tissue revealed.

If a contrast media such as iodine is restricted to one location, the calculation permits deriving and displaying an image showing where the iodine is located. Sequences of exposures can be achieved at relatively high rates. This involves the use of rotating filters placed between the source and the patient, interrupting the x-ray beam at a rate dependent on the speed of rotation. The limiting performance is determined by the speed with which the digital image acquisition system can accept the images and the exposure conditions imposed on the x-ray tube. Thus, imaging of the iodine flow in the arteries, for example, could be managed at rapid rates while that of iodine in the thyroid could be a relatively slow process.

An example of a slow process is that used in the "hold your breath" time required for imaging the airways of the throat. This involves using gaseous xenon as the contrast agent. The gas is relatively radiodense compared to air and has an atomic number similar to barium and iodine. The images shown in Figures 60 and 61 were acquired when a volunteer inhaled xenon gas and retained it for about ten seconds (Ref. 57). Using a dual energy subtraction technique, it was possible to see proximal airways including the larynx and trachea. Extraneous soft tissue shadows were subtracted leaving only the xenon filled portion of the airways and adjacent bone. Bone subtraction can also be done leaving virtually no shadows but those of the xenon containing airways. The larynx when visualized in this way is seen with images that compare favorably with those obtained using laryngographic contrast agents. The use of xenon presents a problem in that it can be accompanied by a mild anesthetic effect. This may be overcome by a gas such as krypton which is free of any secondary effects.

Another example of energy subtraction was in imaging of the gall bladder as shown in Figure 62. The contrast media was an iodine compound taken orally (Ref. 58). Dual energy subtraction produced a distinctly superior image to those obtained directly and a significant advantage over the routine everyday technique. Thirty minutes after the first image was obtained, the volunteer ate a meal and the procedure repeated. The gall bladder had contracted in the interim and the digital subtraction technique produced a superior image. Soft tissue subtraction rather than bone subtraction produced the better enhanced image. The research of Brody and Macovski provide in depth studies of energy subtraction for other procedures in diagnostic radiology (Refs. 59,60).

Mistretta pioneered modern research in energy subtraction. He was, in particular, the first to demonstrate the potential use for a photoelectronic imaging device to manage the steps for acquiring images from exposures taken at different x-ray tube energies. That accomplishment, however, had wider impact in that it was the first demonstration to the community of researchers in medical imaging that a photoelectronic device could capture and transmit medical images in a remarkably efficient manner.

(3) Image Processing

The systems that have evolved for angiography began with simple subtraction, combined with windowing and image enhancement. They featured also the ability to modify data using logarithmic, square root and linear transfer functions. In addition, there was the ability to discriminate against motion artifacts by shifting one image relative to another by a fraction of a pixel at a time, rotating images and rubber sheeting (Ref. 61). Various types of filtration have been included to provide size and edge enhancement. Videodensitometry is another feature that is available to the user. Clearly, the obvious easier tasks have been incorporated in manufactured systems. There remains the carrying out of subtle research dealing with subjects such as texture and growth patterns associated with specific disease and automated keys to assist the radiologist in improving his performance.

The early history of image processing for medicine covers a variety of sources and topics. Videodensitometry dates back to the early 1960's and is exemplified by the research of Heintzen, Silverman, Wood and others (Refs. 62-79). Most of this work was directed toward studies of the heart and associated circulatory structure. An activity to be singled out for its relationship to the growth of subtraction angiography evolved from research at the University of Hamburg (Refs. 80-83). It appears to have grown from studies of scintigraphy undertaken in the early 1970's and moved in the direction of angiography. An interactive system was reported by 1977, for the clinical application of angiodensitometry. They demonstrated a technique called "computer angiography" which made use of a photoelectronic digital system by 1978. The thrust of this work was to make use of temporal information obtained throughout a procedure into a single image in a format identified as "functional" imaging.

Image processing of the conventional radiograph is also being explored for improvements in diagnosis. An example of a simple procedure is shown in Figure 63, which is a radiograph of an adult chest. The radiograph was digitized and fed to an interactive display. The contrast was windowed and fed through the full range of density in the radiograph as exemplified by

Figure 64. At one level, a calcification was revealed that was not apparent from prior direct examination. A video densitometric trace provided further information on the calcification in terms of location and relative density profile as shown in Figure 65. The extent to which this type of procedure will be used in the future depends in large part on the emergence of a large area x-ray video sensor with spatial resolution, dynamic range and DQE comparable to that obtained from film-screen performance. The technology is available for digitizing and processing conventional film radiographs. However, there has not been any trend toward movement in this direction. This probably can be attributed to several factors including cost, time and the lack of any definitive study that reveals significant improvement in diagnostic performance. Ongoing psychophysical research programs are clearly making headway, and one can be optimistic that their findings will be available soon (Ref. 84).

MECHANISM	DEVICES	APPLICATIONS	NOTES
E-M RADIATION			
Gammas Rays	Gamma Cameras	Nuclear Medicine	
	Rectilinear Cameras	" "	
	Coded Apertures	" "	
	Emission Computed Tomography	" "	
X-Rays	Screen Film	Diagnostic Radiology	
	CT Scanners	" "	
	Photoelectronic Imaging Devices (Intensifiers, Video Tubes, Solid State Arrays)	" "	Fluoroscopy Intravenous Angiography Scatter Free Imaging Linear Tomography Digital Radiology
UV Visible Infrared	Photoelectronic Imaging Devices (Intensifiers, Video Tubes, Solid State Arrays, Laser Scanners, Photodetectors)	Medicine Surgery Ophthalmology Pathology	Endoscopy Microscopy Thermography Diaphanog. Fundoscopy
Radio	Nuclear Magnetic Resonance	Radiology	
ULTRASOUND	A Mode B Mode M Mode Doppler	Abdominal Ultrasonography Obstetrics Echocardiography Neurology Ophthalmology Vascular Disease Fetal Heart Tone	B Mode Static Real Time Phased Array Linear Array
PARTICLES Protons Heavy Ions	Accelerators "	Therapy with Imaging	Sensors Films Plastics
Neutrons	Reactors	Imaging	Sensors Screens w/film & video

TABLE I

Sources of Radiation for Imaging, Corresponding Imaging Devices & Their Applications

PERFORMANCE DATA OF CAMERA TUBES

Tube Type	Super Orthicon	Vidicon Resistron	Plumbicon Leddicon	Si-Vidicon	Chalnicon Pasecon	Saticon	Newvicon
Sensor material	S 10.S 20	Sb_2S_3	PbO(PbO-S)	Si-Diodes	CdSe	SeAsTe	ZnSe-ZnCdTe
Diameter (mm)	75, 115	38,25,18	30,25,18	25,18	25,18	25,18	25,18
Raster Size (mm x mm)	24x32	9.6x12.8	9.6x12.8	9.6x12.8	9.6x12.8	9.6x12.8	9.6x12.8
Luminous Flux for White Light (mlm)	0.5/0.2	0.5...200	0.5	0.2	0.08	≈0.2	≈0.1
Responsivity in visible Portion of Spectrum (µA/lm)	50.150	variable	400	900	1500	350	1200
Signal Current (µA)	10	0.2	0.2	0.2	0.2	0.2	0.2
Dark Current (nA)	-----	15...20	<1	7...15	<1	<1	6
Lag, Signal after 60 ms (%)	2	15...25	1...2	5...8	10	2...3	10
Linearity (γ)	1...0.6	0.85...0.65	1	1	1	1	1
Modulation depth at 5 MHz (%)	60...80	50...70	45...60	40	50...70	60	55
Signal-to-Noise Ratio (dB)							
Without rectification	35...40	45	45...47	45	>45	>45	>45
With rectification	42...48	57	54...56	55	55	55	55
Spurious signals							
In Black (%)	±3	±3	±1	±3	±1	±1	±1
In White (%)	±10	±10	±10	±10	±10	±10	±10

TABLE II

Performance Data of several video camera tubes
(From Reference 23)

	14 in. 36 cm	10 in. 25 cm	6 in. 16 cm
X-ray absorption of input screen at 7 mm HVL-Al (%)	65	65	65
Limiting resolution (lp/mm)	3.6	4.0	5.0
Conversion factor (cd/m^2/mr s)	150	75	30
Contrast ratio (10% area)	18:1	25:1	36:1
Pincushion distortion (%)	6	3	0
Output diameter (mm)	35	35	35

TABLE III

Performance From a Tri-Mode "Philips 14" X-Ray Image Intensifier

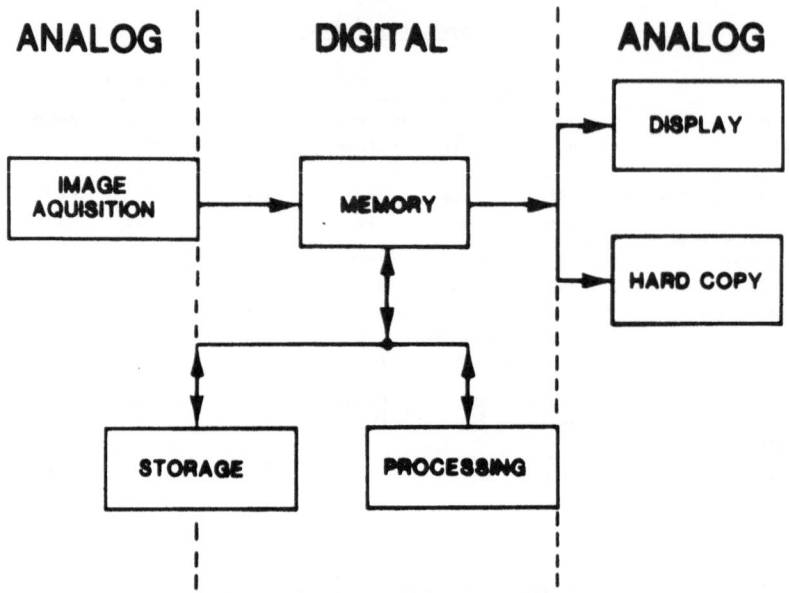

Fig.1 Photoelectronic - Imaging System
(from ref. 1)

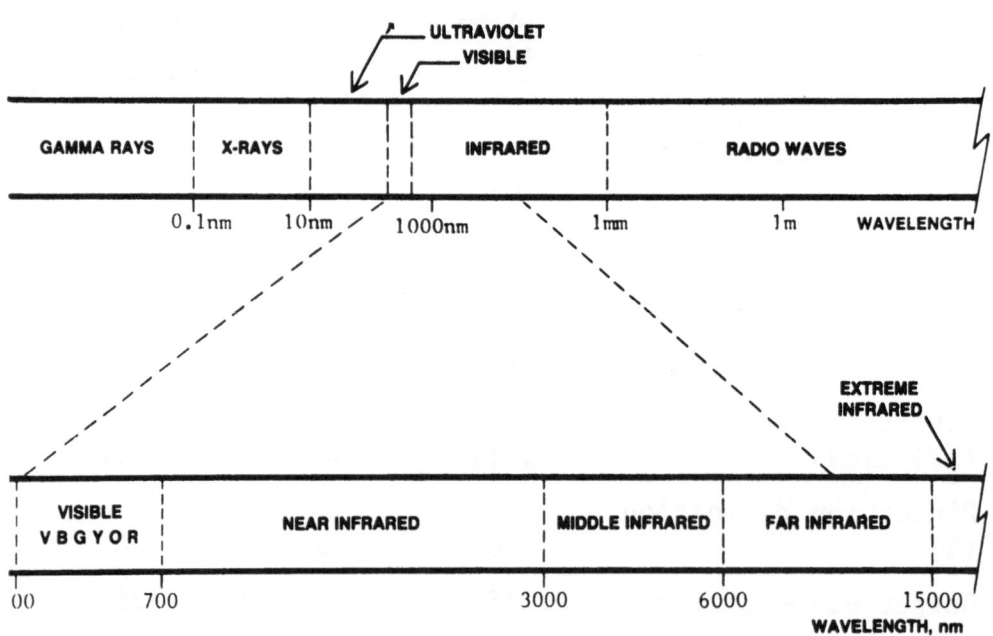

Fig.2 The Electromagnetic Radiation Spectrum

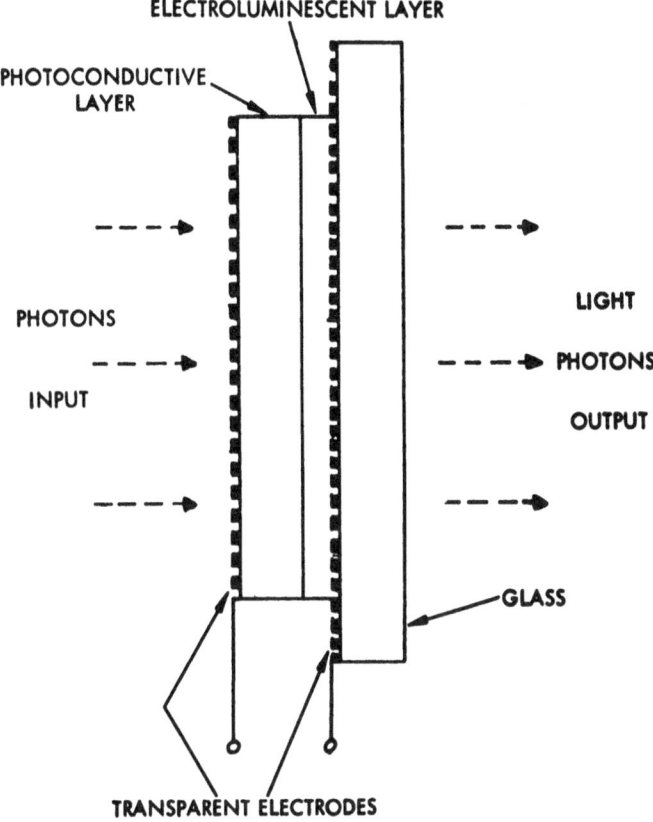

ELECTROLUMINESCENT LAYER

PHOTOCONDUCTIVE
LAYER

PHOTONS

INPUT

LIGHT

PHOTONS

OUTPUT

GLASS

TRANSPARENT ELECTRODES

Fig.3 The Photoconductive - Electroluminescent Light Amplifier

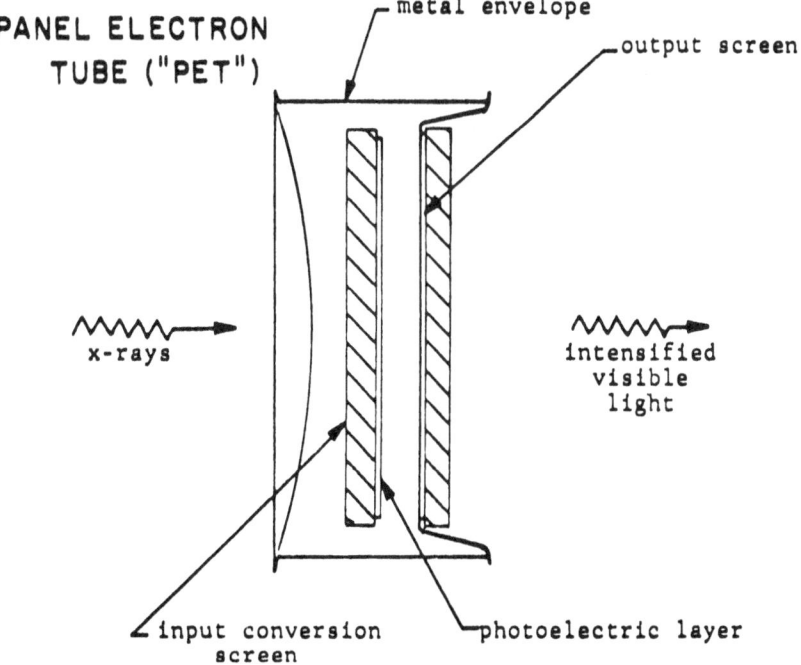

PANEL ELECTRON
TUBE ("PET")

metal envelope

output screen

x-rays

intensified
visible
light

input conversion
screen

photoelectric layer

Fig.4 The Proximity Focussed Intensifier (from ref.21)

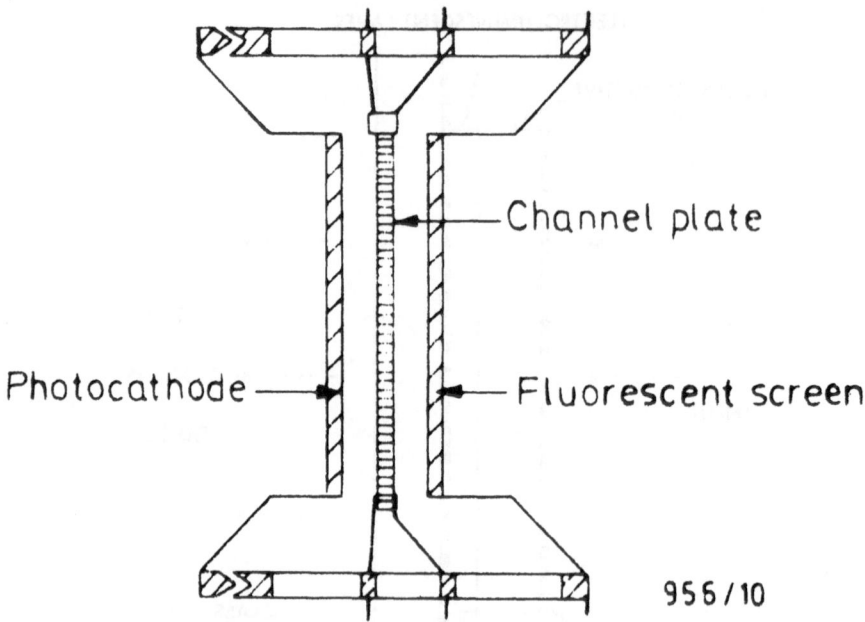

Fig.5 The Channel Multiplier Intensifier
(from Amperex Tech.Info.033 by R.T. Holmshaw)

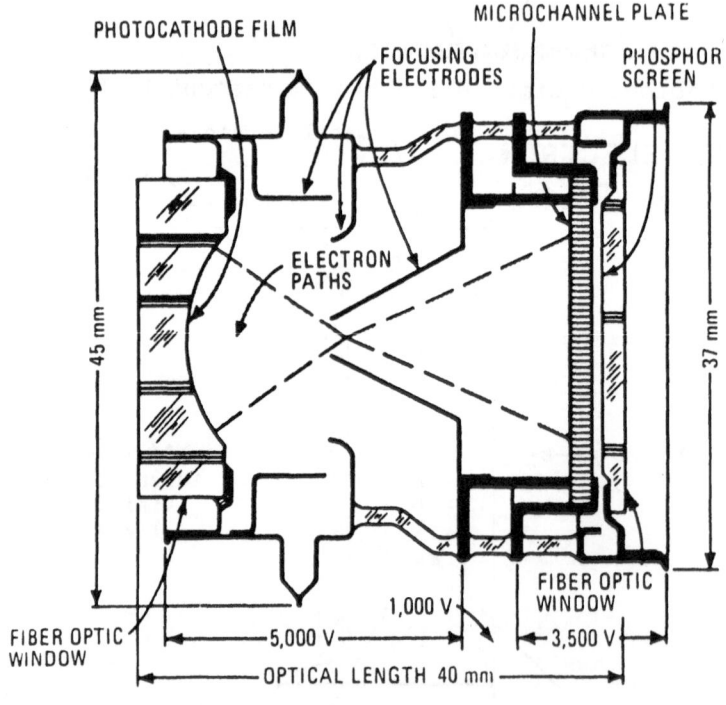

Fig.6 The Low Light Level Inverter Channel Plate Intensifer
(see ref.6 and 15)

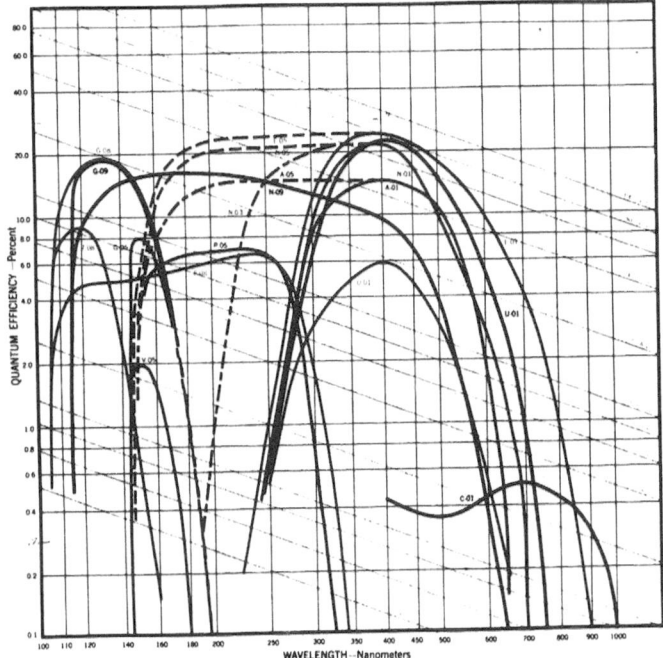

Fig.7 A Typical Variety of Photoelectron Emitter Spectral
Responses (from EMR Brochure on Multiplier Photocathodes)

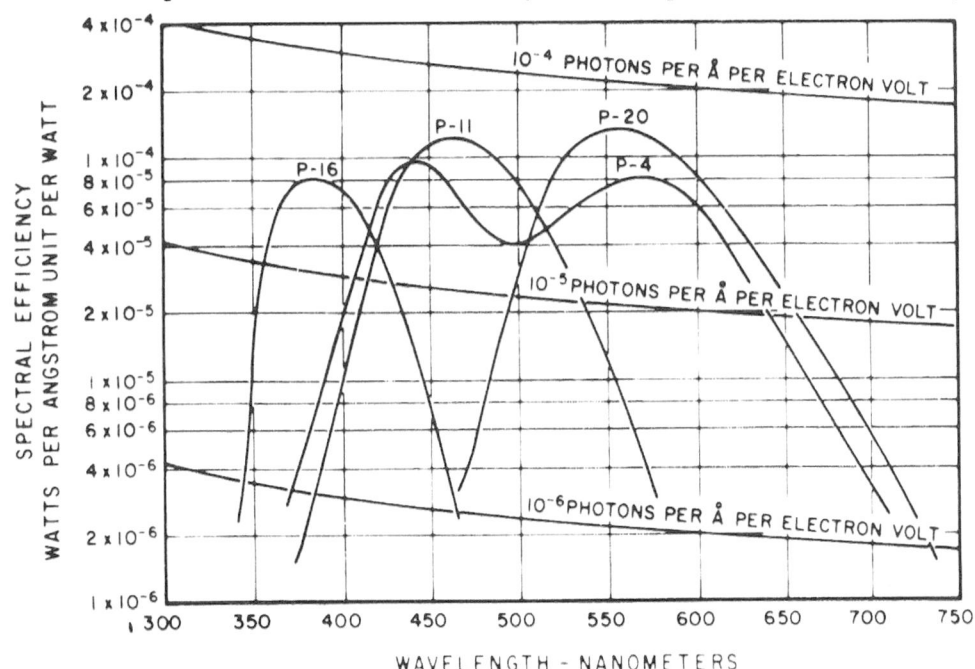

Fig.8 The Efficiency of Several Cathodoluminescent Phosphors
(from ITT Brochure on Special Purpose Photosensitive
Devices)

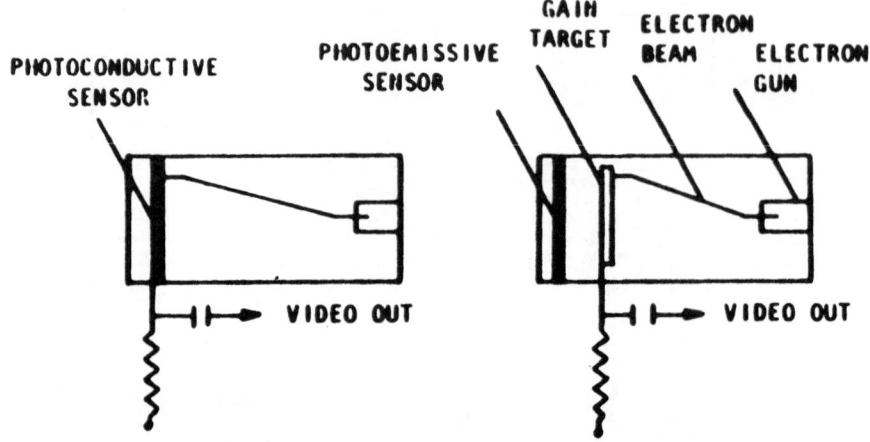

Fig.9 Video Tube with No-Gain: The Vidicon Type TV Tubes
      (see ref. 6 and 15)

Fig.10 Video Tube with Gain: The TV Tubes Which Contain a Target
       To Provide Gain, e.g. the Image Isocon, SEC and SIT Tubes
       (see ref. 6 and 15)

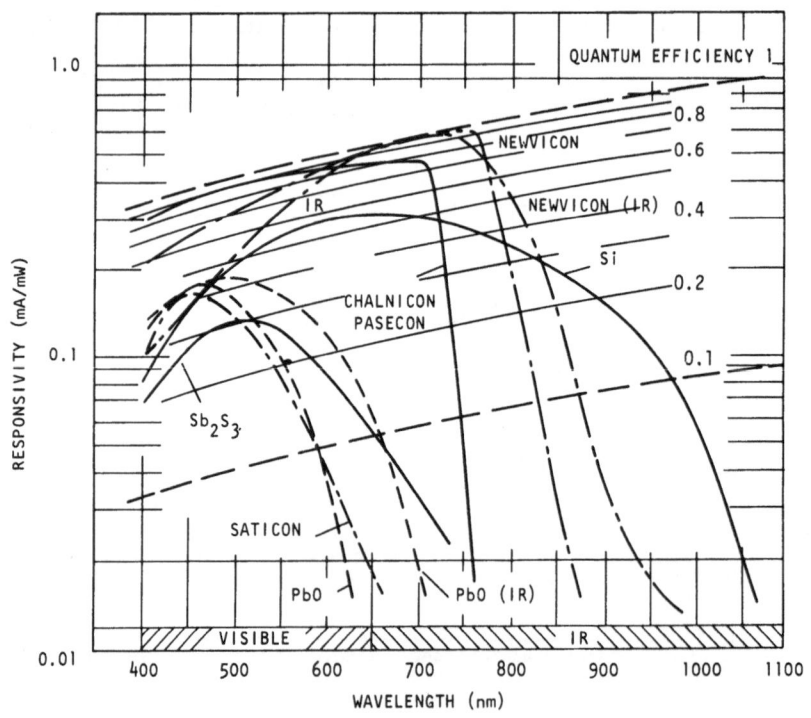

Fig.11 Spectral Response of Vidicon Type Tubes
       (from ref.23)

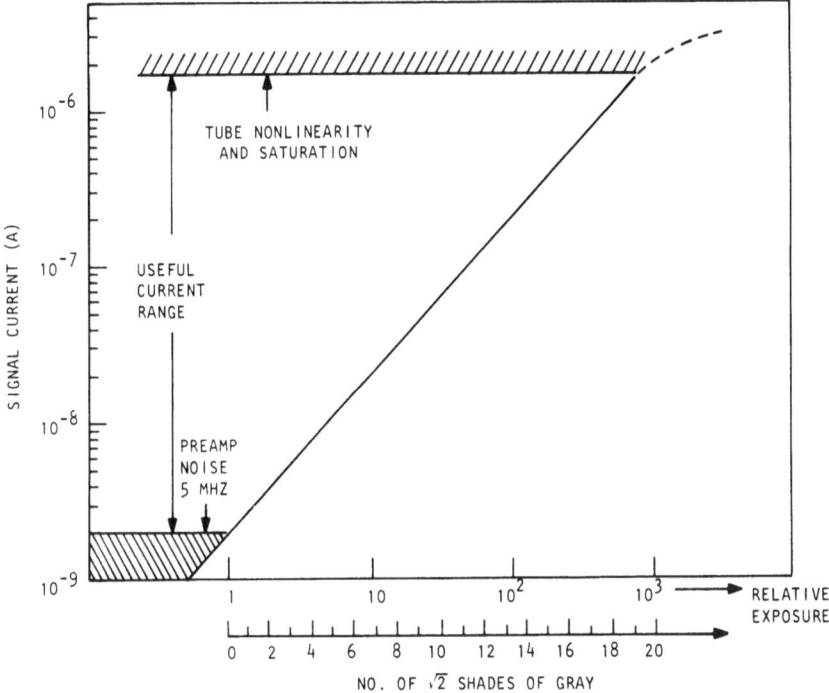

ig.12 Signal Transfer Curve for a Linear Video Tube -
Plumbicon 45XQ (see Roehrig in ref. 1)

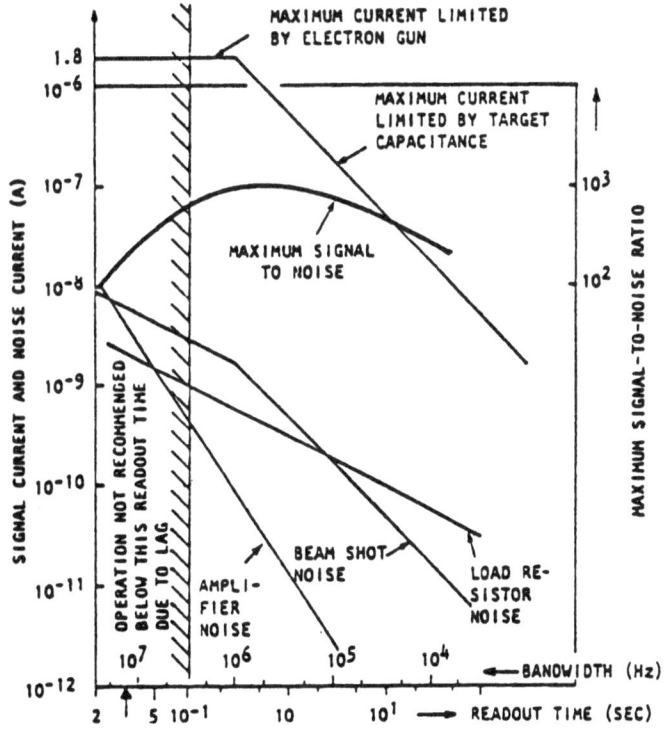

Fig.13 Dependence of Signal and Noise Currents on Scan Rates and
Band-width in a Video Camera

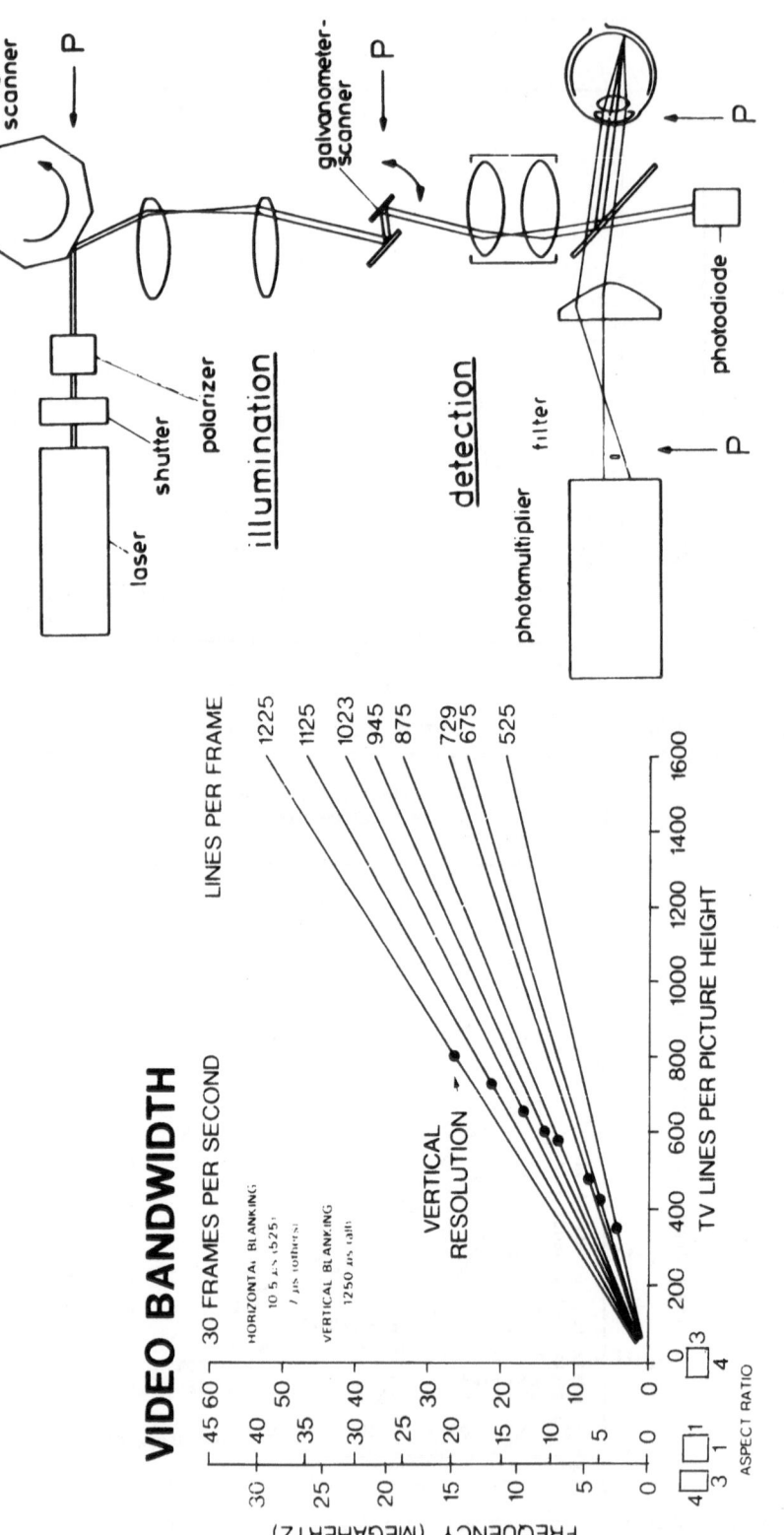

Fig. 15 A Laser Beam Raster Camera (from a Brochure of Heidelberg Instruments)

Fig. 14 The Interrelationship between Video Bandwith Raster Line Number, Aspect Ratio and Spatial Resolution

ILLUMINATING FIBERS
OPTICS
CONE OF ILLUMINATION
SCOPE
POTTED & POLISHED FIBERS
REFLECTOR
LAMP
LIGHT GUIDE

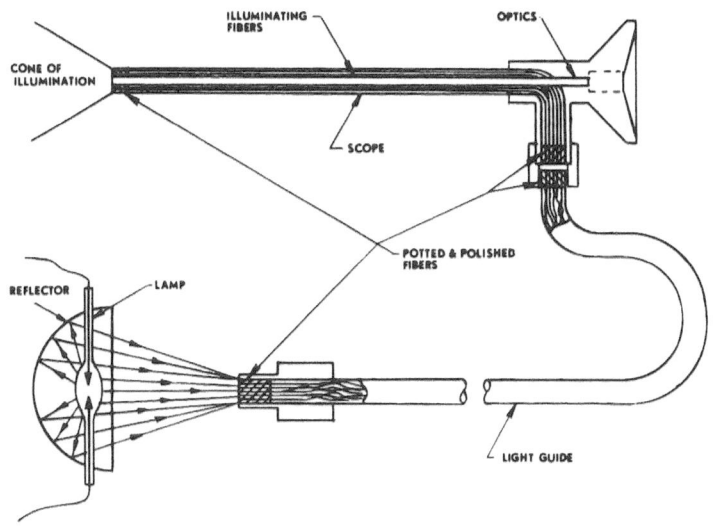

From W. S. Andrus, Photonics Spectra
October 1982

Fig.16 Optical Structure of an Endoscope

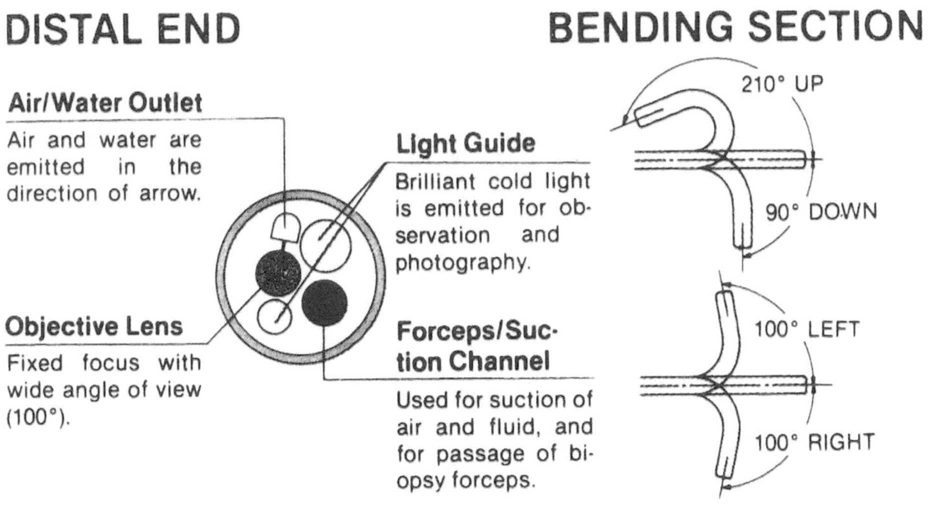

DISTAL END
BENDING SECTION
Air/Water Outlet
Air and water are emitted in the direction of arrow.
Light Guide
Brilliant cold light is emitted for observation and photography.
Objective Lens
Fixed focus with wide angle of view (100°).
Forceps/Suction Channel
Used for suction of air and fluid, and for passage of biopsy forceps.
210° UP
90° DOWN
100° LEFT
100° RIGHT

Fig.17 The Distal End and Bending Section of an Endoscope (from a Brochure of the Olympus Optical Co., Tokyo)

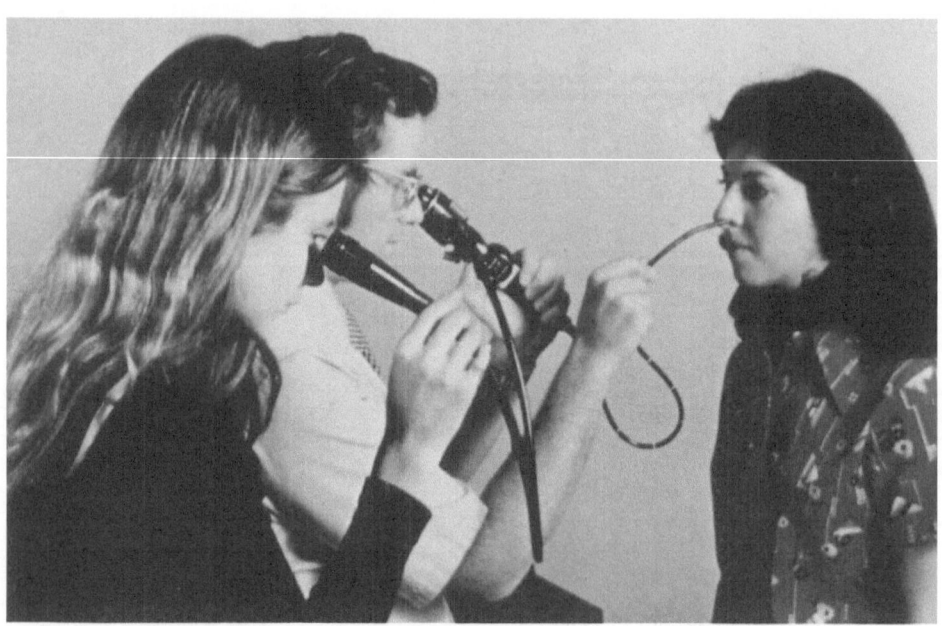

Fig.18 The Endoscope With a Teaching Aide

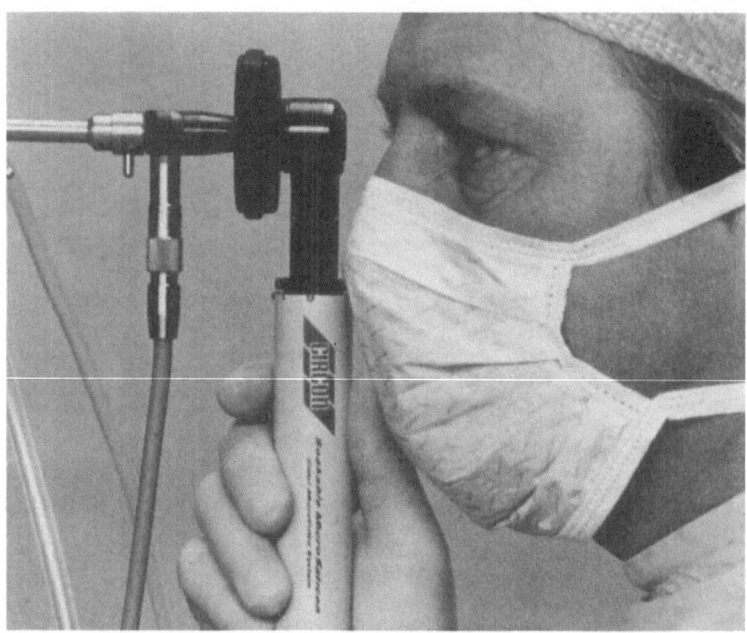

Fig.19 The Endoscope With a Beamsplitter for Coupling a Video Camera
(from the Circon Corp.)

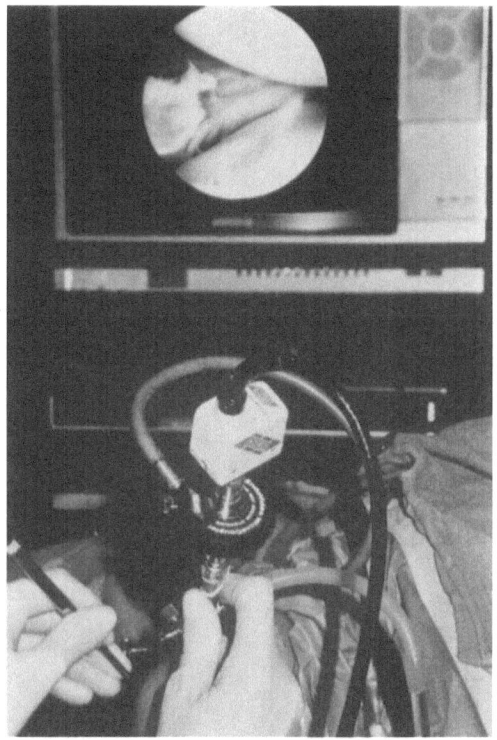

Fig.20 The Endoscope With a Video Camera as Used in Surgery
(from the Circon Corp.)

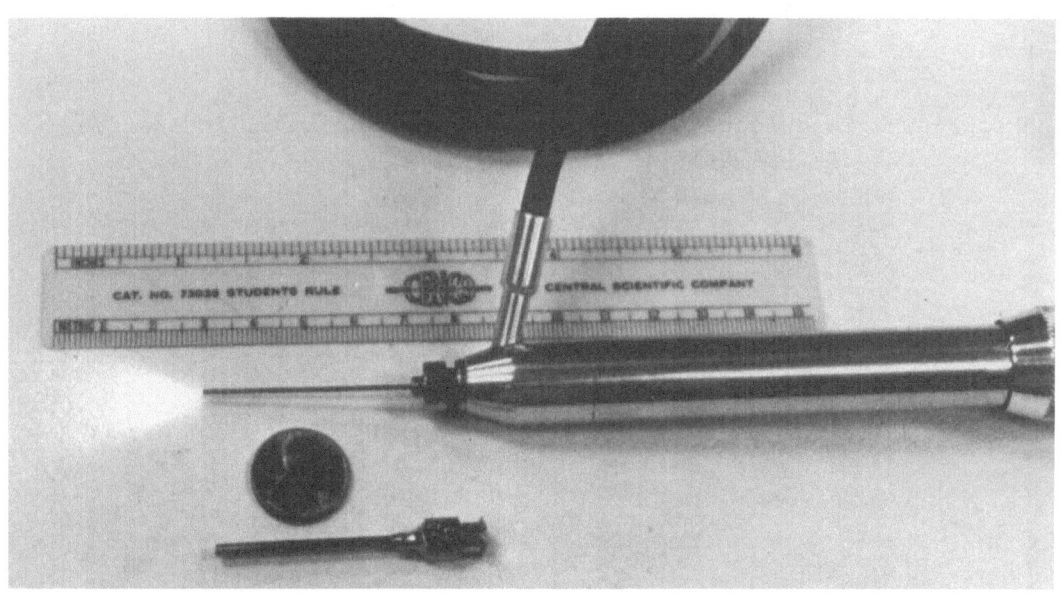

Fig.22 The Hypodermic Fiberoptic Endoscope
(permission of M. Epstein, Ph.D.)

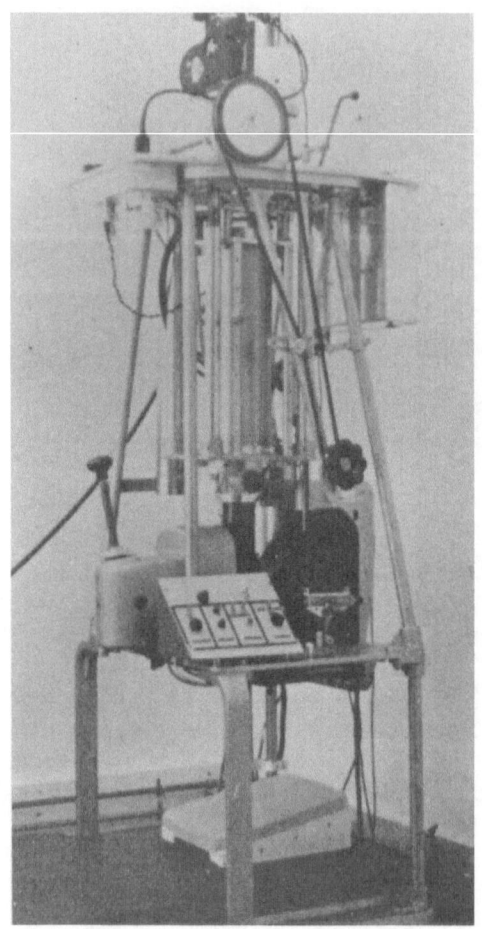

Fig.24 The Funduscope With an Early Low Light Level Intensifier
     Used to Demonstrate Cine-Angiography

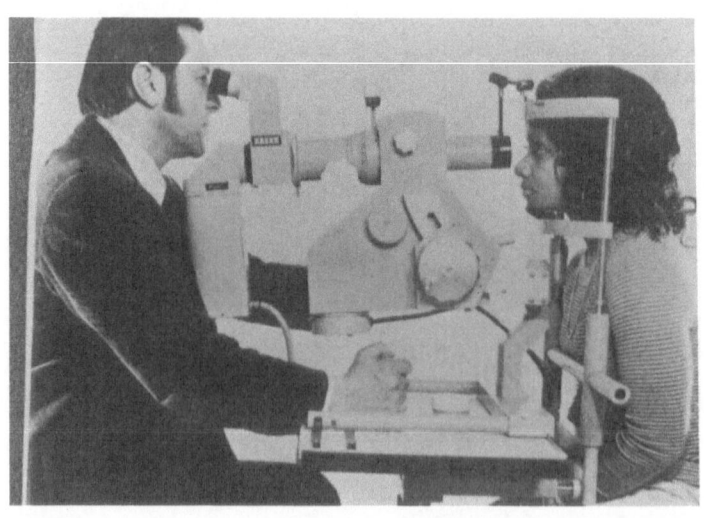

Fig.25 The Modern Funduscope Encorporating a Video Camera

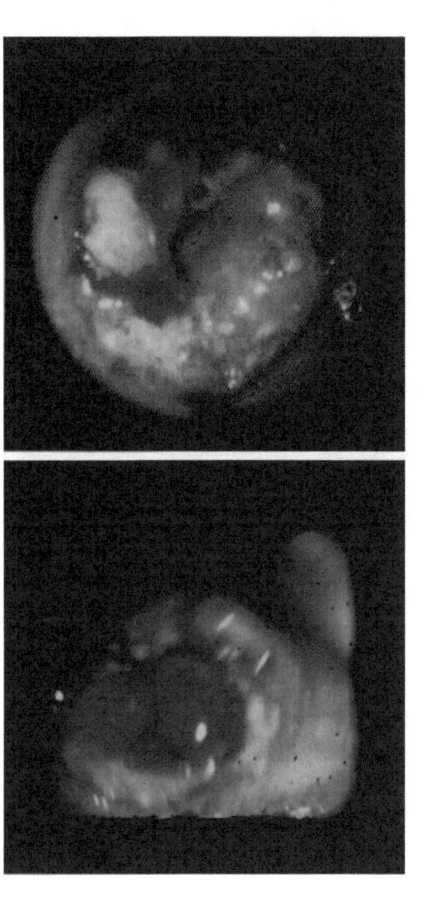

Fig.21 Images Obtained Through an Endoscope(permission of J. Levine, M.D.)

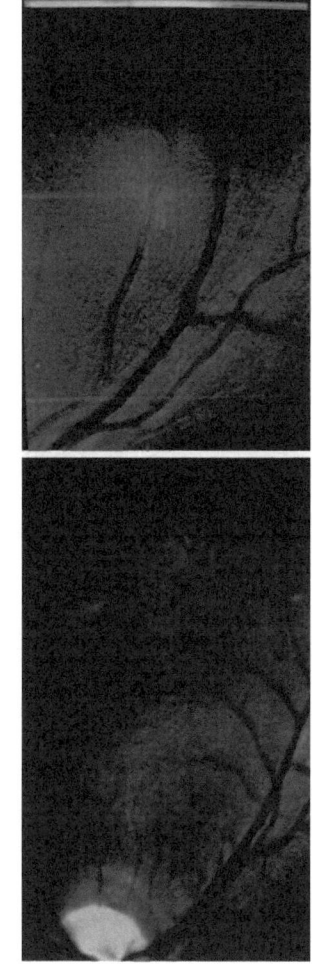

Fig.23 Retinal Images Obtained With a Laser Beam Scanner Camera (permission of B.Katz, M.D.)

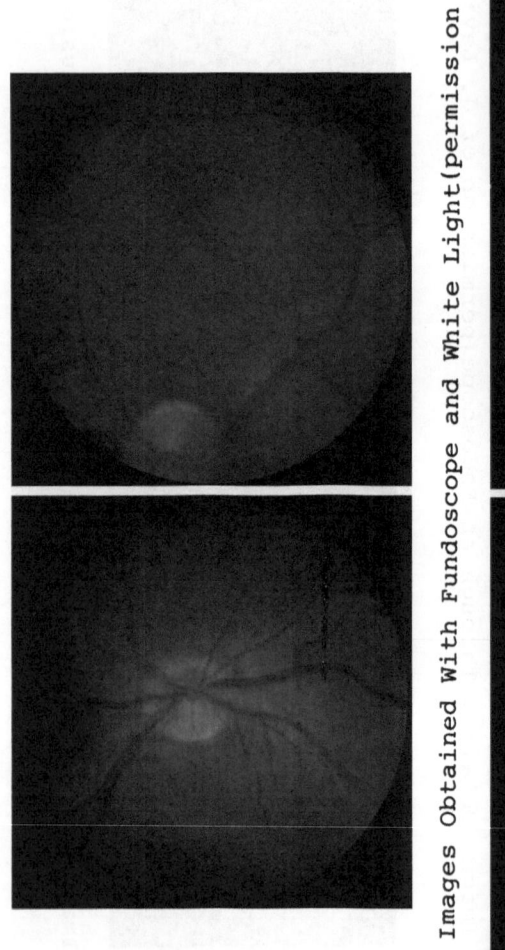

Fig.26 Photographic Images Obtained With Fundoscope and White Light(permission of B. Katz, M.D.)

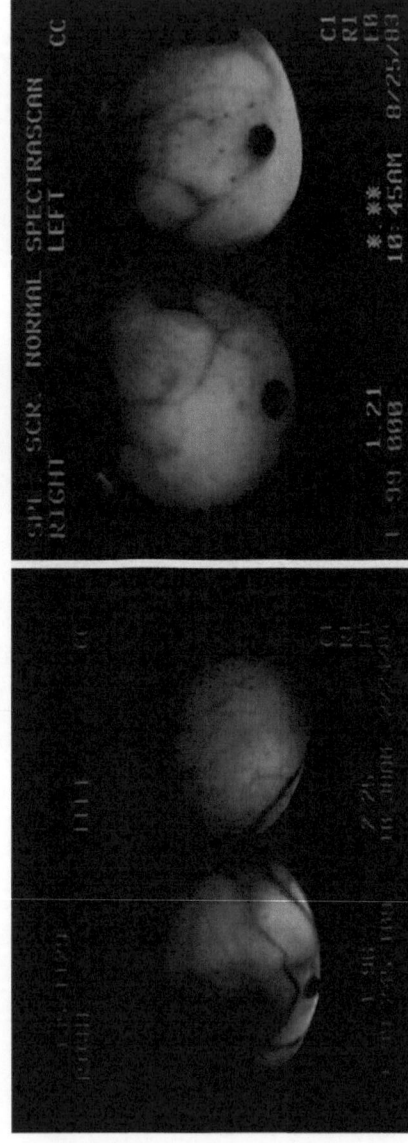

Fig. 30 Diaphanographic images

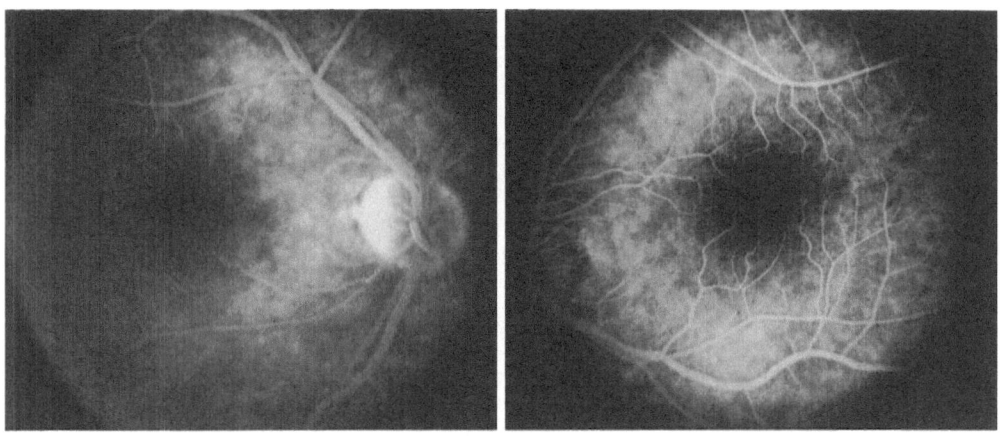

Fig.27 Photographic Image Obtained With Funduscope and Exposure
With Blue Light for Fluoroangiography
(permission of B.Katz, M.D.)

Fig. 28 A Clinical System Used in Diaphanography
(provided by Spectrascan Corp.)

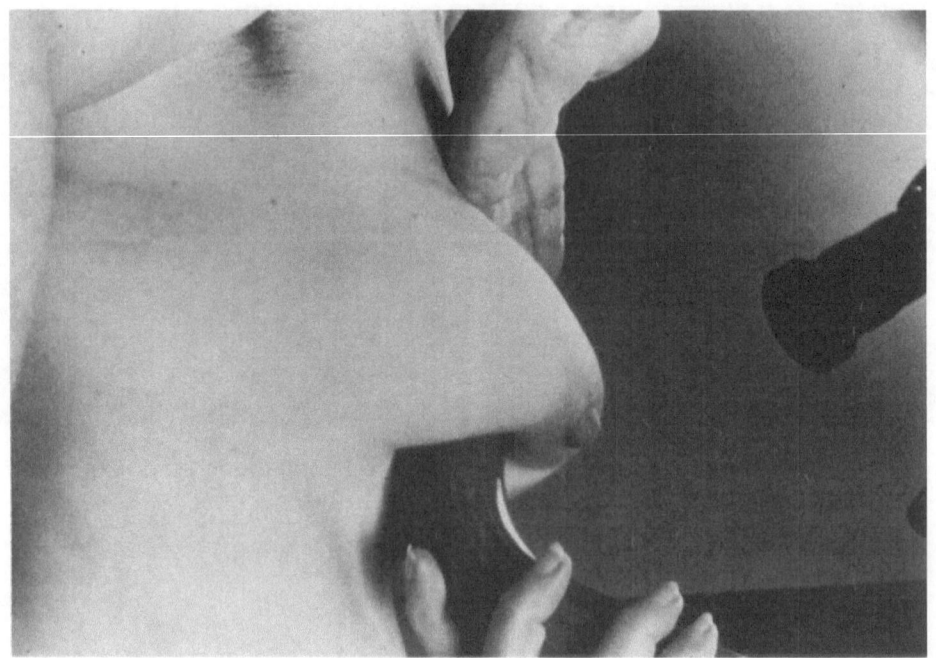

Fig.29 Positioning the Light Probe Against the Breast
(provided by the Spectrascan Corp.)

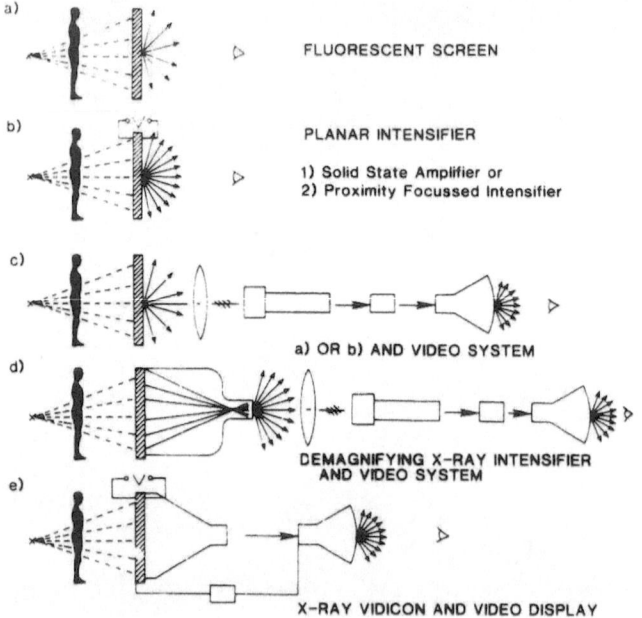

Fig.31 A Variety of Image Acquisition Devices and Systems for
Diagnostic Radiology
(from ref. 2)

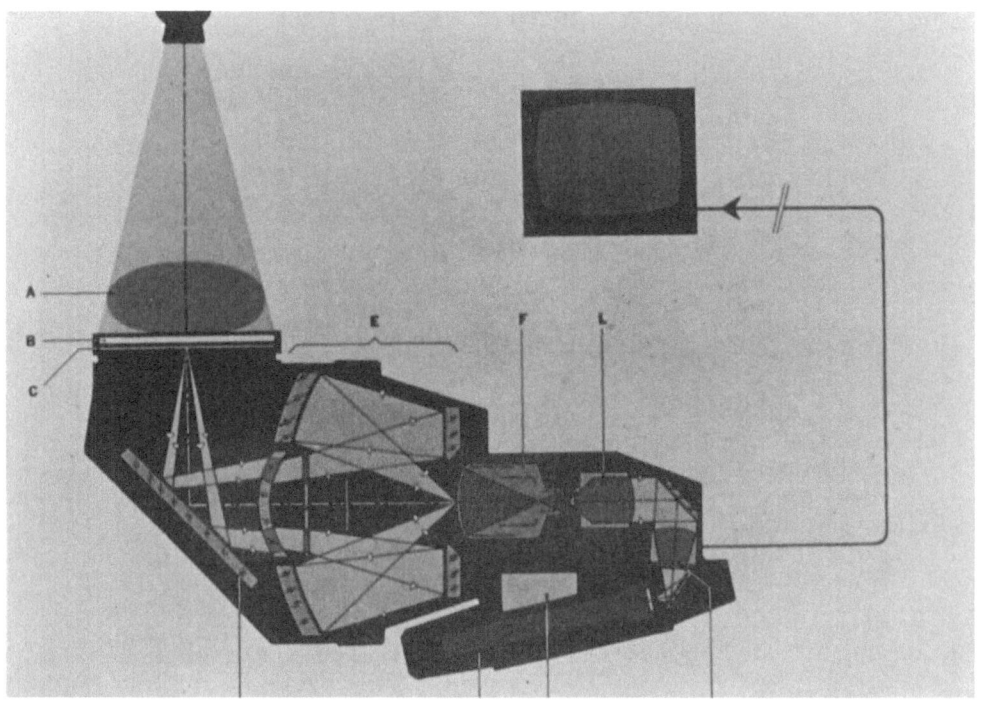

Fig.32 The Delcalix X-Ray Camera
(from DeOude Delft Corp.)

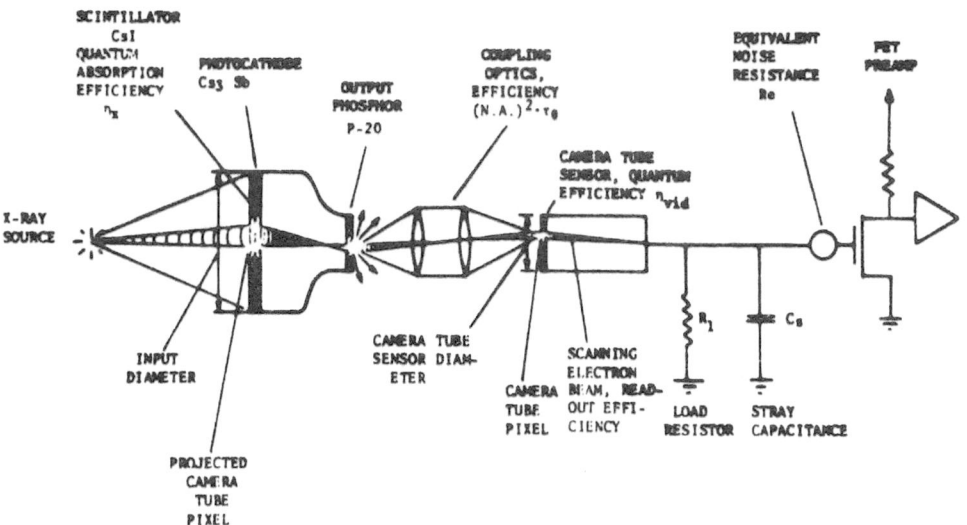

Fig.33 The Critical Components of a Modern Fluoroscopic System
(from Nudelman and Roehrig in ref. 1)

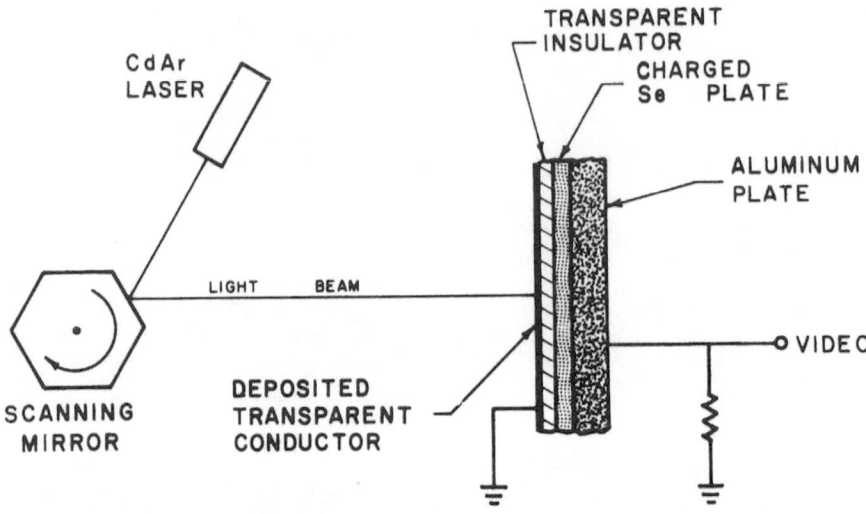

Fig.34
Read-out of an Exposed Selenium Plate With a Raster Scanning
Laser
(from ref. 35)

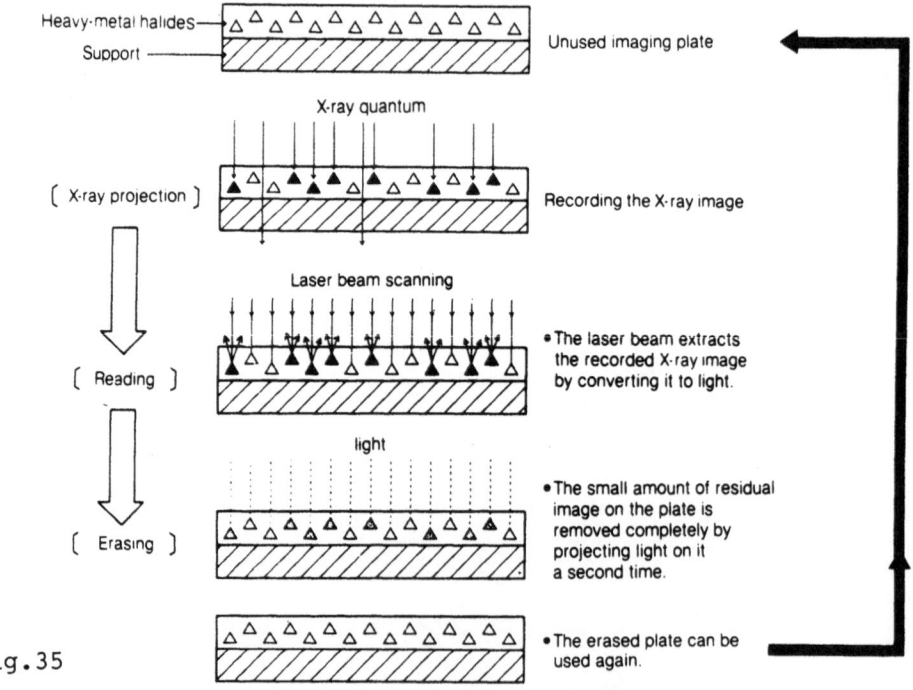

Fig.35

The Fuji Metal-Haloid X-Ray Plate: Photoelectronic Mechanism
for Operation (from ref. 36)

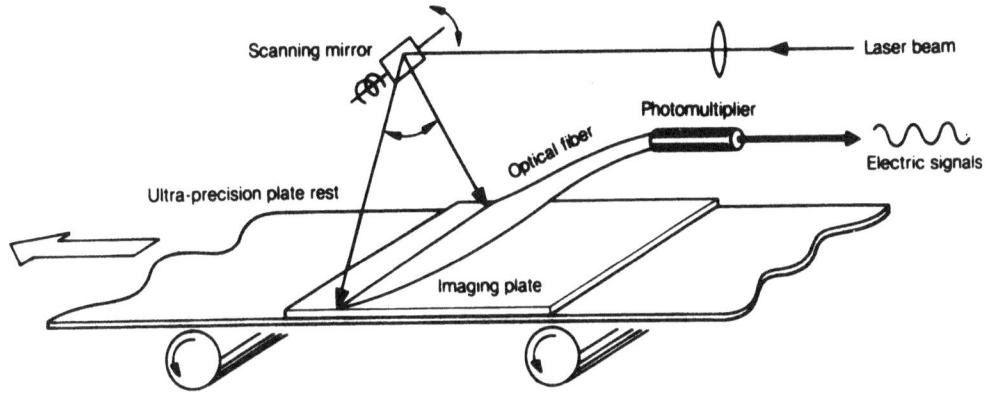

Fig.36

Read/out of the Metal-Haloid Plate With a Laser Line Scanner
(from ref. 36)

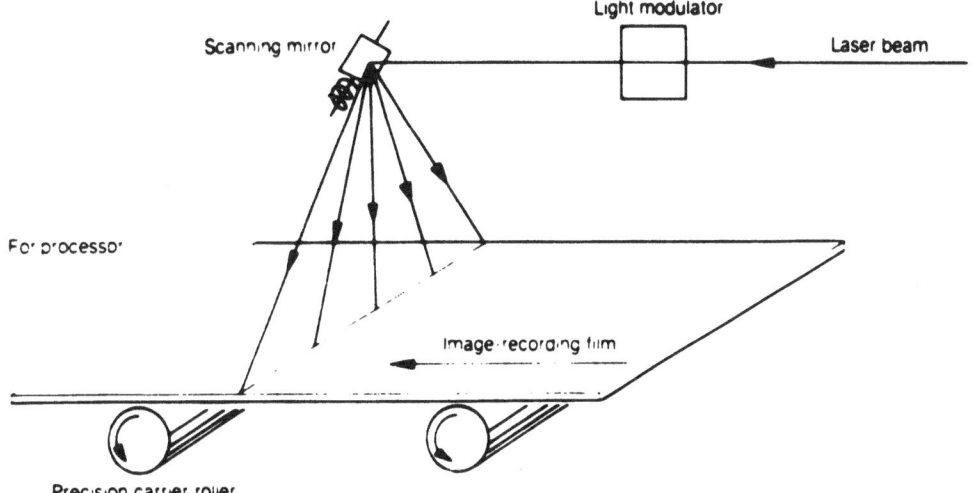

Fig.37

Recording on Film With a Laser Line Scanner From an Image
Stored on the Exposed Metal-Haloid Plate
(from ref. 36)

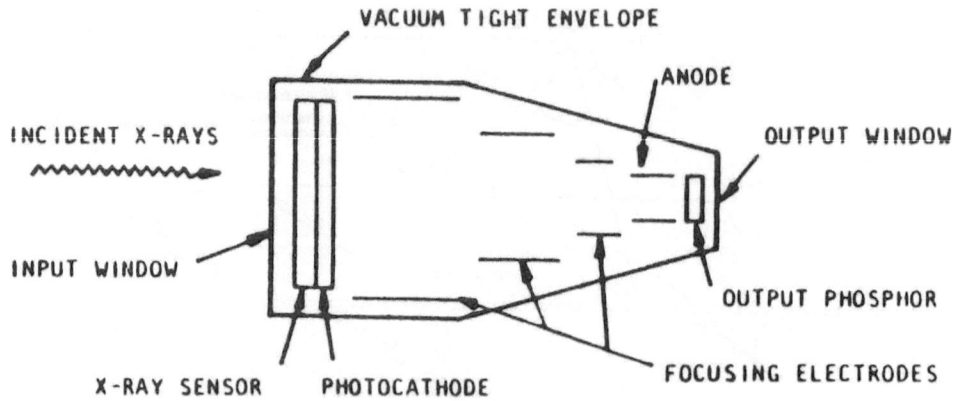

Fig.38 Basic Structure of an X-Ray Image Intensifier

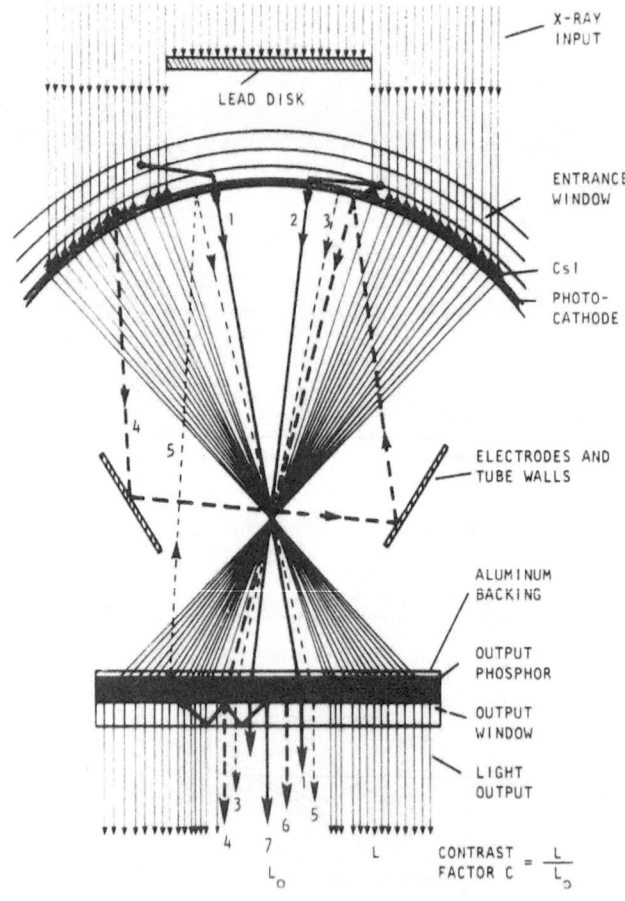

Fig.39
Schematic Illustrating Factors Influencing the Contrast Factor
of an X-Ray Image Intensifier
(permission of H. Roehrig)

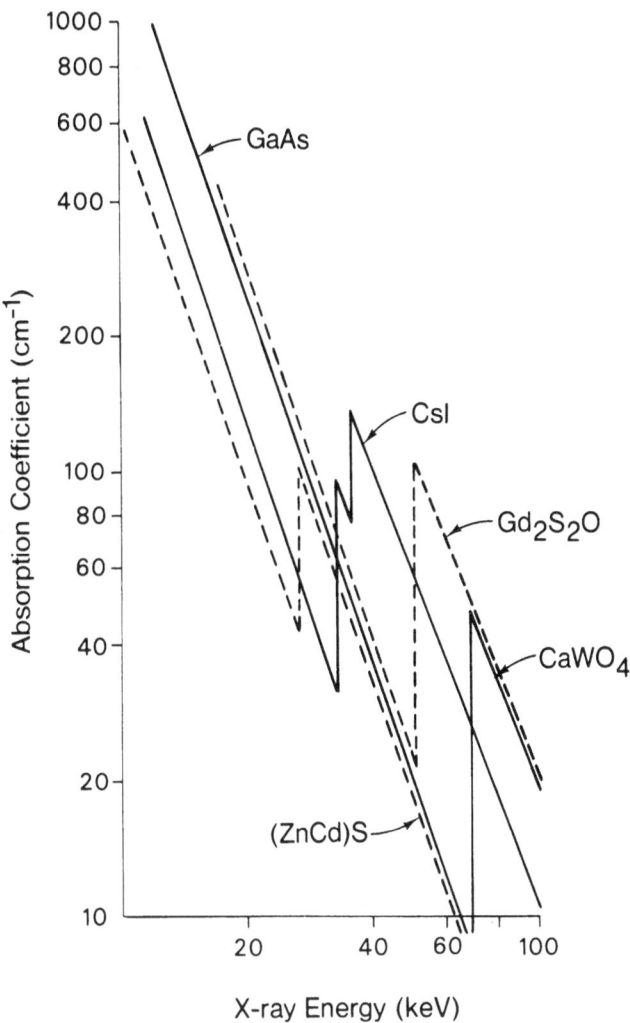

Fig.40 Absorption Characteristics of Several X-Ray Sensors

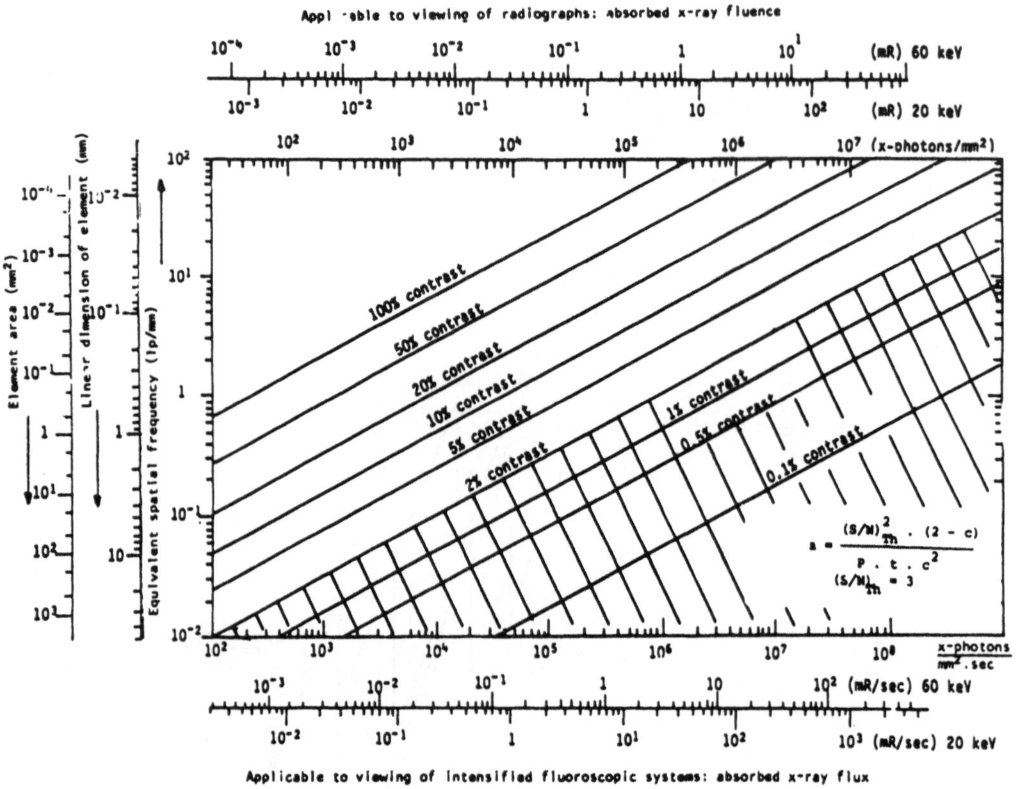

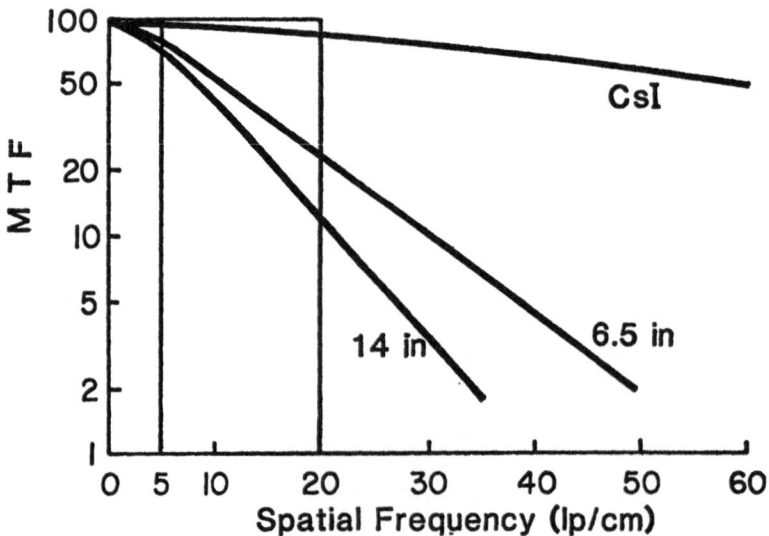

Fig.41 Quantum Limited Spatial Resolution as a Function of
Absorbed Dose and/or Dose Rate for Different Contrasts

Fig.42 The Modulation Transfer Function of an X-Ray Image
Intensifier

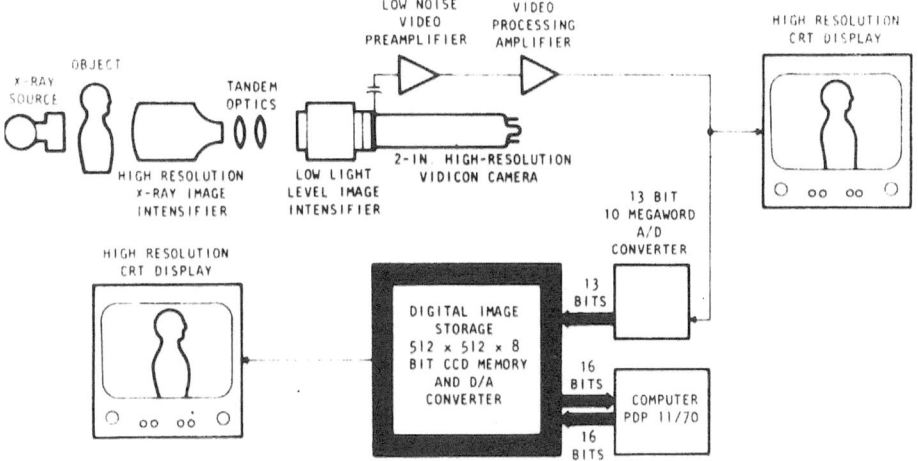

Fig.43 Photoelectronic-Digital Data Image Acquisition System
(from ref. 41)

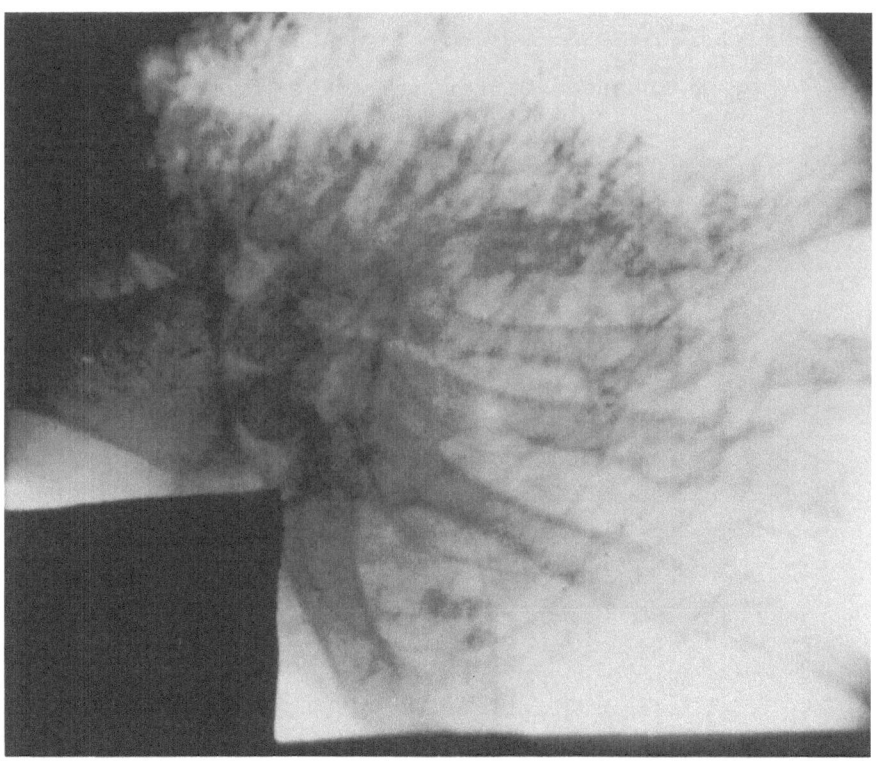

Fig.44 Subtraction Phantom With Contrast Media
(from ref.41)

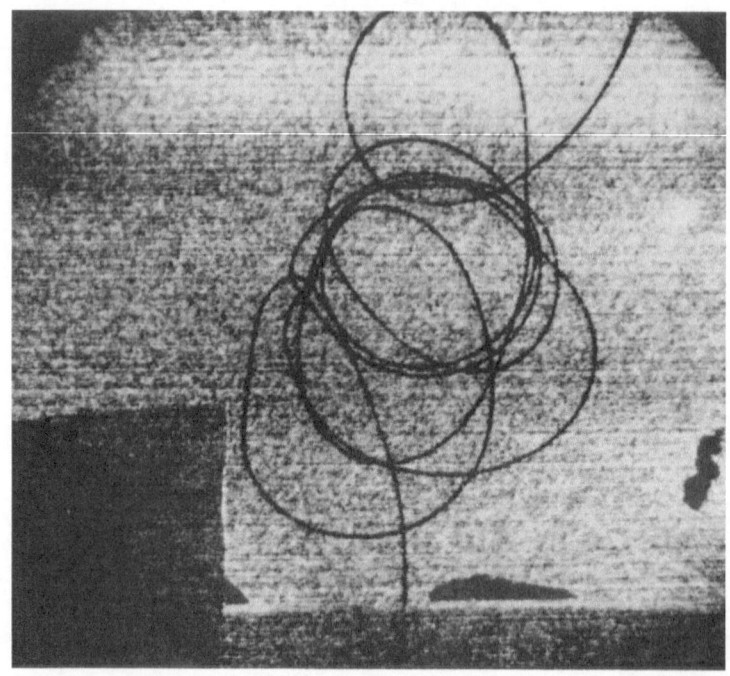

Fig.45 Enhanced Difference Image (from ref.41)

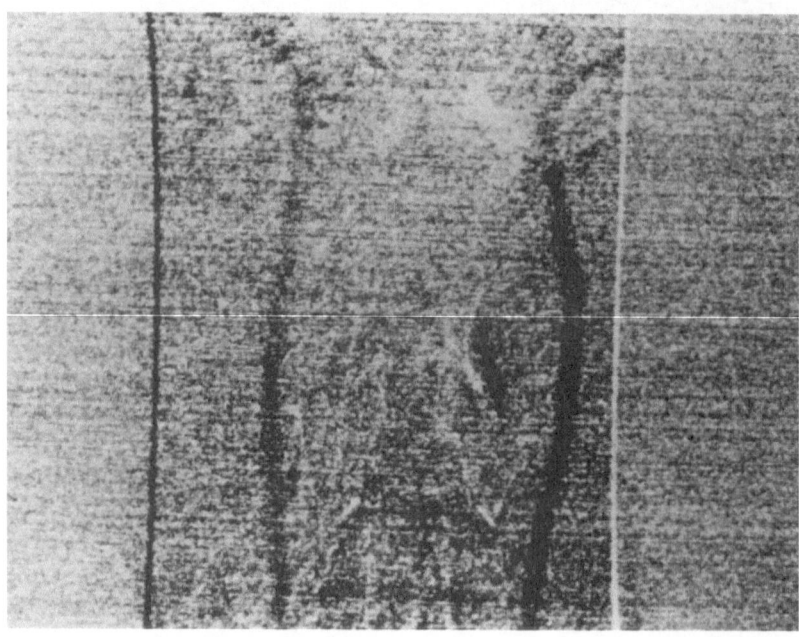

Fig.46 First In-vivo Intravenous Subtraction Image of a Dog's
Corotid Arteries (from Ref.41)

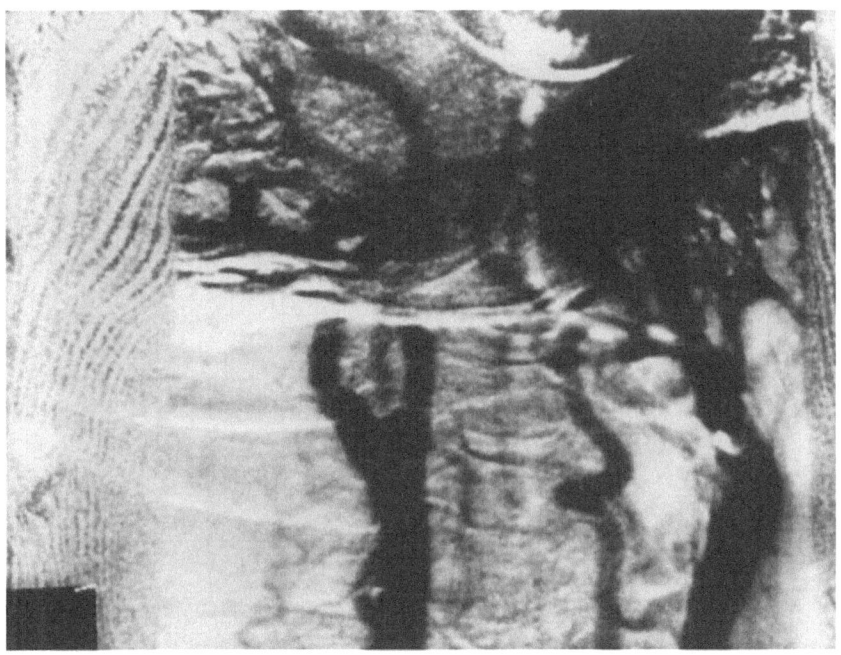

Fig.47 First In-vivo Intravenous Subtraction Images of a Human's
Corotid Arteries(from ref.44-46)

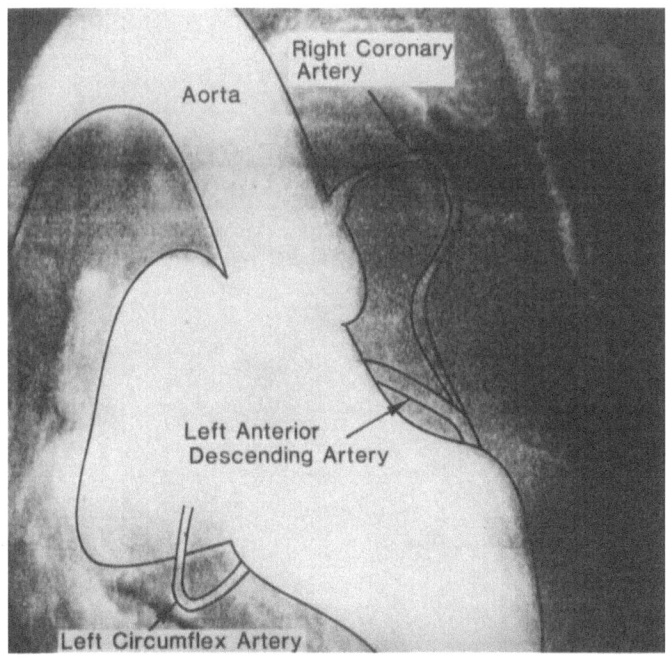

Fig.48 Intravenous Angiographic Image of a Dog's Heart Showing
Several Coronary Arteries (from Ref.39,40)

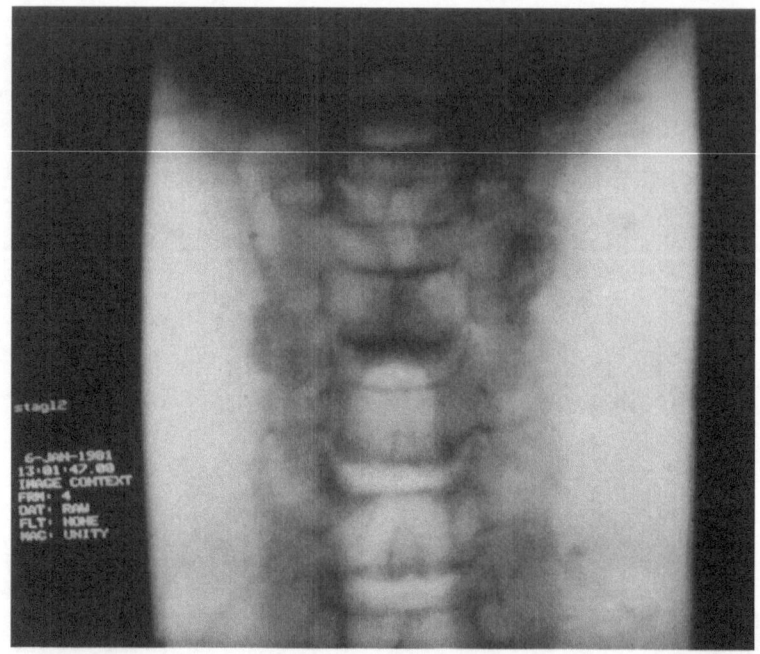

Fig.49 Radiograph of Neck Prior To Arrival of Contrast Media
(permission of T.W. Ovitt, M.D.)

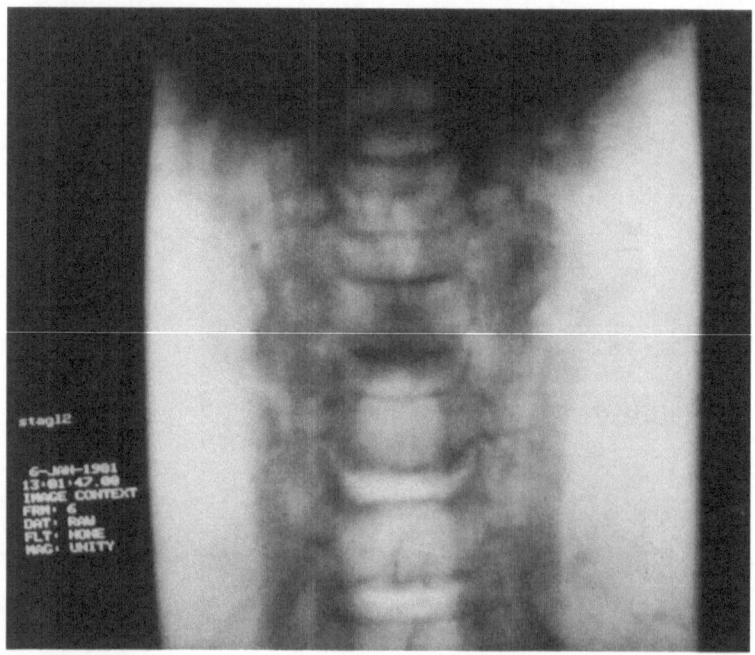

Fig.50 Radiograph of Neck After Arrival of Contrast Media
(permission of T.W. Ovitt, M.D.)

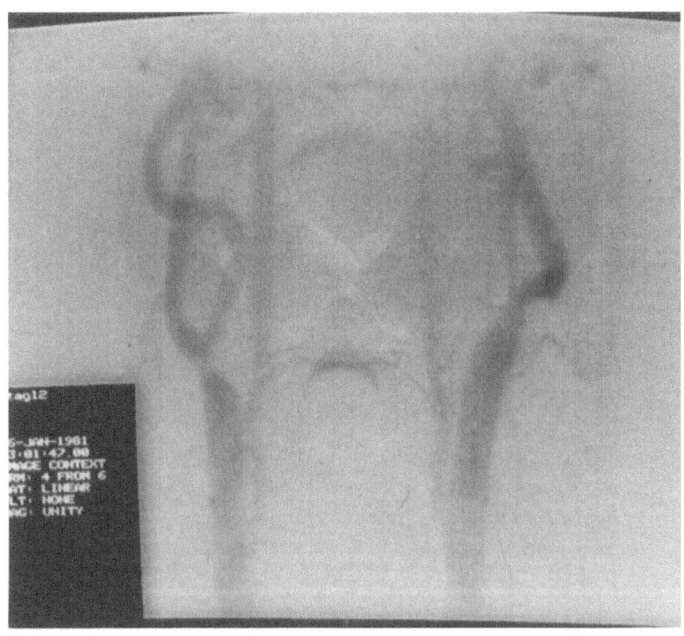

Fig.51 The Low Contrast Difference Image
(permission of T.W. Ovitt, M.D.)

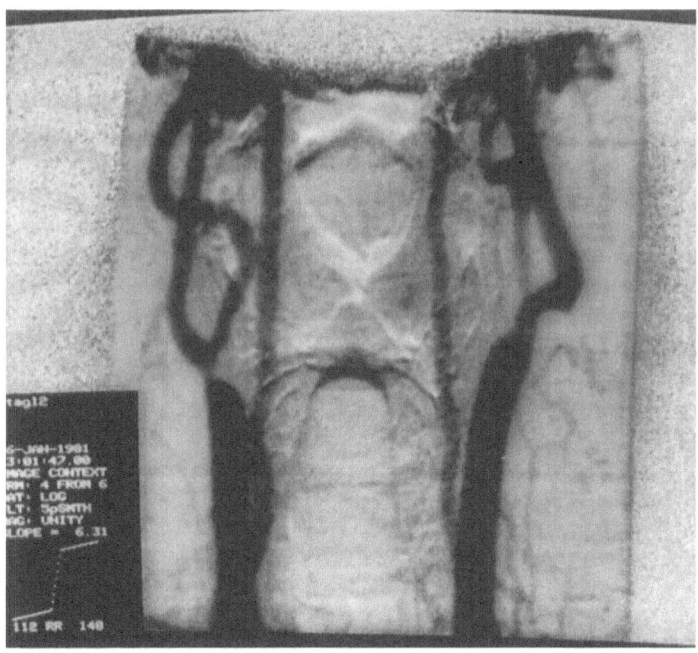

Fig.52 The Enhanced Difference Image
(permission of T.W. Ovitt, M.D.)

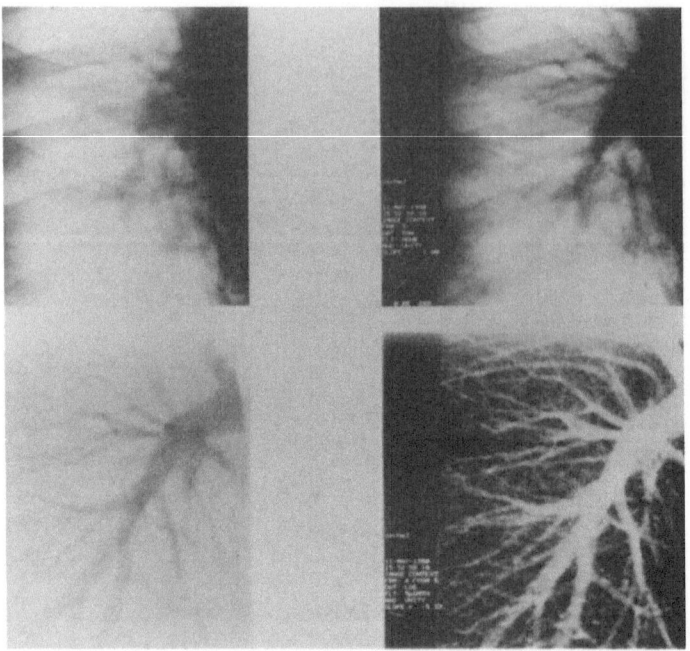

Fig.53 Sequence of Images Obtained in Intravenous Angiography
of the Lungs (with permission of G.D. Pond, M.D.)

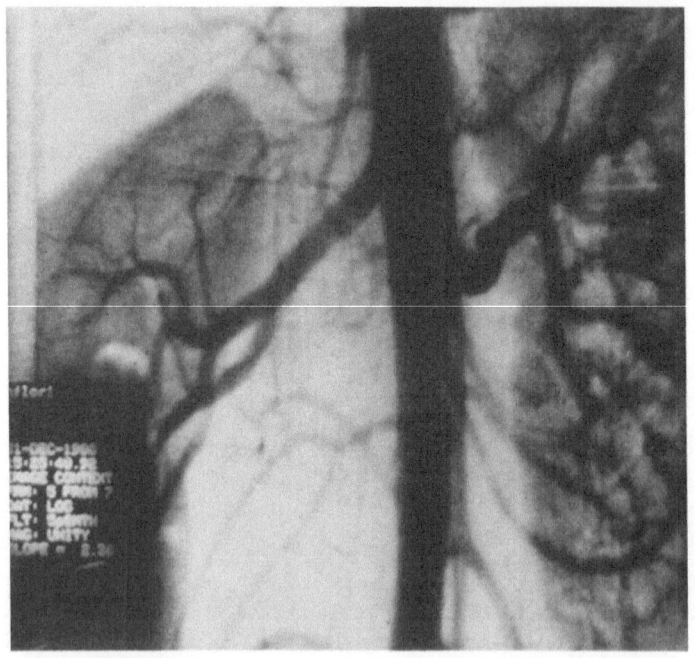

Fig.54 Intravenous Angiogram of the Renal Arteries
(with permission of B.J. Hillman, M.D.)

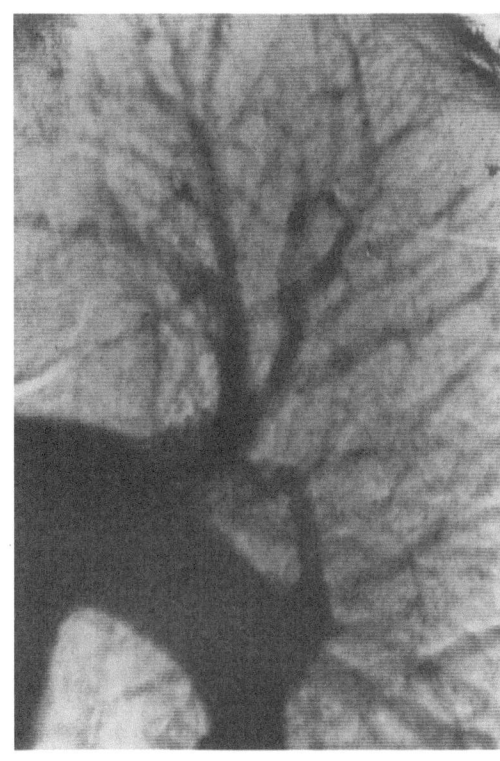

Fig.55 Intravenous Angiogram
       of the Lungs (with
       permission of G.D.
       Pond, M.D.)

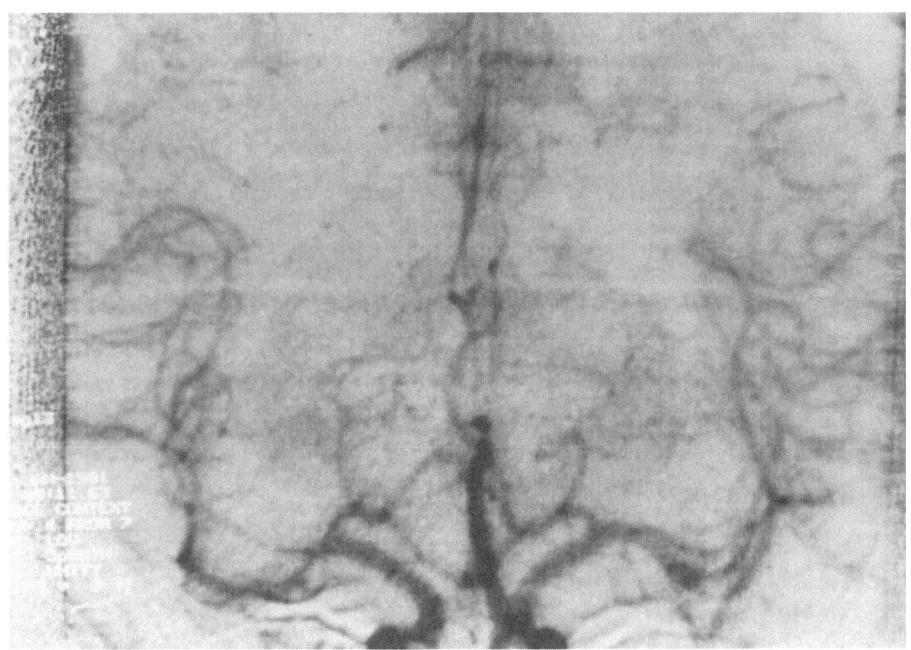

Fig.56 Intravenous Angiogram of Head - Arterial Phase
       (with permission of J.Smith, M.D.)

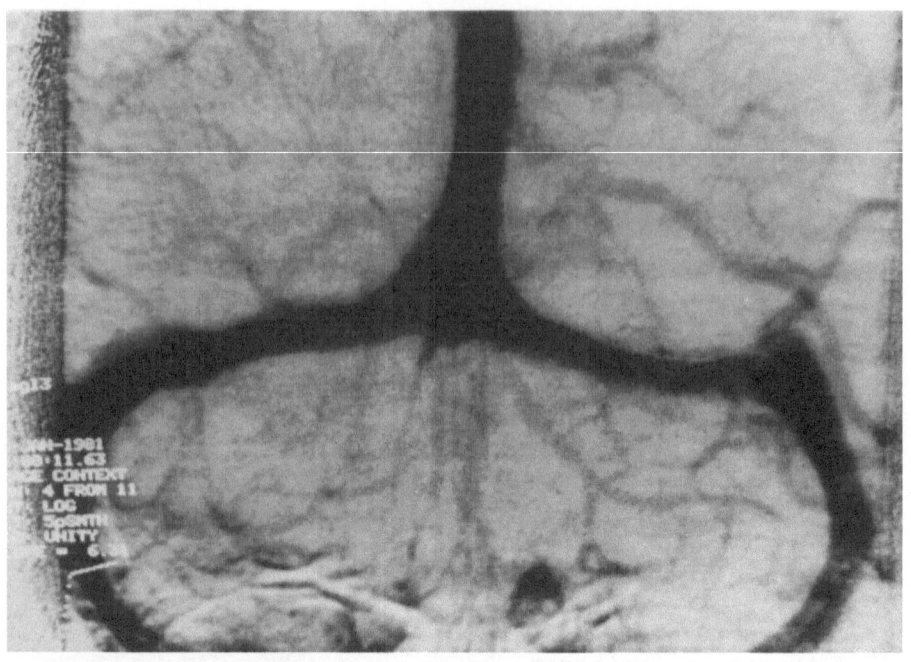

Fig.57 Intravenous Angiogram of Head - Venous Phase
(with permission of J. Smith, M.D.)

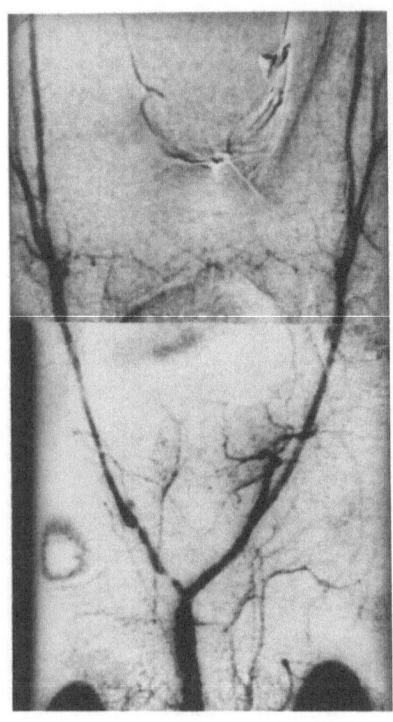

Fig.58 Intravenous Angiogram of the
Iliac and Femoral Arteries
(with permission of G.D.
Pond, M.D.)

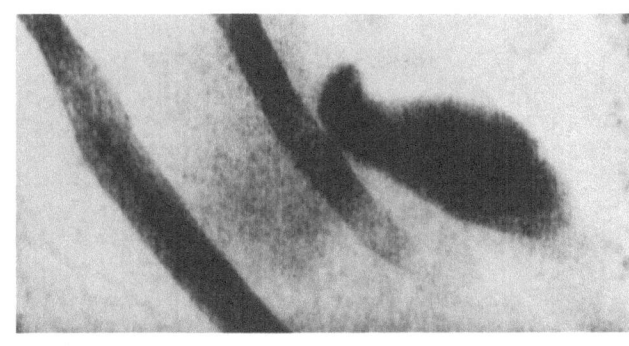

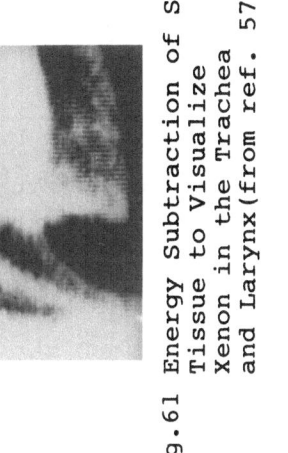

Fig.62 Energy Subtraction
to Visualize Iodine
in the Gall Bladder
(from ref. 58)

Fig.61 Energy Subtraction of Soft
Tissue to Visualize
Xenon in the Trachea
and Larynx(from ref. 57)

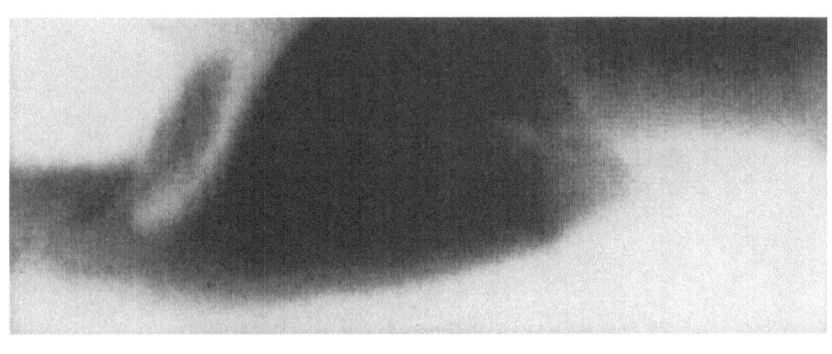

Fig.60 Energy Subtraction
of Bone to Visualize
Xenon in the Larynx
(from ref. 57)

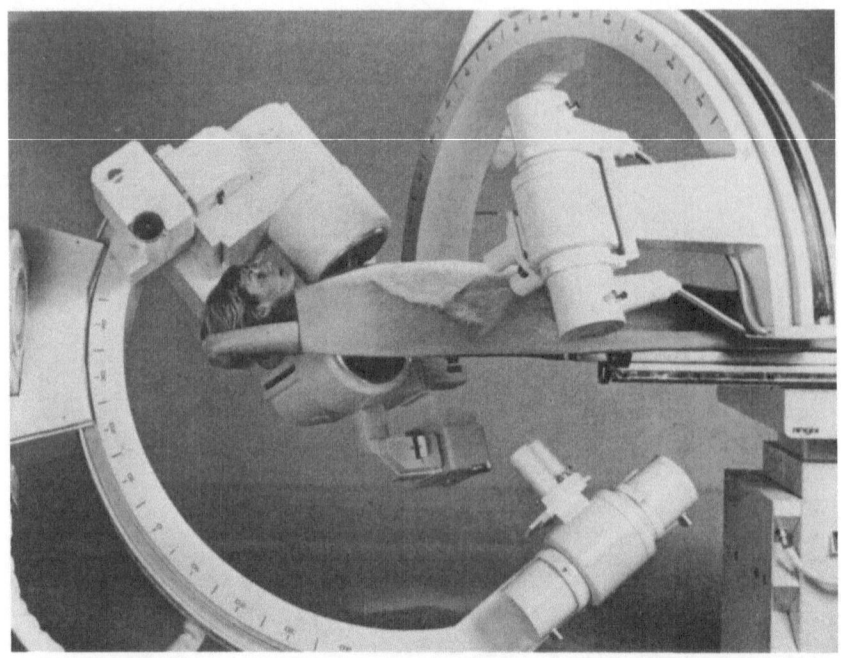

Fig.59 Nature of Equipment for Biplane Angiography
(from the Thomson CGR Corp.)

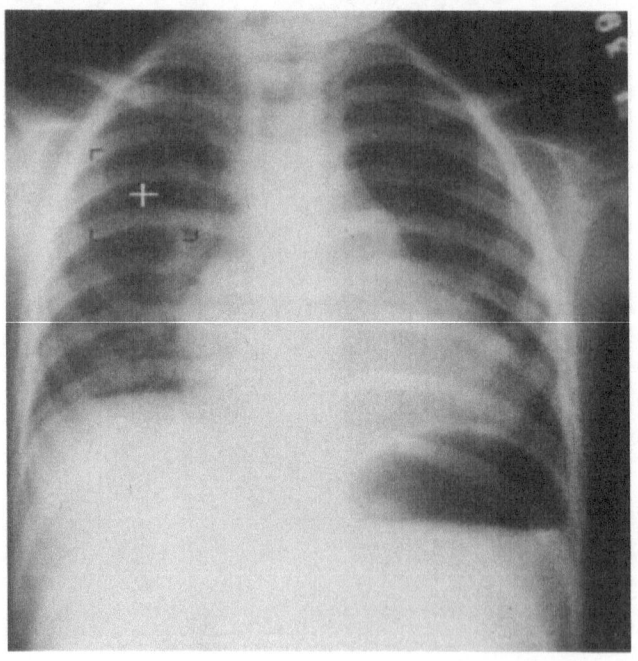

.Fig.63 A Radiograph of the Chest

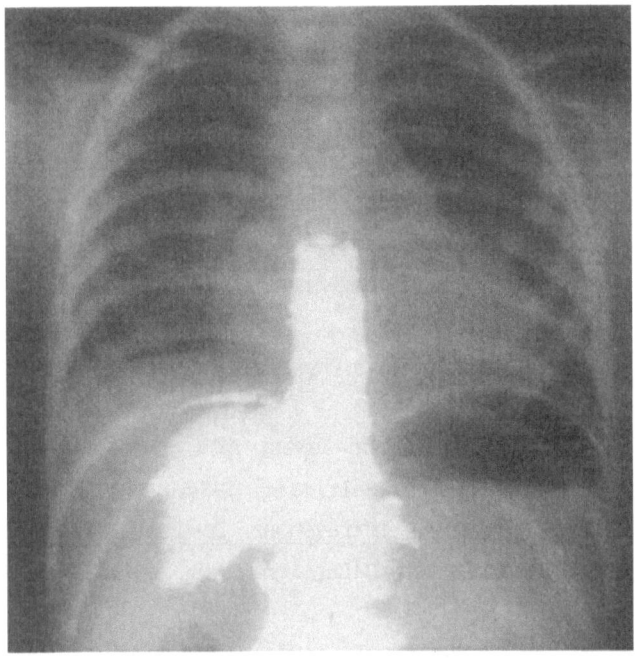

Fig.64 Contrast Enhancement With Density "Windowing" to Reveal a
Calcification (with permission of G. Seeley)

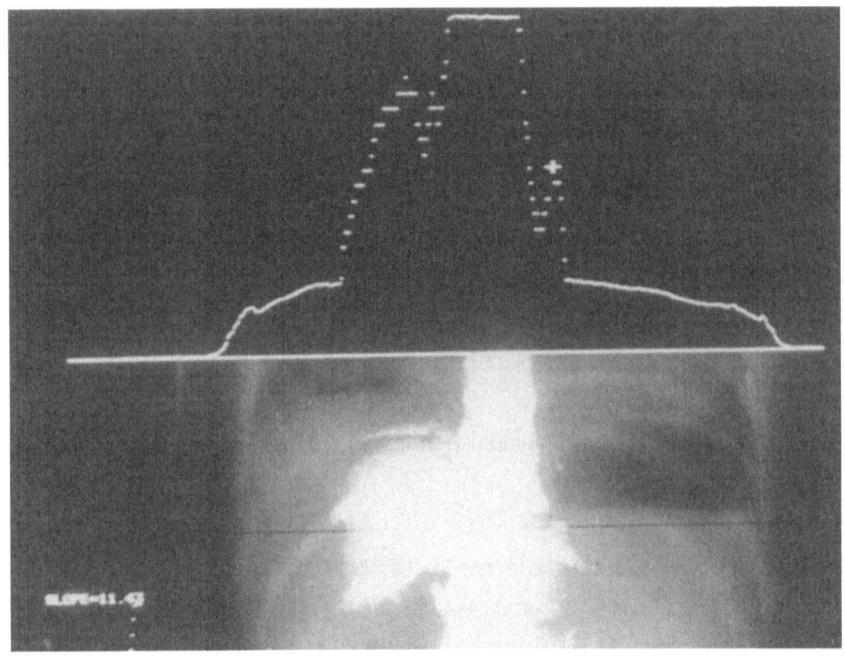

Fig.65 A Videodensitometric Trace Profiling the Calcification
(with permission of G. Seeley)

REFERENCES

1. Nudelman, S. and Roehrig, H., Photoelectronic-Digital Imaging for Diagnostic Radiology in Three Dimensional Imaging Methods, in Medicine & Biology, Vol. I, R. Robb, Editor, CRC Press, Cleveland, In press.

2. Heintzen, P.H. and Bennecke, R., Eds., Digital Imaging in Cardiovascular Radiology, Thieme-Stratton, New York, 1983.

3. Harrison, R.M. and Isherwood, I., Eds., Digital Imaging - Physical and Clinical Aspects, Institute of Physical Science and Medicine, London, 1984.

4. Anderson, J., Higgins, C., and James, A.E., Eds. The Digital Imaging Process, Williams & Wilkins, Baltimore, 1984.

5. Marc-Vergnes, J.P., Ed., Intravenous Digital Angiography, Societe D'Edition De l'Association D'Enseignement Medical Des Hopitaux De Parin, in press.

6. McLean, T. P. and Schagen, P., Eds., Electronic Imaging, Academic Press, New York, 1979.

7. Picture Archiving and Communications Systems for Medical Applications, I. Proc. SPIE, 318, 1983; II, Proc. SPIE, 418, 1983.

8. Digital radiography, Proc. SPIE, 314, 1981.

9. International Workshop on Physics and Engineering in Medical Imaging (IEEE-Berlin), Proc. SPIE, 372, 1982.

10. 1st International Symposium on Medical Imaging and Image Interpretation (IEEE-Berlin), Proc. SPIE, 375, 1982.

11. Application of Optical Instrumentation in Medicine. VI, Proc. SPIE, 127, 1977.

12. Non-invasive Cardiovascular Measurements, Proc. SPIE, 167, 1978.

13. Recent and Future Developments in Medical Imaging. II, Proc. SPIE, 206, 1979.

14. Application of Optical Instrumentation in Medicine. VIII, Proc. SPIE, 233, 1980.

15. Biberman, L., and Nudelman, S., Eds., "Photoelectronic Imaging Devices", Vols. I & II, Plenum Press, New York, 1971.

16. German Patent #968667, April 4, 1952.

17. Australian Patent #157101, April 9, 1952.

18. French Patent #1079964, April 18, 1952.

19. U.S.A. Patent #2650310, October 10, 1952.

20. Cusano, D.A., "Radiation Controlled Electroluminescence and Light Amplification in Phosphor Films," Phys. Rev., 98, 546-547, 1955.

21. Wang, S.P., Robbins, C.D., Bates, C.W., Jr., "A Novel X-ray Image Intensifier Proximity Focused Tube," Proc. SPIE, 127, 188, 1977.

22. Weimer, P.K., Forgue, S.V., and Goodrich, R.R., "The Vidicon-Photoconductive Camera Tube," RCA Rev. 12, 306-313, September 1951.

23. Heimann, B. and Heimann, W., "Fernsekamera Roehren Eigenschaften und Anwendungen, Fernseh und Kinotechnik, 32, 1, 1978.

24. Shack, R., Baker, R., Buchroeder, R., Hillman, D. and Bartels, P.H., "Ultrafast laser scanner microscope," J. Histochem. Cytochem. 27, 160-173 (1980).

25. Bille and Klingbeil, U., "Laser-scanning ophthalmoscope with active focus control," Proc. of the 6th International Conference on Pattern Recognition, Munich, Germany, 126, Oct. 19, 1982.

26. Nudelman, S., Fisher, H.D., Frost, M.M., Capp, M.P. and Ovitt, T.W., "A study of photoelectronic digital radiology. 1. The photoelectronic-digital radiology department," Proc. IEEE, 70, 700, 1982.

27. Nudelman, S., Healy, J. and Capp, M.P., "A study of photoelectronic radiology. 2. Cost analysis of a photoelectronic-digital versus a film-based system for radiology," Proc. IEEE, 70, 715, 1982.

28. Nudelman, S., Frost, M.M., Nevin, W.S. and Crowell, M.H., "Videotaping from the Lecturescope during fiberoptic bronchoscopy", Proc. of the 2nd European Electro-Optics Markets and Technology Conference, Montreux, Switzerland, 433, April 2-5, 1974.

29. Epstein, M., "Hypodermic fiberscope," Opt. Eng., 13, 139-142, Mar./Apr. 1974.

30. Delori, F.C., Airey, R.W., Dollery, C.T., Kohner, E.M. and Bulpitt, J., "Image Intensifier Cine-angiography," Adv. Electrn. Electron, Phys., 33B, 1089-1099, 1972.

31. Cutler, M., "Transillumination as an aid to diagnosis of breast lesions," Surgery, Gynecology, and Obstetrics 48, 721-727, 1929.

32. Ohlsson, B., "Diaphanography-A new method for investigation of the breast," World Journal of Surgery, 4, 701-707, 1980.

33. Carlson, E., "Transillumination light scanning," Diagnostic Imaging, April, 1982.

34. Marshall, V., Williams, D. and Smith, K., "Light scanning: Evaluation of a new modality to screen for breast cancer," Presentation at Radiological Society of North America, Nov. 1982.

35. Hevezi, J.A., Zermeno, A., Marsh, L., and Ong, P., A new approach to xeroradiographic imaging, SPSE 30th Annual Conference Advance Printing of Papers Summaries, Jacobson, L., Eds., Society of Photographic Scientists and Engineers, Washington, D.C., 1978, 59.

36. Sonoda, M., Takano, M., Miyahana, J., and Kato, H., Computed radiography utilizing scanner laser stimulated luminescence, Radiology, 148, 833, 1983.

37. Bates, C.W., X-ray intensification employing an external CsI(Na) input and scintillator, Appl. Opt. Suppl., 12(5), 938, 1973.

38. Ovitt, T.W., and Nudelman, S., Improved Instrumentation and Techniques for Non-Invasive Detection, Characterization and Quantification of Atherosclerosis for Research and Diagnostic Application, Q. Rep. to the National Institute of Health (NHLBI), Contract N01-HV-5-2969, Bethesda, Md., September 22 to December 21, 1976.

39. Ovitt, T.W. and Nudelman, S., Development and Evaluation of Instrument Systems for Non-Invasive Detection, Characterization and Quantification of Atherosclerotic Lesions, response to Rep HNLBI-HV-77-7, Bethesda, Md., submitted April 7, 1977.

40. Ovitt, T.W., and Nudelman, S., Improved Instrumentation and Techniques for Non-Invasive Detection, Characterization and Quantification of Atherosclerosis for Research and Diagnostic Applications, Final Rep. to the National Institute of Health (NHLBI), Contract N01-HV-5-2969, Bethesda, Md., May 1977.

41. Frost, M.M., Fisher, H.D., Nudelman, S., and Roehrig, H., A digital video acquisition system for extraction of subvisual information in diagnostic medical imaging, Proc. SPIE, 127, 208, 1977.

42. Ovitt, T.W., Nudelman, S., Fisher, H.D., and Frost, M.M., Computer assisted video subtraction for intravenous angiography, presented to RSNA-AAPM, 63rd Scientific Assembly and Annual Meeting, Chicago, 1977.

43. Nudelman, S., Photoelectronic imaging devices (PEID) for diagnostic radiology (abstr.), Tokyo Symp. 1977 on Photo and Electro Imaging, SPIE and SPSE 13-1, Tokyo, 1978.

44. Christenson, P.C., Ovitt, T.W., Fisher, H.D., Frost, M.M., Nudelman, S., and Roehrig, H., Intravenous angiography using digital video subtraction: Intravenous cervicocerebrovascular angiography, 1st Arizona clinical paper presented at the Annual Meeting American Society Neuroradiology, Toronto, Canada, 1979.

45. Ovitt, T.W., Christenson, P.C., Fisher, H.D., Frost M.M., Nudelman, S., Roehrig, H., and Seeley, G., Intravenous angiography using digital video subtraction: X-ray imaging system, Am. J. Neuroradiol., 1, 387, 1980.

46. Christenson, P.C., Ovitt, T.W., Fisher, H.D. Frost, M.M., Nudelman S., and Roehrig, H., Intravenous angiography using digital subtraction: intravenous cervico-cerebrovascular angiography, Am. J. Neuroradiol., 1, 379, 1980.

47. Brennecke, R., Brown, T.K., Bursch, J., and Heintzen, P.H., Digital processing of video-angiocardiographic image series using a minicomputer, Comput, Cardiol., No. 76CH1160-1C, 255, 1976.

48. Brennecke, R., Brown, T.K. Bursch, J., and Heintzen, P.H., Computerized video image preporocessing, in Digital Image Processing, Nigel, H.H., Ed., Springer-Verlag, New York, 1977, 244.

49. Brennecke, R., Hahne, H.F., Moldenhauer, K., Bursch, J.H., and Heintzen, P.H., Improved digital real time processing and storage techniques with applications to intravenous contrast angiography, Comput. Cardiol., No 78CH1391-2C, 191, 1978.

50. Mistretta, C.A., Ort, M.G., Kelcz, F., Absorption edge fluoroscopy using quasi-monoenergetic X-ray beam, Invest. Radiol., 8, 402, 1973.

51. Mistretta, C.A., A multiple image subtraction technique for enhancing low contrast, periodic objects, Invest. Radiol., 8, 43, 1973.

52. Ort, M.G., Mistretta, C.A., and Kelcz, F., An improved technique for enhancing small periodic contrast changes in television fluoroscopy, Opt. Eng., 12, 169, 1973.

53. Kruger, R., Lancaster, J., Mistretta, C., et al., Current results in real time computerized fluoroscopy and radiography, presented to RSNA-AAPM, 63rd Scientific Assembly and Annual Meeting, Chicago, 1977.

54. Kruger, R.A., Mistretta, C.A., and Lancaster, J., A digital video image processor for real-time X-Ray subtractioon imaging, Opt. Eng., 17, 652, 1978.

55. Kruger, R.A., Mistretta, C.A., Houk, T.L., Riederer, G.J., Shaw, C.G., Goodsitt, M.M., Crummy, A.B., Zwiebel, W., Lancaster, J.C., Rowe, G.G., and Flemming, D., Computerized fluoroscopy in real time for noninvasive visualization of the cardiovascular system, Radiology, 130, 49, 1979.

56. Crummy, A.B., Strother, C.M., Sackett, J.F., Ergun, D.L., Shaw, C.G., Kruger, R.A., Mistretta, C.A., Turnipseed, W.D., Lieberman, R.P., Nyerowitz, P.D. and Ruxicka, F.F., Computerized fluoroscopy: digital subtraction for intravenous angiocardiography and arteriography, Am. J. Radiol., 135, 1131, 1980.

57. Roehrig, H., Fu, T.Y., Nudelman, S., Pond, G.D., Hunter, T.B., and Bjelland, J.C., Dual energy imaging of airways using xenon as a contrast agent, paper presented at the 68th Scientific Assembly and Annual Meeting of the RSNA, Chicago, 1982.

58. Hunter, T.B., Pond, G.D., and Roehrig, H., Gastrointestinal radiology using dual energy imaging, presented at the Annu. Meet. Assoc. University Radiologists, Mobile, Ala., March 1983.

59. Brody, W.R., Cassel, D.M., Sommer, F.G., et al., Dual energy projection radiography: initial clinical experience, Am. J. Roentgenol., 137, 201, 1981.

60. Macovski, A., Alvarez, R., Lehmann, L.A., Roth, E., and Brody, W.R., Iodine imaging using three energy spectra, Proc. SPIE, 314, 140, 1981.

61. Fisher, H.D., Nudelman, S., Ovitt, T.W., Capp, M.P., Frost, M.M., Ouimette, D., and Roehrig, H., Photoelectronic-digital radiology: development and evaluation leading to intravenous angiography, Proc. SPIE, 273, 227, 1981.

62. Heintzen, P., Simple method for recording of radiopaque dilution curves during angiocardiography, Am. Heart J., 69, 720, 1965.

63. Heintzen, P., Bursch Jl, Osypka, P., and Moldenhauer, K., Rontgenologische Kontrastmittelmischungen zur Untersuchung der Herz und Kreislauffunktion, Elektromedizin, 12, 967, 1981.

64. Heintzen, P., Bursch, J., Osypka, P., and Moldenhauer, K., Rontgenologische Kontrastmitteldictemessungen zur Untersuchung der Herz und Kreislauffunktion, Elektromedizin, 12, 145, 1967.

65. Silverman, N.R., Clinical videodensitometry, Am. J. Roentgenol., 114, 840, 1972.

66. Silverman, N.R., Intaglietta, M., and Tompkins, W.R., A videodensitometer for blood flow measurement, Br. J. Radiol., 46, 594, 1973.

67. Intaglietta, M., Silverman, N.R., and Tompkins, W.R., Capillary flow velocity measurements in vivo and in situ by television methods, Microvasc. Res., 10, 165, 1975.

68. Heintzen, P.H., Malerczyk, V., Pilarczyk, and Scheel, K.W., On-line processing of the video-image for left ventricular volume determination, Comput. Biomed. Res., 4, 44,197.

69. Wood, E.H., Sturm, R.E., Sanders, J.J., Data processing in cardiovascular physiology with particular reference to roentgen videodensitometry, Mayo Clin. Proc., 39, 849, 1964.

70. Tsakris, A.G., Donald, D.E., and Sturm, R.E., Ejection fraction and internal dimensions of left ventricle determined by biplane videometry, Fed. Proc. Fed. Am. Soc. Exp. Biol., 28(4), 1358, 1969.

71. Sturm, R.E. and Wood, E.H., Roentgen image-intensifier television recording system for dynamic measurements of roentgen density for circulatory studies, Roentgen-, Cine- and Videodensitometry. Fundamentals and Applications for Blood Flow and Heart Volume Determination, Heintzen, P.H., Ed., Thieme, Stuttgart, 1971,23.

72. Von Bernuth, G., Tsakiris, A.G., and Wood, E.H., Effects of variations in the strength of left ventricular contraction on aortic valve closure in the dog, Circ. Res., 28, 705, 1971.

73. Smith, H.C., Frye, R. L., Donald, D.E., Roentgen videodensitometric measure of coronary flow, determination from simultaneous indicator-dilution curves at selected sites in the coronary circulation and in coronary artery-saphenous vein grafts, Mayo Clin. Proc., 46,800, 1971.

74. Smith, H.C., Frye, R.L., Wood, E.H., et al., Sequential measurement of saphenous vein graft flows and dimensions, Circulation, Suppl. 2, Abstract 22, 1972.

75. Greenleaf, J.F., Ritman, E.L., Wood, E.H., et al., Dynamic computer generated displays for study of the human left ventricle, Proc. SPIE, 35, 131, 1972.

76. Ritman, E.L., Johnson, S.A., Sturm, R.E., and Wood, E.H., The television camera in dynamic videoangiography, Radiology, 107, 417, 1973.

77. Ritman, E.L., Sturm, R.E., and Wood, E.H., Biplane roentgen videometric system for dynamic (60/sec) studies of the shape and size of circulatory structures, particularly the left ventricle, Am. J. Cardiol., 32, 180, 1973.

78. Robb, R.A., Johnson, S.A., Greenleaf, J.F., et al., An operator-interactive computer-controlled system for high fidelity digitization and analysis of biomedical images, Proc. SPIE, 40, 11, 1973.

79. Wood, E.H., New horizons for study of the cardiopulmonary and circulatory systems, Chest, 69, 394, 1976.

80. Hohne, K.H. and Pfeiffer, G., The role of the physician-computer interaction in the acquisition and interpretation of scintigraphic data, Meth. Inform. Med., 13, 65, 1974.

81. Hohne, K.H., Nicolae, G.C., Pfeiffer, G., Dix, W.R., Ebenritter, W., Nowak, D., Boehm, M., Sonne, B., and Buecheler, E., An interactive system for clinical application of angiodensitometry, Informatik Fachb. Band 8, Digitale Bildverarbeitung, Springer-Verlag, Berlin, 1977, 234.

82. Hohne, K. H., Boehm, M., Erbe, W., Nicolae, G.C., Pfeiffer, G., and Sonne, B., Computer angiography - a new tool for X-ray functional diagnostics, Med. Progr. Technol., 6, 23, 1978.

83. Hohne, K.H., Boehm, M., Erbe, W., Nicolae, G.C., Pfeiffer, G., Sonne, B., and Buecheler, B., Die Messung und differenzierte bildliche Dartsellung der Nierendurchkblutung mit der Computer-Angiographie, Fortschr. Rontgenstr., 129, 667, 1978.

84. Seeley, G.W., Stempski, M., Roehrig, H., Nudelman, S., & Capp, M.P., Psychophysical comparison of a video display system to film by using bone fracture images, Proc. First International Symposium on Medical Imaging & Image Interpretation (ISMIII), Pub. IEEE, 212-216, Oct. 1982.

INTEGRATED HOSPITAL INFORMATION SYSTEMS.

A.R. Bakker,

BAZIS, Leiden University Hospital

The Netherlands.

ABSTRACT

After some considerations on data and their use in hospitals the concept Integrated Hospital Information System is described together with its potential advantages. Essential elements in this concept are:

- storage of data of various type within a central database, together with their interrelationships,
- conversational access to the data at the moment when and the place where users need these,
- data should be presented in a shape geared to the needs of the specific user,
- the system should check the quality of the data supplied as good as possible,
- the system should take care of a part of the coordinating activities in the hospital.

Attention is given to system structure, followed by an overview of applications within a HIS. The notion integration is considered in some detail.
User access rights and other data protection aspects are a specific point of interest.
As a case the Leiden University Hospital Information system is described with data on its use. Costs and trends in costs are considered.

Trends in further development of HIS systems are considered and limiting factors in this development are indicated. It is emphasized that restrictions in the available technology are only of minor importance here.

NATO ASI Series, Vol. F19
Pictorial Information Systems in Medicine
Edited by K. H. Höhne
© Springer-Verlag Berlin Heidelberg 1986

Important topics in this development will be
- standardization of subsystems demarcation
- standardization of data description
- standardization of technical characteristics to allow for comparison of
  systems
- formulation of reference cases for data protection

The data stored in a HIS database are till now of alpha numeric type. Two
questions will be considered:
- possible relations between HIS and PACS
- what lessons can be learned from the HIS experience and can these be of
  value in further PACS developments.
As to the first question a PACS system should logically be integrated into
a HIS allowing in this way access from the same workstation to data both
on the HIS database (e.g. medical record and appointments) and the PACS
database.
Support of internal management of a PACS (e.g. of a multi level storage
structure) by HIS-data on patient admitted and appointments with
outpatient departments.

As to the second question the HIS experiences have shown amongst others
that:
- fast responsetimes
- user friendliness
- high availability
- design of the system for continuous further development
- measurement as guiding tool for further development
  are of great importance.

1. ON DATA IN A HOSPITAL AND THEIR USE.

In a hospital we are confronted with a lot of people and departments who
are -directly or indirectly- involved in achieving the primary goal of a
hospital, i.e. patient care. These persons and departments must have data
at their disposal in good time to perform their jobs. Data which in
general are supplied in some form by other persons/departments.

Because of the fast development of medical science and technology the amount of data to be recorded and the number of people involved in the care for a specific patient is increasing.

The dataflow between these persons and departments is rather complicated and very voluminous.

Moreover management in the various areas at the various levels needs data. Finally, especially in university hospitals, data are necessary for scientific research and education [1]

After these general remarks the characteristics of (the use of) data in a hospital are more specifically described:

- the data are "sensitive" in several respects [1], [2]:
  . errors/deficiencies in the (provision of) data can be fatal for the patient. Example: data on medication
  . the availability of certain data can be critical in time. Example: the result of a pathology-test during the operation: malignant yes/no
  . the connection between data can be critical, example: a certain labtest-result is pathological, except when the patient uses drug x. If both data are available at the same time in their connection there's nothing wrong, otherwise completely wrong conclusions might be drawn
  . data often are of a confidential nature, sometimes in an unexpected way. Example: the registration of a visit to a specific department on a certain day of the week may be sensitive, because on that day a specific disease is being treated.

- A well-defined nomenclature and standardized content and form of the data in use is often missing. Already with apparent simple data as 'department', 'specialism' and 'bedday' there exists a confusion of tongues; with data in the medical area it is sometimes even worse [3]. This fact is a serious complicating factor in the exchange and interpretation of data, within the hospital as well as (may be even stronger) in the outer world. One cause of this problem is the existence of different views on a number of basic data.

For example a medical procedure knows the following views:
. medical practice (medical history)
. financial settlement of the procedure
. (possibly) scientific research and education
. management information
. epidemiology - health care statistics.

The data often have to be stored for a very long time (and in an accessible way) on different grounds:
. in the interest of the patient (medical record).
  The fact that physicians strongly differ in opinion on the desirable period of storing the patient data is interesting.
. legal regulations (Archive Law)
. scientific retrospective research and education.

- The majority of the data has to have a high availability: not only because of the above-mentioned critical-in-time-aspect or generally efficiency; but also because the non- availability of certain data for a considerable time might disturb the whole functioning of the hospital [5].
- It is possible that in department A (of the hospital) data about a patient are incomplete in the sense that in another department B, also involved in the treatment of the patient, data about the patient may be known which are of interest for the treatment of the patient at the former department but are not at its disposal.
  This problem also may exist between hospitals A en B; the need is probably weaker or at least less manifest, the solution is more difficult.

2. THE CONCEPT HIS AND ITS POSITION IN THE HOSPITAL

The basic concept of a HIS might be described naively as follows:
  to supply authorized workers in the field at their working location with all available information (of high quality) immediately at the moment they need it in a ordered way geared to the specific need and wishes of the individual [3],[4],[5].

A HIS is aiming at:

- a more efficient use of the restricted resources
  available for the health service.
- a qualitative improvement in services to patients
- providing support for research projects
- providing support of education

A HIS might lead to quantitative and qualitative improvements in the following ways:

. registration-once- of data with qualitative improvement by checking the input in itself (form and content) and in relation with data already known (consistency). The fact that more people have an interest in a precise registration can also improve the quality.

Example: the checking of the input of a labtestresult

(is it a correct number, is it not out of certain predefined bounds) and in relation with results already known for this patient (is the increase or decrease in time of this result acceptable -the so- called deltacheck)

. savings in the registration of data and their transfer to other persons; copying can be avoided, the errors in copying also. It also can save the patient from answering many times the same questions.

. duplication of diagnostic tests (e.g.lab tests) sometimes might be avoided by better communication; leading to less burden of the patient and cost saving.

. fast accessibility of the data for authorized persons; this can have a positive influence on patient-care. By the lasting accessibility data can become more complete (patient-related data as well as overall-data).

Example: the fast access to medical history data can be of vital importance for the correct treatment of an emergency-patient.

. standardizing influences

Example: registration of many items in a uniform way and often as a reference to a nationally or even internationally accepted standard, e.g. diagnosis as a reference to the ICD-9-standard of the World Health Organisation.

. by the above-mentioned points the exchange of data for operational purposes may be facilitated. Also the availability of data for scientific research and education, and for management-information can be improved. This fact itself can lead to new improvements, in quantitative and qualitative respect.

Example: when data are not readily accessible and not standardized a scientific retrospective investigation of for example a relation between a certain technique of a cardiovascular operation and certain complications becomes an enormous if not impossible task.

. improvement of the coordination of the various activities of the hospital.

Example: an integrated appointment system, where the utilisation of the hospital-facilities can be balanced against the inconveniences for the patient.

. the use of data in relation with each other is made possible. For the patient possibly dangerous situations can be signalled in an early stage.

Example: - the prescribed drugs in relation with allergies or other drugs administered

- but also: the signalling in an early stage of an infection in the hospital.

. support or control of processes

Examples: - a large amount of the whole process in the clinical chemistry laboratory can take place under control of a part of the HIS

- but it is also conceivable that the medical decision-process is directly supported by a HIS.

Although there are some examples of this kind of application in some restricted areas a certain modesty on this point seems suitable.

There is still a long way to go before this can be done on a large scale.

Structure and characteristics of a HIS.

Although there exist different approaches in attemps to realize a HIS, a common core of principles can be recognized.

Central in the system is a **database** holding:

- patient data, both medical and administrative data
- data on the facilities and capacities of the hospital
- reference files
- working files

Users (of a variety of disciplines) communicate with the system by means of conversational terminals. These terminals should be available at the working location and easy to use. Since a rather complete HIS will offer support in some form for almost all functions in the hospital ultimately at least 90% of the personnel can be expected to need some form of access to the system.

As a consequence one of the major characteristics of a HIS is the large volume of workload of conversational transactions to be processed. This combined with a large variety of types of transactions.

apart from conversational terminals also satellite computers (e.g. those used in laboratories for data acquisition purposes) will be hooked up to the system.

Centralization versus decentralization
▄▄

Although there exists agreement to a large extend about the basic structure of a HIS (a logically central database accessible by authorized users from their working location) there is a lot of debate on computer technical set up. Basically two philosophies are competing:

a. a centralized set-up, with all data physically
   located in a database in a professional
   computercentre,
b. a decentralized set-up where the database is
   spread over more computersystems installed at
   different locations.

Within approach b. a distinction can be made in a set-up where the various decentralized systems have a hierarchical relation and the situation where they have not.

Although practical systems might show solutions that deviate slightly from these basic models, it is in general not difficult to assign a system to one of this classes. It is remarkable to notice that there exists already for many years a debate on this topic. Each of the solutions has its convinced advocates . This is the more remarkable because in a highly technological question one is inclined to expect that fact and figures would be decisive, we find however that emotional arguments and intuition play a dominant part.

Arguments used in favor of a centralized set-up are:

- economy of scale in the equipment cost, based on the observation that
  for major system-components like disc storage and CPU capacity Grosch
  law (for double the price, 4 fold capacity) still holds true for the
  size of systems we have in mind,
- a professional computercentre, that can guarantee a reliable service of
  the system and a good data protection,
- a set-up that has shown to be feasable,

Against this set-up is argued that:

- the professional computercentre tends to be burocratic
- modifications of the system become more difficult
- users have difficulties to rely on a remote facility where they do not
  have direct control of.

As arguments in favor of a decentralized approach are mentioned:

- the system can be build up gradually; system capacity is easier to
  expand,
- more identification of the user with the system,
- less vunerable to hardware failures.

As arguments against are given:

- difficult to maintain a coordinated evolution of the system,
- difficult (and expensive) to maintain a reasonable level of data
  protection,
- set-up implies a lot of system overhead to keep the various parts of the
  distributed database consistent (especially after system failures),
- No evidence that this alternative approach would lead to a cheaper
  system.

It is not the place here to make a choice between the 2 positions, the
topic is mentioned because it illustrates that even in high technology
projects as HIS or PACS it can be extremely difficult to come to firm
conclusions.

3. INTEGRATION

From this overview it can be concluded that a main characteristic of a HIS must be "integrated". There is a lot to say about the rather vague concept "integrated".

Generally spoken the concept "integrated" means that the parts are coherent, the whole system is more than the sum of the parts.

For a HIS it means:[6]

. data-integration, i.e. data are registered once and afterwards (many times) used by every authorized person who needs them in various presentations and combinations

. functional-integration, i.e. authorized users can flexibly switch from one function of the HIS to another function

. integration of technology (software and hardware), i.e. users must be able to have access via coherent technical facilities (e.g. one terminal network) to those data and functions within the HIS they need for their work, and the technical facilities must be "qualified" for this use (for example short response time and high availibility)

A part of the advantages listed in paragraph 2 can also be achieved by non-integrated, isolated applications (for example fast input checking of the input in itself). The major part of the advantages however can only be achieved by an integrated HIS.

One of the major menaces for an integrated HIS nowadays is the appearance of many micro-computer-based isolated systems which are commercially available at relatively low costs. With these systems it is often possible to gain (limited) success at a short term at low costs and with a small implementation effort.

However, these systems often need or record data which are also needed elsewhere in the hospital (for example patient-masterindex, in-patient-file etc.) and before long there will emerge inconsistent patientfiles in the hospital, each of poor quality.

The integration of these isolated applications afterwards usually is a very costly and sometimes impossible effort.

This is aggravated by the fact that there is very poor standardisation in the area of health care information systems; linking these isolated systems is a heavy task, not so much due to hardware incompatibilities, as well due to incompatibilities in the datadescription [3].

Although, the application of a micro-computer can be very useful; it has to be done intelligently; the necessity and possibility of integration should be regarded before decision making.

It should be mentioned here that integration may also be a disadvantage e.g. when relations between HIS functions because so intimate that further development becomes hardly possible because modification of one program has as a consequence modification of many other programs. A good database mgt system is an essential tool to reduce this disadvantage.

4. APPLICATIONS OF A HIS

A HIS is meant to support a wide variety of activities in the hospital, ranging from meal supply to accounts receivable, from diagnosis registration to medication or clinical chemistry. To keep the system manageable it is in general divided in a number of subsystems or application packages that have a formal interface with each other and with the database. One of the problems in the HIS area is that no uniformity exists as to the division in subsystems. Nevertheless it seems useful to give some indication of the applications that can be supported by a HIS nowadays. The following list is not complete but is meant to give an impression of current practice:

- patient registration (master index)
- patient location (admission/discharge; outpatient visits)
- appointment scheduling
- clinical chemistry and hematology
- microbiology
- infection statistics
- pathology
- radiology
- radiotherapy
- pharmacy
- bloodbank
- ECG-registration
- EEG-registration
- meal supply
- medication

- diagnosis registration
- patient correspondence (discharge letters)
- operation registration
- nursing support system
- order entry
- document lending system
- total activity registration
- invoicing
- accounts receivable
- accounts payable
- general ledger
- budget
- diagnosis related groups
- personnel
- pay-roll

- computer assisted instruction
- workschedule planning
- text processing
- general utilities for registration and statistics
- patient oriented retrieval in the database (medical record)
- database analyses

In figure 1 the relations between applications are presented in a scheme. Linked to the patient master index we see in the upper half of the figure subsystems directly supporting patient care. Apart from the direct results these subsystems in general also produce input to the total activity registration in the middle where all activities are registered to yield insight which workload is generated by whom. This registration is a excellent base for derivation of management information.
In the lower half administrative and management support functions are indicated yielding apart from their direct output also input to the general ledger.
On the left subsystems are indicated that give an overall access to data in the database.

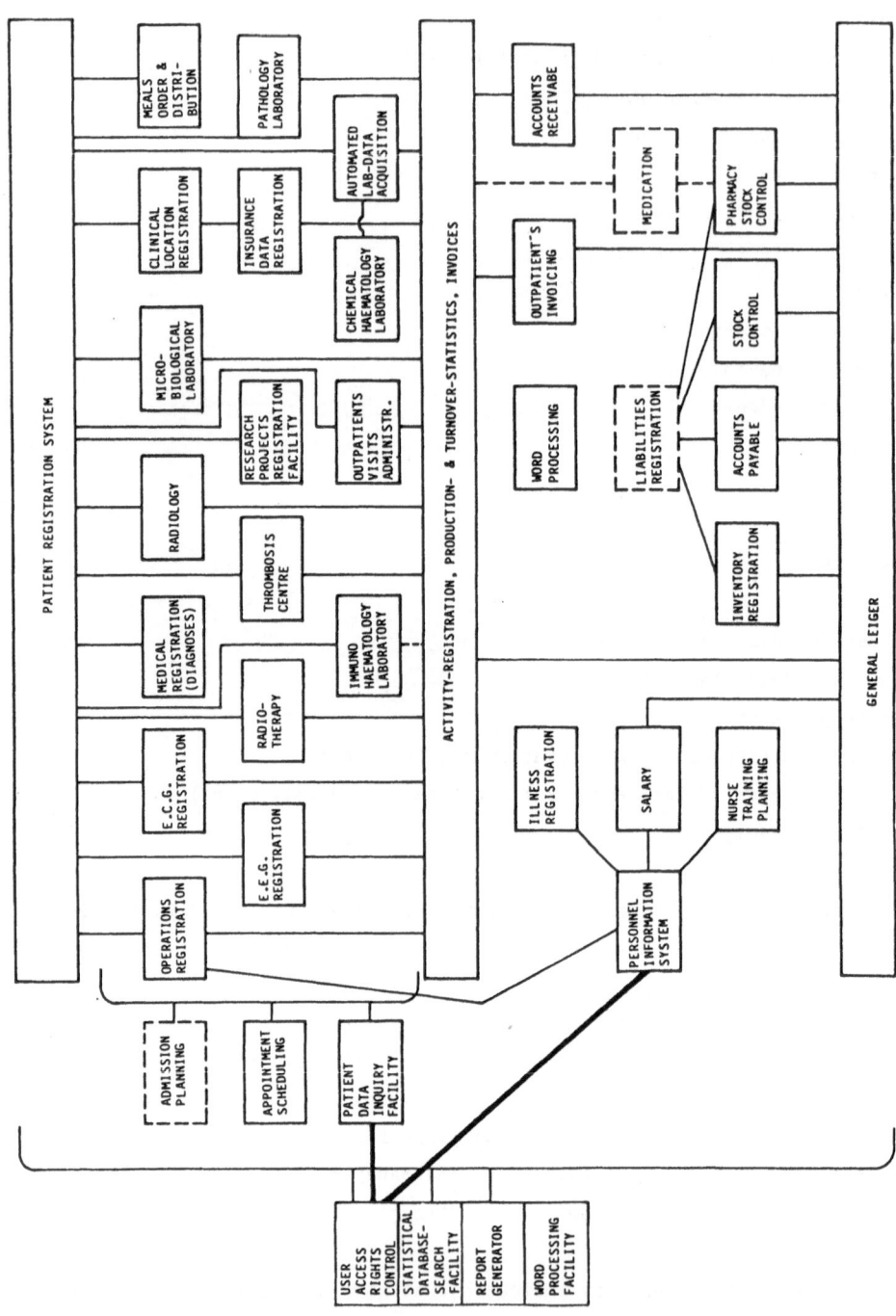

Fig. 1. Overview of relations
between HIS subsystems

In general it can be stated that main emphasis in a HIS is at present on:
- support of communication within the hospital
- support of departmental activities
- administration

The support of direct clinical activities in relation to both physician and nurse is limited, apart from communication and archiving. A considerable further development activity can be expected here, the interest in nursing support by a HIS is growing rapidly. Support on medical decision making (expert systems, protocols) is still in its infancy, apart from certain incidental cases (e.g. dosage of anticoagulants) application on a significant scale is not expected within five years from now.

5. PERFORMANCE ASPECTS.

The development of a HIS, together with its introduction in the hospital organisation is a big adventure where it is impossible to foresee the way practical applications would look like and their requirements as to systems resources. This characteristic of HIS projects lead to many unpleasant surprises that not seldom implied the abandonment of a project. If a system is meant to play an important part in the functioning of the hospital it should be a reliable, flexible and handy tool. Surprises were amongst others reported on:
- availability of the services of the system (too many interruptions)
- responsiveness of the system
- shortage of storage capacity and terminals
- inflexibility of applications once started that made further development of the HIS difficult

Since resource requirements are rather difficult to predict measurements on both system's and user's behaviour, together with extrapolations can yield essential data for guiding further development and implementation.

Data derived from the measurements sometimes combined with simulation techniques can reveal potential bottlenecks at a moment there is still an opportunity to avoid reduction of operational service level.

Within the Leiden HIS the following types of measurements are carried out:

. static measurements on system's parameters, these measurements are written to a file on a frequent basis (every 4 minutes).

Measurements comprise:

number of input characters, number of input messages, number of output characters, percentage of CPU utilisation, number of interrupts, number of disc accesses, percentage of disc load for the various components of the disc subsystem, queue lengths, number of various types of database actions and their resource consumption.

. Journalling of all incoming messages and database mutations (primarily to allow for recovery), yielding valuable data for analysis of user behaviour and for simulation.

. Analysis of contents and growth of the database together with counts of various types of actions on different subfiles.

. Dynamic measurement facilities; at points of interest within the system software probes can be installed yielding periodic counts or histograms. These probes are in general installed at points that are suspected to be potential bottlenecks. Moreover probes can be installed at points where data can be collected that are needed as input for design calculations.

Performance considerations leading to timely avoidance of bottlenecks in general are not spectacular although essential for achievement of adequate performance. Two examples [8],[9] of performance considerations and their effects are presented here briefly as an illustration.

A software disc cache
━━━━━━━━━━━━━━━━━━━━━━

When at a certain moment measurements indicated that the load of the disc subsystem was becoming the limiting factor in further expansion of the HIS-throughput it was amongst others suggested to implement a software disc cache (keeping a number of disc records in main memory according to some algoritm) in analogy to the widely used cache on main memory. Although the locality of disc adressing probably will be lower than that of main memory adressing the savings of a disc cache hit would be much larger as compared to that for a main memory cache. So a relatively low hitrate of e.g. 30 to 50% is already very attractive.

The following figures for a typical short disc access illustrate this:

  seek time (heads positioning) 20 m sec.,

  rotational delay 8 m sec.,

  transfer time 1 m sec.

  transfer from a disc cache (in main memory), about .5 m sec.

To support design considerations on this software disc cache the logging function was extended with recording of all disc accesses. The logged data served as input to a simulation program that predicted disc cache performance as function of size of cache buffer. The effect of different cache replacement algorithms was considered .

It was decided on beforehand to have all write accesses directly carried out (write through) to symplify recovery in case of hardware faillure. Moreover in principle only short disc accesses (one sector) were cached no program-swap actions or consecutive read actions. Results of some algorithms are presented in figure 2. The figure shows that even for a rather small cache buffer the predicted effect is very attractive.

Comparison with the unreachable ideal replacement scheme (based on lookahead on the log tape) indicated that performance of the second chance algorithm that is illustrated in figure 3 is reasonable.

In this algorithm disc blocks recently referenced are stored in a circular buffer. To each block a tag is assigned that may have the values 2,1 or 0. When reading a block in the buffer its tag is set at 2. The same is done when a block already in the buffer is referenced. By looking for space to bring a new block in the cache, the buffer is scanned cyclicly and tags of blocks being scanned, are reduced by 1 till a tag-value 0 is found. That block is deleted and replaced.

Using a look ahead feature the performance could be improved slighly further. When consecutive disc locations turn out to be in cache at the same time, this feature implies that also some following disc sectors are read into the cache buffer automatically.

Based on these simulation results a software disc cache (based on the second chance algorithm) was implemented in the Leiden HIS. At present with a workload that has increased by more than a factor 2 and a much more broader scope of applications the cache performance is 42 % in the dayshift and 50% outside the dayshift for a cachebuffer of 500 kbytes.

120

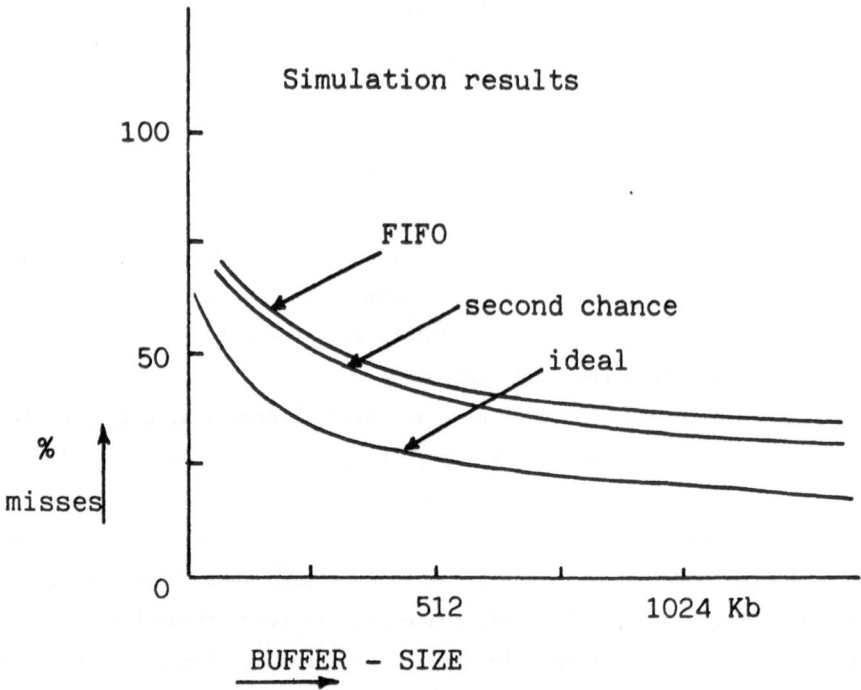

Fig. 2   Simulation results as a function of
buffer size for various replacement algorithms

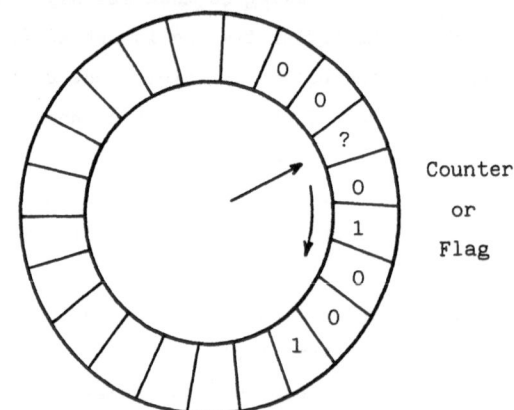

Fig. 3; Principle of second
chance algorithm

Automatic reallocation of cylinders in a disc subsystem

━━━

Installation of a new disc subsystem, that initially did not yield the
expected performance, stimulated disc performance analysis. The analysis
revealed the reasons for the poor performance and the specific problem
could be solved in cooperation with the hardware manufacturer. Besides
that our attention was drawn again to the load characteristics of a disc
spindle where a large percentage of the spindle busy time is consumed by
head positioning.

The average positioning time (for random disc accesses) for the disc
subsystem considered is given by the manufacturer to be about 29 m sec. By
careful positioning of files on the discs the operational average can be
consideribly lower (in our case about 20 m sec.). However the
reallocation of files is an ad hoc solution that combines as disadvantages
that it is time- consuming and error-prone. It was suggested to let the
system itself take care of reallocation of disc cylinders based on
characteristics of disc utilisation. The basic idea is to maintain the
logical structure of the disc and perform within the software discdriver a
mapping of logical cylindernumbers on physical cylindernumbers.

To investigate the feasability of this idea a simulation was carried out
based on logged data about all disc accesses during a day. It was proposed
to base the mapping algoritm on the total number of accesses per cylinder
in the dayshift (the dynamic character of this utilisation being
neglected), and sort the logical cylinders as to decreasing intensity of
utilisation. The cylinder most frequently used would be positioned in the
middle of the disc and the following cylinders alternating at the next
free cylinder to the right and to the left.

In formula:

$$j = [(N+1) / 2] + (-1)^i * [i /2]$$

N is the number of cylinders per disc pack
i = sequence number of a cylinder after sorting as to frequency of
utilisation
[ ] is the 'entier' function.

The simulation revealed that starting from the frequency distribution measured over a certain day the reallocation as indicated here would reduce average seek times by about one third. This means that the total elapse time of a short disc access would be reduced by about 20%.

The question remained in how far the frequency distribution showed considerable variations from one day to the other. For that purpose the frequency distribution found at a certain day was also applied for following days. Some results are shown in figure 4. Here' for a typical disc drive simulated average positioning time is shown at several days using the frequency distribution of preceeding days.

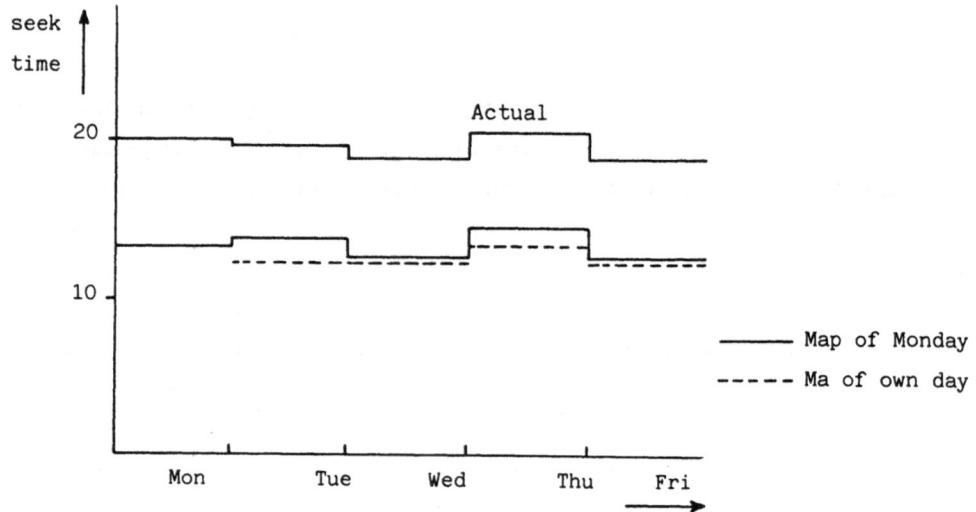

Fig. 4   Results of cylindermapping simulation
using maps of previous day

The performance of the cylinder mapping algoritm shows only minor differences if a previous frequency distribution is used as compared to that of the current day. However on a longer term a gradual shift of the distribution can be expected. In our case physical reallocation of cylinders can be performed each night when safe-copies are produced online.

The saving of 20% of spindle utilisation was considered to be attractive but not essential at that time. It was decided to implement this feature with a low priority. At this moment programming is ready and implementation is expected before the end of this year.

Conclusion

Performance analysis has proven to be a valuable tool in the design and construction of an integrated hospital information system. For PACS systems it is expected that this statement will hold even stronger.

6. DATA PROTECTION ASPECTS

In relation to a HIS data protection has to deal with: [2],[9]
usage integrity,
data/program integrity and
availability.
To achieve a reasonable protection level measures, in hardware, software and organization, have to be implemented. It is important to select a well-balanced set of measures since a chain is as weak as the weakest link.

Usage integrity

Since data stored in a HIS-database are at least partly of a confidential nature, for privacy reasons, access should be limited to those who need the data in carrying out their function in the hospital. After definition of the accessrights measures should be implemented to have any access checked on authorization with an acceptable burden for the user.
Almost everywhere as a basic rule is accepted that all hospital staff involved in the care for a certain patient should have access to the part of the patients data in so far he needs the specific type of data; e.g. no financial data, nursing data or diagnosis for a laboratory technician; no operation history for a ward nurse.
As to the access rights for data of previous cases handled in the same institution, (possibly by another specialism), less agreement exists.

The MEDICAL CONFIDENTIALITY of the data OBTAINED according to the following diagram must be fully guaranteed when USING these data

Authorization Table:
Authorization Arrangement for ACCESS TO PATIENT DATA

MAY, 1978

	TYPE OF DATA	PERSONS OR ORGANIZATIONS WITHIN THE AZL MAKING ENQUIRIES								
		TREATMENT SPEC.			CENTR. LAB	EDUCATION		SCIENCE		ADMINISTRATION AND ORGANIZATION
		doctors	nursing staff	admin. staff	Heads of Dept.	doctors resp. for treatm.	doctors not resp. for treatm.	doctors resp. for treatm.	doctors not resp. for treatm.	
1	administrative data	+	+	+	+	+	(±)	+	(±)	+
2	location history	+	+	+	+	+	(±)	+	(±)	+
3	diagnosis lists per patient	+	+	+	+	+	(±)	+	(±)	-
4	operation history per patient	+	+	+	[±]	+	[±]	+	[±]	-
5	diagnosis surveys with names	+	-	-	[±]	+	[±]	[±]	[±]	in conj.with Hd.Med.Reg. report to treat.spec.or Head of treat.dept.
6	lab. results per patient	+	+	+	+	+	(±)	+	(±)	-
7	lab. surveys, without names	[±]	-	-	+	[±]	[±]	[±]	[±]	[±]
8	PA results per pat. (file component)	see under 9 + 10: files for own specialism, files for other specialism								
9	PA conc. per pat. (diagnosis list component)	+	+	+	+	+	(±)	+	(±)	-
10	X-ray photographs and results	+	+	+	-	+	[±]	+	[±]	-
11	ECG per patient	+	+	+	[±]	+	[±]	+	[±]	-
12	files for own specialism	+	+	+	not app.	+	(±)	+	(±)	not app.
13	files for other specialism	(±)	-	-	[±]	(±)	[±]	(±)	[±]	-
14	copies of discharge letters (own spec.)	+	+	+	not app.	+	(±)	+	[±]	not app.
15	copies of discharge letters (other spec.)	+	-	-	[±]	(±)	[±]	(±)	[±]	-
16	emergency break-in	+	+	-	-	-	-	-	-	-

LEGEND AUTHORIZATION: + granted; - not granted; (±) only with consent of doctor responsible for treatment; [±] only with consent of the Head of department responsible for treatment or the Head of laboratory

Fig. 5 Authorization scheme for access to patient data, Leyden University Hospital

In figure 5 the basic scheme for user access rights in Leiden University Hospital is presented. It is hardly believable that it took several years to arrive at this very simple and rough scheme that is applicable for both computerized and non-computerized data. Till the moment of introduction of the HIS obviously people never had considered the question of access rights for the data seriously.

The concept of individual access rights implies that any user, before being granted access to the data has to be identified. Often an individual password or magnetic card is used for identification.

In a hospital one should have provisions for a rapid access to data in case of an emergency. In one of the systems this is realised as follows: In general access to the patient data is only granted to those users who fulfill two conditions:
- they have access to the required type of data, because they need that type of data in carrying out their job (e.g. laboratory technicians have access to laboratory data only)
- within the system there is a relation recorded between he patient and the department of the user (recent admission or appointment).
Some users however have the right to bypass the second check. If they do so they have to type in the reason for violation (a system requires a minimum length of this message) after which access is granted. However a message is printed in the office of the medical records officer that the violation took place with indication of the moment it occurred, the patient concerned and the user. The medical record officer checks whether the violation was justified. In practice violations almost always have to deal with emergency situations indeed.

Data/program integrity
━━━━━━━━━━━━━━━━━━━━━

Measures should be implemented to guarantee the quality of the data recorded (and the programs). Threats to this quality may be of both intentional and unintentional nature. As to the quality of data one should bear in mind that it is a common human feature to look more critically at results of other people than at own results. Therefore when working with pooled data the contents of shared files (e.g. patient master index) should be of a significantly better quality than that of local departmental files.

Basically one should take care of consistency checks on the data both within a record and in relation to other database records. The quality of the data should be checked both at the moment of input and on a periodic base.

Equipment measures should safeguard the operation of the computer. Safecopies of the database should be available in several generations and all mutations on the database should be logged on a journalling tape to enable recovery after a system crash.

The system should operate in a professional environment where capable personnel is available to take care in case of system trouble.

Quality assurance of new developed or modified software is an important issue; relatively small program errors might lead to serious pollutions of the database that are very dificult to correct.

Availability

Since a HIS is an essential tool in the hospital its services should be available when users need these. Although no full non-stop requirement can be justified an availability of 99% (or higher) is required.

This requirement leads to the need to duplicate the equipment at least partly, but this is not sufficient. Environmental provisions like fire protection and air conditioning are necessary in addition. Software should allow for (fast) recovery after system failure and procedures should give a reasonable guarantee of adequate reaction in crisis situations (stand-by procedures, written instructions).

7. AN EXAMPLE OF A HIS

As an example of a HIS the Leyden University Hospital Information System is outlined here briefly. This example was chosen just because the author knows it system best, it is emphasized that there exist various other interesting systems.

Since the concept HIS was considered to be attractive but the realization difficult, the HIS at Leyden was started in 1972 as a government sponsored experiment with as goal to achieve knowledge and experience in this field. The project was not aiming at a complete HIS but at the realization of sufficient functions to allow for a judgement on the technical feasability of the concept and its usefulness for the hospital. As a consequence the experimental system had to be introduced on a hospital wide scale.

The project was the start of a HIS development that lead to an operational system in 20 hospitals in the Netherlands with together over 11,000 acute beds. The development is going on since 1972 and accelerated every year. With the decreasing prices of equipment and their increasing performance a HIS has demonstrated to be a valuable tool for a wide variety of activities in the hospitals.

The HIS is aiming at the obejectives as indicated earlier in this paper and covers at present almost all application areas indicated in chapter 4 and some more.

Some characteristic features are listed here:
. the system is organized in a number of subsystems (application packages) that are the link between users and the database,
. from the start on users were heavily involved in the specification of subsystems,
. there exists a formal interface between the application programs and the database to allow for flexibility and expandability,
. the system is highly conversational with a present hundreds of terminals (both VDU and printing terminals) throughout the hospital,
. the technical structure is centralized with a duplicated central computer accommodated in the database and serving all terminals. It is not excluded that for some specific functions satellite computers are attached (e.g. for data acquisition in the clinical laboratories).
. one of the two computers carries out all production work and the second has three tasks:
  — to serve as back-up in case of faillure of the production computer,
  — to serve as a tool for software development (that is strictly separated from production)
  — to serve as a tool for quality control of software.

. the system is operational   24 hr/day, 7 days/wk (as the hospital is),

. the system software, both operating system, database software and
  teleprocessing software is an own development,

. all application software is written in a high level programming
  language, FORTRAN, with a gradual introduction of PASCAL now.

Before the end of the sponsored period already two other university
hospitals in the Netherlands decided to implement the system and a
cooperative structure was set up [10] for further development of the
system and support of the organizations using it.

Central in this cooperation is a foundation (called BAZIS) that is
responsible for development of both application and system software, for
second line support and for coordination (e.g. of hardware
configurations). The board of the foundation consists of representative of
participants. Participants are large hospitals or groups of cooperating
hospitals that base their information processing on the HIS. Each
participant has its own duplicated computer system and its own data
processing department that is responsible for introduction of the system
and operational aspects. Moreover these local data processing departments
participate in the total BAZIS development effort. For applications that
are judged to be only of importance for one of the participants these
local data processing departments have their own development activity.

That the HIS is growing stedily is illustrated by figure 6, where the
number of terminals installed by some HIS participants is indicated as a
function of time. Although Leiden (indicated by AZL) is still the largest
the speed of introduction at the other hospitals has been higher than that
at Leiden.

Figure 7 shows the number of imput messages per day (in the dayshift,
between 7.30 a.m and 5 p.m). A message is defind here as a string of
characters terminated by a cariage return symbol, after which the HIS has
to react. We see that not only the number of terminals installed as given
in figure 5 is growing, but also the intensity of use, at leiden with as
an average 20% every year.

In figure 8, the intensity of use at leiden is presented as a seasonal
curve over the year for the successive years. The figure shows clearly the
steady increase of workload and part of that a clear significant effect.

129

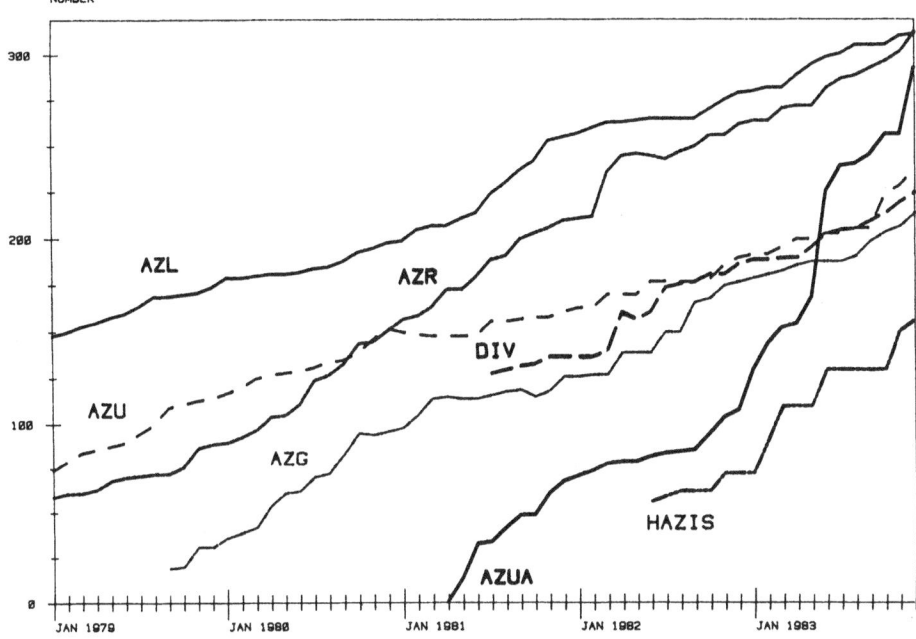

Fig. 6 Overview of number of connected HIS terminals

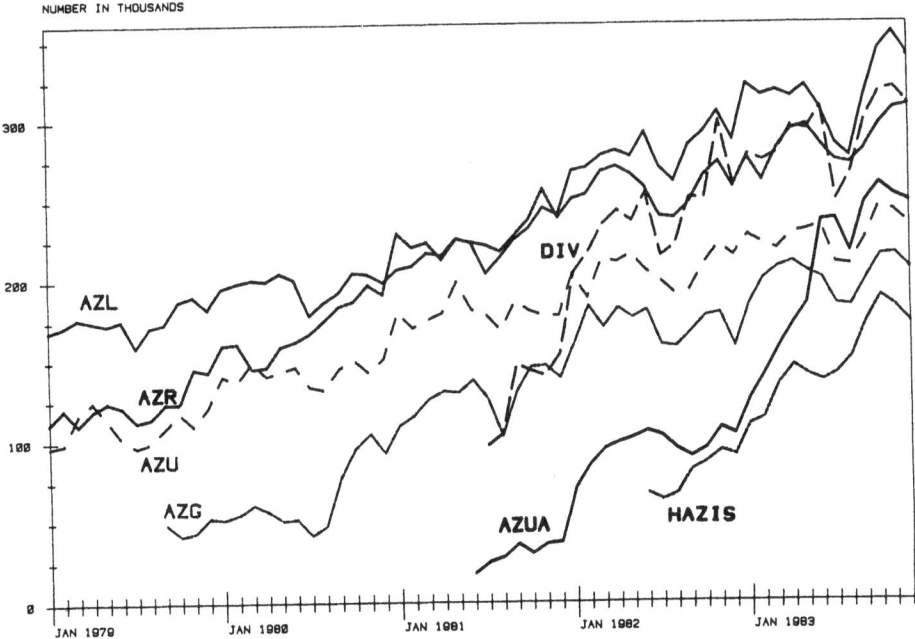

Fig. 7 Overview of number of message per day in day-shift

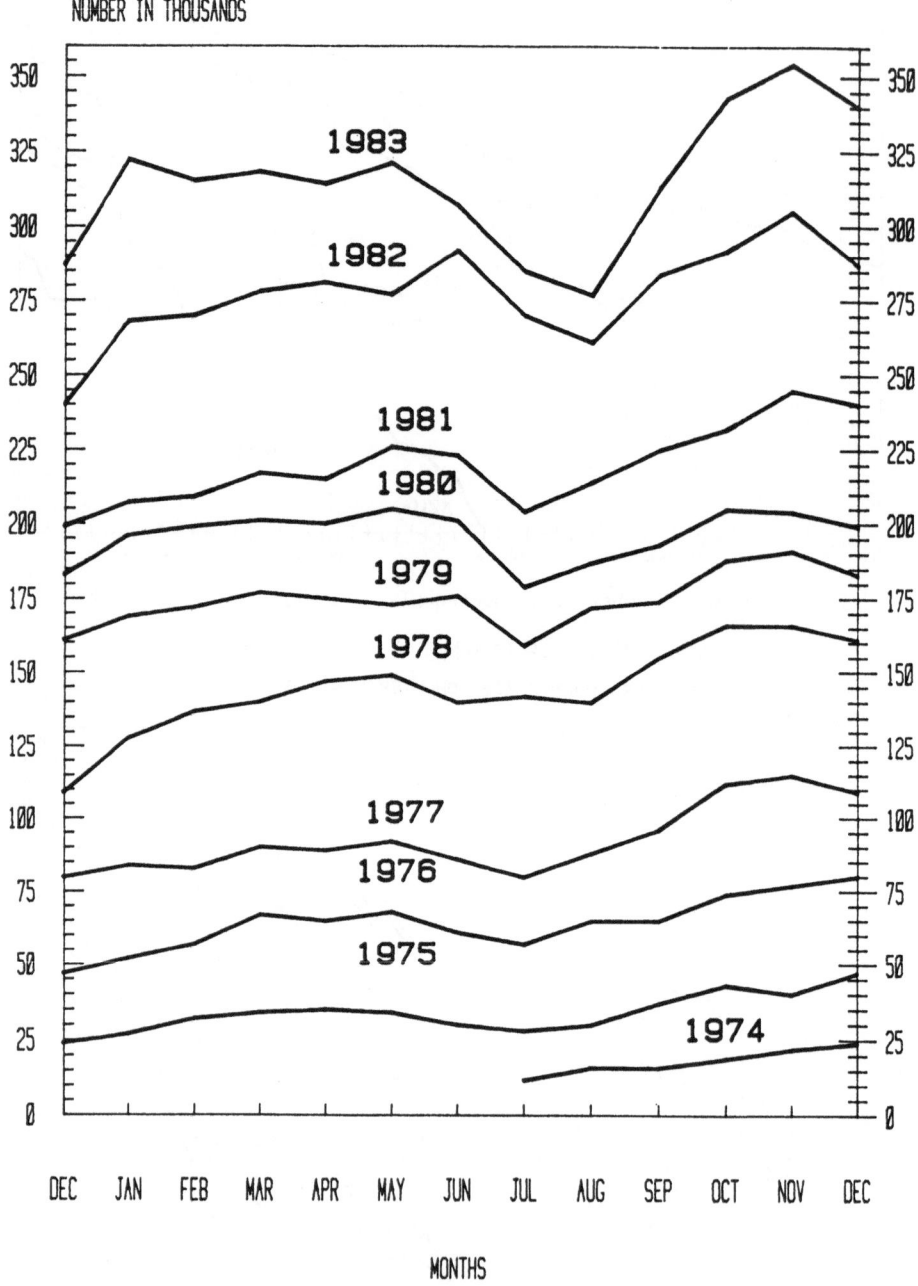

Fig. 8 Number of messages per day of ZIS Leyden
platted against the month of the year.

8. COSTS OF A HOSPITAL INFORMATION SYSTEM

In this period of economic recession and budget restrictions an increased interest in costs and benefits of dataprocessing in the hospital can be noticed.

Around Leiden University Hospital Information System a cooperation for further development of the system exists. Within this cooperation interest in cost figures is increasing. Since cost components are not uniquely identified a model was defined in 1981 for allocation of all costs concerned. For each hospital costs could be allocated to the various aspects of use of a HIS (development, implementation and operation) and specified for the different application packages within the system [11]. This model has been applied now at Leiden University Hospital for the years 1980, 1981, 1982 and 1983. Some of the figures found are presented here.

The costs contain all expenditures for :

personnel cost: software development staff, staff for implementation, operation of central hardware and the activities needed for decentralized hardware (e.g. terminals and some satelite computers for laboratories).

overhead cost: the expenditures for dataprocessingmanagement and overhead of hospital management and staff-departments.

hardware costs: depreciation of both central and decentral equipment in 5 years, maintenance and expenditures for materials. Costs for computers as part of special equipment e.g. CT, SMAC are excluded.

As for development costs, they are considered as investsments with a depreciation in five years; costs are shared on an equal base by all participants.

Cost figures

For 4 years (1980-1983) this method was used for calculating the costs of the HIS at Leiden. Some figures are presented here:

Related to the total expenditures of the hospital, the costs for dataprocessing are reasonably stable,(see table 2), with an increase in intensity of ZIS use by about 20% each year . The total budget of the hospital amounts to about 250 million guilders. The d.p. budget amounts to about 3.3 million guilders, which corresponds to about 1.3% of the total budget.

	1980	1981	1982	1983
Total hospital budget (million Dfl.)	225.5	238.0	251.3	255.8
Dataprocessing budget	3.28	3.31	3.57	3.28
Dataprocessing as % of total budget	1.46%	1.39%	1.42%	1.28%

Table 2 Budget and cost data of Leiden University Hospital.

The expenditures are specified in table 3 for the different kinds of costs. The hardware cost component is given seperately in the last column.

	development	implementation and support	operation	hardware component
1980	34.1 %	21.1 %	44.8 %	26.0 %
1981	34.6 %	20.4 %	45.0 %	30.9 %
1982	34.4 %	20.3 %	45.3 %	22.6 %
1983	32.5 %	22.7 %	44.8 %	23.2 %

Table 3. Specification of cost categories as percentage of dataprocessing budget at Leiden University Hospital.

For the cost of operation the distribution over cost components yields the following result:

cost for <u>staff</u>: (29.4 %), staff-overhead and office (8.7 %),

<u>hardware</u>: depreciation (31.3 %), maintenance (14.0 %), materials: (8.9%), housing and energy: 2.5 %,

<u>system software license</u>: (5.2 %).

In our situation the system-load, measured as the number of transactions per day, increases each year with about 20% , while expenditures are at the same level.

The trend in cost of operation for a set of 'standard applications' is analysed. These standard applications concern patient registration and location, laboratories, billing, general ledger, medical registration and authorisation of insurance coverage. The results are given in table 4.

First the cost per year are considered. An average annual decrease is found of about 10% . When interpreting these cost figures one should realise that the workload for this package of applications is still increasing. Correction for this effect leads to a significantly higher annual cost decrease of about 18%. When as an additional effect inflation is taken into account the annual price decrease becomes even higher.

	1980	1981	1982	1983
decrease straight		13 %	7.8 %	8.1 %
corrected for more intensive use of the packages		21.5%	17.3%	17,1
corrected for inflation		28,2%	23.6%	20,3

Table 4. Decrease of costs of operation of 'standard' set of applications.

For some large laboratories the total dp costs per patient sample have been analysed. The results are presented in table 5. where in general a steady decrease is found.

	1980	1981	1982	1983
Chemistry	1.53	1.40	1.09	0,88
Hematology	1.93	1.76	1.46	1,19
Microbiology	1.25	1.20	0.97	0,82
Pathology	-	2.86	1.48	1,40

Tabel 5. Laboratory costs per patient-sample in Dfl., Leiden University Hospital.

Discussion

Although the intensity of use of the HIS shows an annual increase of 20% the total cost of the system at Leiden University Hospital is not increasing The total cost in 1983 amount to about 1.3% of the total hospital budget. For the coming 5 years this percentage is expected not to grow beyond 1.5% with a further growth of the workload of 20% each year. Cost for a PACS system are not included in these considerations.

This implies that the cost per transaction for operational applications is decreasing as an average every year with about 20%. Applications where cost and benefits are almost in balance at present can be expected to yield a significant positive outcome in the years to come.

Within the total cost the categories :
   a. application software development and maintenance,
   b. implementation,
   c. equipment cost,
   d. operation of the computer centre,
are of about equal size. There is no indication for a significant change in the distribution of the cost over these categories.

In Leiden University Hospital Information System special attention has been given to limitation of equipment cost. By means of special developed system software it is possible to run the huge workload on a medium size computer (PDP 11/70). Asynchronous terminals are used and special micro computer based concentrators are applied to reduce the communication cost. In this environment hardware costs amount to about 25 % of total cost. This percentage is stable although significant replacement (and extension) of equipment has been realized in the period analysed. It is often assumed that hardware cost are of decreasing importance when using hospital information systems. Our figures lead us to the conclusion that this assumption is incorrect and that the increasing demand for capacity caused by the ever increasing workload is in balance with the price performance improvement of equipment.

Central computer equipment at Leiden University Hospital is fully duplicated. This duplication has three main advantages :
a. a very high availability can be achieved (99.5 % round-the-clock).
b. the backup computer can be used for quality assurance of new developed and modified application software.
c. development of software can be fully seperated from the production environment.

Analysis of cost figures shows that duplication of central computer equipment only amounts for about 5 % to the total data processing costs.

Favourite trends in costs are caused by:
. growth of the cooperation; the costs spread over more participants
. better price/performance of new equipment
. growth of the intensity of use (economy of scale)

9. BENEFITS AND EVALUATION.

As mentioned a HIS is aiming at significant improvement in the information handling within the hospital, with positive effects for patient care, utilisation of resources, research and education.

In view of the considerable costs involved both in development and in implementation and operation of such systems the question lays at hand whether the positive effects achieved justify the costs.

Two approaches exist:

- a global evaluation
- cost benefit analysis per application area.

Let me state on beforehand that in both approaches the amount of published results is small.

As to the first approach there is the well-known case of Mountview hospital El-Camino (Technicon System) [12]; and besides that the one of Leyden University Hospital Information System [13]. Both evaluations were carried out by independent consultants.

Evaluation turned out to be difficult for three reasons:

- relevant parameters were difficult to determine,
- when parameters were identified it was difficult to accurately detect their values before and (to a smaller extend) after the introduction of the system,
- when changes in the parameter values were found it was difficult to detect whether these were (at least partly) caused by other effects than introduction of the HIS.

Although the few evaluations reported on in literature yielded positive outcomes they were probably not sufficiently convincing (or too costly) to make evaluation a normal part of a project.

Where evaluation of an integrated system is a large time consuming activity it could be expected that on the scale of an application subsystem more material would be available. This may be true but even here no uniformity exists and difficult questions remain to be answered.

It remains difficult to determine to what extend the various effects found can be attributed to introduction of the system. E.g. one could suppose that the simple fact an advanced information system was being used had considerable influence of the user population (some users leaving because they didn't like such a system, other users being attracted by the new technology).

It is extremely disappointing that till now no reliable tool seems to be available to show clearly whether all our efforts to develop advanced information systems were worthwile and did achieve their goals to a reasonable level. The choice for an integrated information system remains till now a matter of intuition, taste or even fashion. It is my strong conviction that we should pay much more attention to the problem of evaluation and cost benefit analysis if we intend to continue with our developments. In a climate of reduction of budgets for health care such an effort is vital. Of course we should not only look at the short term benefits but also long range benefits should be taken into account.

10. LESSONS TO BE LEARNED.

What can we learn from experiences in HIS development and implementation?
- To facilitate productive cooperation between various groups in the field, it is essential to have a common terminology, data definition and a reference structure. In the HIS area this aspect got insufficient attention, leading to a situation where exchange of software is practically impossible. Also the exchange of specifications and technical solutions hardly gets a chance.
  In view of the large amount of manpower required to construct a HIS, this leads to an enormous waste of intellect and money.
  Perhaps in PACS development there is still a possibility to avoid such a confusion of tongues.
- A HIS, when intended to be really used in the hospital as an operational tool should have a level of performance that satisfies the users; the designer/manufacturer is inclined to be more tolerant as to deficiences of the system since he understands these and has good hope for improvement. 'Unfortunately' it is the user who decides on real use of the system

One should not introduce a system before an acceptable performance level can be reached; 'experimental' introduction of an immature system at the 'testsite' may easily disturb the climate, making further development and introduction significantly more difficult.

- Development of HIS systems in general was financed by special grants. The typical period of sponsoring being 4 tot 5 years. Many projects failed completely because they were not planned for continuity. If at the end of the sponsored period no organisation is prepared to pay for further development and operational costs, the project is in serious trouble. Already with a significant reduction of available funds (e.g. by 30%) key staff people will leave, contributing in that way to the rapid termination of the project.

In planning for continuity also relations with industry should be considered since industry is the primary channel for distribution of technological projects.

- Early HIS projects were at the frontier of available technology (both hardware and software). This holds, especially true for number of terminals, workload, database structure and storage requirements. Some technological developments went faster than anticipated, others slower. Although technological limitations have been a significant factor in HIS development, as an average performance improvement of hardware has been according to expectation, this opposed to development in software technology that seriously lag behind.

It turned out to be of paramount importance to monitor carefully technological development both as to performance and as to price.

- Since both the external workload (user behaviour) and internal workload of a system are difficult to predict maesurement facilities to monitor the behaviour are essential. Performance analysis has demonstrated to be a very valuable tool in guiding further development and configuration management of a HIS.

Because of the similarities between early HIS projects and present PACS projects it is expected that this statement also holds true for the latter.

- In evaluation of HIS systems three main problems were met:

  a. the previous situation is described insuffiently,

  b during the introduction of the complex system (which often takes several years) also other forces were acting on the hospital. One cannot simply compare situations before and after introduction of the system. Elimination of other effects is quite complicated.

c. Comparison with similar institutions that do not use the system is also rather difficult since that assumes other differences to be insignificant. One could imagine that management that is prepared to introduce advanced technology as to information processing will also be inclined to introduce other modern techniques. The opposite reasoning could also be produced stating that management that has to pay much attention to the introduction of modern information processing technology might neglect other technological developments.

Without giving a recipe for successful evaluation we learned the lession that if one intends to do an evaluation one should plan this already in the early stages of the project.

- Data protection is an important aspect of a HIS. It turned out that rules for user accessrights hardly exist in the hospitals. It took quite some time to come to a reasonbly clear definition of user accessrights. This discussion only for a minor part had to do with patient privacy it had more relation with professional responsibilities.

Availability of services of a HIS in an hospital should be very high. If the systems intends to be an essential tool in the hospital and one expects people to adjust there working procedures, then the services of the system should be available when users need these. In most large scale hospital information systems this led to the conclusion that computer equipment should be duplicated to a large extend.

For PACS systems availability can be expected to be of equal importance as that of a HIS.

In a HIS the complete database can be copied daily for security reasons. For PACS systems the procedure for safe-guarding image data deserves our special attention.

- For an experimental project that is partly being used for production purposes the seperation of development and production should be regid. A formal acceptance test of 'improved' programs is essential.

11. ON THE RELATIONS BETWEEN A HIS AND A PACS.

Present HIS implementations mainly deal with alpha-numeric data (sometimes also with graphs, e.g. ECG). The size of an elementary record in such HIS systems is restricted to several thousands of bytes (e.g. pathology report, radiology report or a discharge letter). The average size is even much smaller. When picture processing is considered in relation to a HIS one should realize that:

- the amount of data per picture is several orders of magnitude greater than that of HIS records which leads to very heavy requirements as to storage, transfer and display capacity.

- Images in a hospital are no selfcontained entities but belong logically to the medical record where also results of other diagnostic tests, treatment and patient health status are stored. For users of images access to other data from the medical record is a requirement especially for data on patient identification and location, results of diagnostic tests, diagnosis and therapy.

It probably is attractive to offer a facility to image users for communication with the HIS medical record by means of the same workstation.

Even more important is a linkage of the HIS database and a PACS database for logistical reasons. The storage management of large numbers of images is a difficult task where probably several levels in the storage hierarchy will have to be defined (e.g. archive, active set and working set). Data form the HIS database on appointments and admittance /discharge of patients can be a extremely valuable tool in this storage management where images with a high probability to be used in the near future can be activated on beforehand. Initially one can think of activating images of patients recently admitted or with appointments with specific physicians. As a refinement other data from the medical record might be taken into account as well.

12. FURTHER HIS DEVELOPMENTS AND PROBLEMS INVOLVED.

After some experiences with development of information systems of a limited scope, the concept of a 'total hospital information system' came up in the late sixties. Since that time we have seen many attempts to realize systems of considerable scope. Most of them failed, some became a (partial) success.
Although limitations in the available technology were till now probably a major reason for the disappointing experiences, other aspects need our attention as well since these might become the major limitations in the foreseeable future[3]

The early developments were started with inadequate technical facilities as to hardware, like background memories, directly accessable memories, data communication equipment and as to software, like operating system, teleprocessing and recover facilities. The availability of the systems was too low because the high cost of hardware prevented duplication.

User satisfaction in general was low because of poor performance as to responsetime, user friendliness and availability. In the early years systems that were succesful were only aiming at a limited portion of the whole scope that should be covered in the ideal situation. In first instantce mainly financial systems were realised or departmental systems. Later on communication systems were constructed covering only a portion of the total information handling in the organization concerned. The medical decision-making till now is hardly supported and management support is limited.
Further development of the systems towards being a tool for all disciplines in the hospital is expected, not only as a communication device but also to support medical and managerial decision-making.

It is tried here to identify major factors hampering the achievement of the ideal goals of the systems. It is emphasized that technological progress will not automatically lead to realization of a system close to the ideal. The hampering factors need more attention and an international approach to stimulate a successful development. It is felt that the concept of the integrated information systems is still

very attractive and might yield a major contribution to better health
care systems.

Technological developments

For storage and retrieval of alpha-numeric data the present technical
facilities of the high volume magnetic disc memories (either removable or
of the Winchester type) offer sufficient possibilities to realize the
online storage capacity needed. A further growth in capacity combined
with improvement of the price/performance ratio can be expected. It is
expected that the video disc technology will become attractive for long
term storage of data, but even if this development will not yield what is
expected, sufficient storage facilities are available for a realization
of the 'ideal system'.
Where CPU capacity is still growing significantly, no significant
improvement is expected in the access speed of background memories.
(Apart from specialized equipment with parallel read/write heads;
intended for bulk transfer).
Apart from sophisticated software facilities to allow for improved
throughput of disc subsystems (disc cache either within the operating
system or within an intelligent controller) parallel access via more than
one controller seems to be necessary for the bigger systems. This might
also be solved by composing the system of a number of computers connected
in a network where workfiles are located close to the working location,
in this way distributing not only CPU capacity but also storage access
load.
Data communication hardware can offer the required speed and reliability
as long as alphanumeric data have to be tranferred on a local network.
data communication over public telephonelines still suffers from speed
limitations (till ISDN will be available).

Especially as to database management systems and teleprocessing
facilities the speed of development is definitely lower than in the
hardware area.

Transportability of application software between hard/software systems of
different suppliers is not possible in practice, because of
incompatibility in interfaces to the system software and differences in
the language specifications, especially in the way in which the

languages interfere with the underlying hardware architecture. The use of high level programming languages is well accepted now.

The introduction of parameter driven systems to allow for adaptability to a specific organizational setting may contribute to the transportability of software. However, when this technique is applied to a large extent, setting parameter values in a consistent way becomes a problem in itself and needs more thinking. Nowadays it tends to get the characteristics of assembler programming (low level, limited built-in checks, time-consuming, error-prone).

The introduction of query languages, report generators and general packages for registration of limited scope may reduce programming effort considerably, but efficiency in computer resource consumption in general is rather low.

The number of conversational terminals in advanced hospital information systems amounts nowadays to one per 3 to 10 beds or to one per 8 to 20 workers in the hospital. It is expected that this figure will change in the coming decade, terminals will become as common as tool as telephones. One terminal per 2 workers seems a reasonable limit (equivalent to two terminals per hospital bed).

Structure of the systems
━━━━━━━━━━━━━━━━━━━━━━━━━━━

Health and hospital information systems in general are of a considerable size as to the various functions and to the volume of software necessary. To be able to manage both technical aspects like development and maintenance and organizational aspects the systems are partitioned in subsystems.
Unfortunately there exists no standard on the partitioning in subsystems. Often more or less similar terms are used like: patient registration, clinical laboratories, appointment scheduling, invoicing, medical records, etc.. However, the assignment of specific functions to the various subsystems may differ considerably. In some systems, for example, testordering is considered to be part of the clinical laboratory subsystem, whereas in other systems it is considered to be part of the nursing subsystems.
The reporting of results to the wards is in some systems within the

laboratory subsystem, in other systems, however, it is part of the message switching subsystem, the medical record subsystem or the nursing subsystem. The example shows that even for an application area like clinical chemistry, which most people feel is well-defined, no uniformity in the demarcation of subsystems exists.

Given this lack of uniformity in definition of subsystems it is not surprising that transportation of subsystems from one development group to another hardly occurs. The only practical examples deal with rather isolated applications like radiotherapy planning or non-integrated departmental subsystems. Often it is assumed that the lack of transportation is caused by incompatibilities between the hard/software systems used. Although these incompatibilities are a big problem indeed and hamper transportation it would be too optimistic to expect fruitful transportations as soon as these incompatibilities would be eliminated. This might be illustrated by the fact that also between groups using compatible hard/software systems no transportation takes place in general.

The transportation of application software would be stimulated if a standard partitioning in subsystems were available. For hospital information systems a reference model might be defined. Although such a reference model is not a standard the situation in the field would become much more clear if deviations from the model would be a part of the documentation of a subsystem.
In this respect a project in the Netherlands to set up a hospital information model describing a grouping of functions within the hospital and the related files should be mentioned. It might be one of the steps in the direction of a reference model. International cooperation in this respect seems necessary to reduce the amount of duplication of development efforts.

Data definition

Perhaps the most basic problem as to transportability of application software is the lack of a uniform data definition of the items to be recorded. Even the common notions like admission, discharge, bed-day, ward etc., often turn out to lead to confusion. It can not be hoped that application subsystems of different origin will fit together smoothly as long as such uniform data definition is lacking. There is an urgent need for a standardizing activity on data definition. Since the development of health and hospital information systems is such a huge effort that international cooperation is indispensable this standardizing activity should be set up internationally.

Comparison of systems characteristics

Apart form non-uniformity of terminology and data definition in the medical and hospital administration area the field suffers from nonuniformity in technical terminology. No uniform definition is e.g. available for the common notions: responsetime, transaction, availability. No objective comparison of different systems as to these characteristics is possible.

Since there is no uniform model for cost calculation either, comparison of systems till now remains subjective. Objective comparison as to performance and cost might stimulate further improvement of the systems. There is a clear need for a standardizing effort on this aspect.

Economical aspects

Although it is generally felt that besides qualitative benefits also quantitative benefits (will) emerge from hospital information systems, the number of documented cases with evidence for a positive cost/savingbalance is limited. This may be explained by the economical situation in the past decades, with sufficient funds available for advanced system developments with in general limited attention for quantifiable benefits.

In many countries nowadays the percentage of the GNP that is spent on health care is considered to be too high already and there is a tendency to reduce budgets. In this climate there is a real risk that budgets for information systems development and operation will be cut. However one should realize that health and hospital information systems may contribute significantly to a more efficient functioning of the health care system.

Apart from the risk that budget cuts will reduce development capacity in this field, since some people feel information systems are a luxury, there is a risk that people feel such systems, when cost effective, will be a source of unemployment.

The luxury argument should be confronted with well documented cost effectiveness analyses, which might be stimulated by international coordination.

As to the unemployment argument one should realise that in most countries the percentage of GNP to be spent on health care is fixed or even reducing. Of the health care budget 60-70% is spent on personnel costs, so the total employment in this field can be derived directly from the total budget. Reduction of the budget will directly lead to reduced employment in this field. The cost of the information system is only 0.5 to 2% of the budget of the using organization and within this information systems budget again the percentage of personnel cost is pretty high, nowadays already 50-70% and expected to increase further. So as opposed to what is usually assumed, no significant shift in the spending of the budget between personnel cost and equipment cost will result from the introduction of information systems. As long as average salaries do not differ significantly between dp personnel and health care personnel (incl. doctors) the introduction of these systems will not influence employment figures.

The decision for advanced information systems should be based on its effects on the quality of health care to be achieved within the assigned budget. Unemployment is not effected by this decision, when the total budget is determined by external factors.

Data protection

With the growth of health information systems the data protection aspects gain in importance. This holds true for both usage integrity and for data and program integrity.

The variety of medical data stored in a common database (either set up centrally or in a distributed implementation) may lead to encroachment of the privacy of patients. With the growing number of conversational terminals we are confronted with an increasing risk for illegal access.

The further development and acceptation of these systems might be hampered if no adequate data protection can be realized. A variety of measures of hardware, software and organizational nature is suggested to reduce the risks. Working group 4 of IMIA is especially active in this field of data protection in health information systems.

A sloppy attitude of users of the system is one of the most serious risks. Education of users deserves our attention since a high standard of usage integrity is essential for a wide acceptance of the system and a sound further growth.

An international working group

To stimulate successful further development of health and hospital information systems more international cooperation and coordination is necessary. The IMIA established an international working group (WG 10) with as tasks:

- to set up a reference model for partitioning in subsystems,
- to study aspects of integration together with advantages and disadvantages,
- to establish a reference data definition of data elements used in this field,
- to define a terminology in respect to system characteristics,
- to define a reference model for cost calculation,
- to stimulate comparison of systems and transportation of software,
- to establish in cooperation with working group 4 a reference model for data protection measures.

As a consequence one of the major characteristics of a HIS is the large volume of workload of conversational transactions to be processed. This combined with a large variety of types of transactions.

Apart from conversational terminals also satellite computers (e.g. those used in laboratories for data acquisition purposes) will be hooked up to the system.

References

[1] Bakker, A.R., Mol, J.L. 'Hopital Information Systems'
Effective Health Care, Vol. 1, Nr. 4; Elsevier Biomedical
Press, Amsterdam, 1984

[2] Griesser, G. et al. 'Data protection in health information
systems considerations and guidelines'
North Holland Publishing Company, Amsterdam, New York,
Oxford, 1980 . ISBN 0-444-86052-5

[3] Louwerse, C.P., Bakker, A.R. 'Technical requirements for
hospital information systems' Proc. Seminars Medinfo '83, O.
Fokkens ed., pp. 83-90,
North Holland Publishing Company, Amsterdam, 1983.

[4] Bakker, A.R., 'Hospital Information Systems, risks for
failures and actions to be taken' Proc. IFIP-working
conference on hospital information systems, R.H. Shannon,
ed. pp. 243-248,
North Holland Publ.Comp., Amsterdam, 1979. ISBN
0-444-85341-3

[5] Bakker, A.R., Louwerse, C.P., Kouwenberg, J.M.L.
'Data integrity in an integrated Hospital Information
System, practical experiences.
Proc. Medinfo 83, J.H. van Bemmel et al., eds. pp. 959-962,
North Holland Publishing Company, Amsterdam, 1983.

[6] Leguit, F.A. 'Integration in a Hospital Information System,
a necessity or a luxury'
Proc. Medinfo '80, Lindberg/Kaihara, eds. pp. 58-61
North Holland Publ. Comp., Amsterdam 1980, ISBN
0-444-86029-0

[7] Abel, R. 'A disk cache: a significant improvement to an
online, highly interactive, large database system
implemented on a minicomputer', Proc. Euro IFIP 79, pp.
575-580 ed. P.A. Samet.
North Holland, Publishing Amsterdam, 1979.

[8] Snitker, P.P., Abel, R.B. van Gennip, M.
'Performance improvement by automatic reallocation of
cylinders in a disc sub-system'. Messung, Modellierung und
Bewertung von Rechnesystemen, P.J. KUhn und K.M.Schulz, eds.
Springer Verlag, Berlin Heidelberg New York, 1983.
ISBN 0-387-11990-6.

[9] Griesser, G. et al. 'Data Protection in Health Information Systems where do we stand?'
North Holland
Publishing Company, Amsterdam, New York, Oxford, 1983.ISBN 0444 867 139

[10] Bakker, A.R. 'The development of an integrated and co-operative hospital information system'
In Medical Informatics 1984 Vol.9, no. 2,
Taylor & Francis Ltd London and Philadelphia

[11]Bakker, A.R., Hoogendoorn, C., Zanden, van der H.G.M.
'Cost aspects, cost allocation and cost figures for an integrated Hospital Information System'
Medinfo 1983, J.H. van Bemmel et al., eds. pp. 214-217.
North Holland Publ. Company, Amsterdam,
ISBN 0-444-86525X.

[12]Research report "Evaluation of the Implementation of a Medical Information System in a Central Community Hospital" El Camino, Batelle Columbus, December 1975.

[13]Roeleveld, J.A.G.
'The evaluation of the Hospital Information System (ZIS).
In order to the Minister of Education and Science
The Hague, carried out bij Van de Bunt, Amsterdam, 1978.

ON THE ARCHITECTURE
FOR PICTORIAL INFORMATION SYSTEMS

D. Meyer - Ebrecht
Lehrstuhl für Messtechnik
Rheinisch - Westfälische Technische Hochschule
Aachen, FRG

This Advanced Study Institute on Pictorial Information Systems -- elsewhere
the term Picture Archiving and Communication Systems (PACS) is used, and
medical diagnostics is obviously the most prominent application field at this
time -- was representative for the diversity of system aspects: Pictures of
extremely different nature are generated by a variety of imaging systems;
sophisticated image processing methods can potentially improve the diagnostic
use of pictures; data base and expert systems can help to manage pictorial
and non-pictorial information, and to offer supporting knowledge; novel technologies such as optical disk stores, optical fibre LANs, multi-micropro-
cessors, or real-time picture coders have to be applied due to the massive
data problem -- to mention only a few ... It leaves us with the crucial
questions: how shall this all fit together? And, in view of today's situation
in hospital practice which is already complex enough: how shall this all fit
together and still fit into the organisation of the hospital?

A host of building bricks, and we shall now build up a house. People shall
live in that house and use it for their purpose: This is a situation when
we need the ARCHITECT. However, who is that architect, what is ARCHITECTURE
in the PACS enviroment?

This paper will deal with the question "what is architecture". It will introduce the method of modeling as a key to a formal description. An approach to a solution of the controversy between the user's model and the
engineer's model will be proposed. The power of decomposition strategies
will be elucidated by means of a generic systems architecture approach.
Based on this, vital problem areas are being isolated and practical solutions will be described. Those are a distributed approach for the storage
subsystem, a solution for the compatibility problem within the transmission subsystem, and an approach to unification within the processing subsystems. Rather than "blue-sky-research", the ideas and solutions presented
here are the results of about ten years industrial research on this subject,
which followed the loop on the next page, starting from various entries and
leaving it at different exits.

NATO ASI Series, Vol. F19
Pictorial Information Systems in Medicine
Edited by K. H. Höhne
© Springer-Verlag Berlin Heidelberg 1986

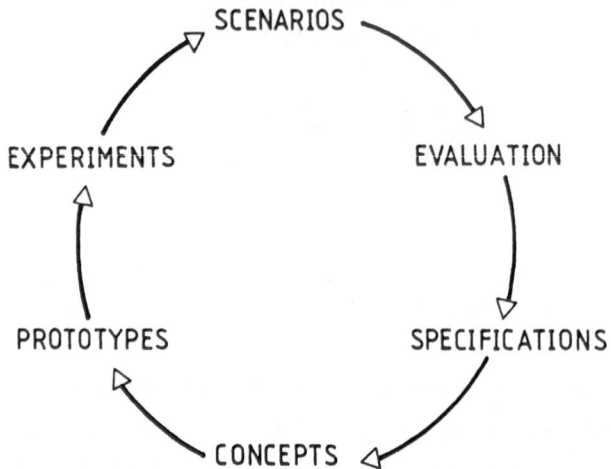

During this research work we have been highly motivated by a series of "aha"-events -- and deeply frustrated by the typical engineering attitude of those who were responsible for product development: "Interesting what those guys say about architecture, but now let's forget this nonsense and go proto-typing!"

> 99 We build systems like the
> Wright Brothers built airplanes ...
> ... Build them, push them off a
> cliff, and start over again! 99

You can do it this way as long as you have time enough, money enough, and as long as you do no harm nature or human beings. However, world has become more complex since that time, particularly because technologies and man-built systems are increasingly interacting with the human individual and with human society. Therefore, when a novel complex system is started to be developed -- like a PACS for medical application -- a thourough conceptual phase is re-quired which covers ALL aspects of the behaviour of the envisaged system and its interactions with its potential users. This is what we call systems architecture.

> 99 Architecture -- art or science? 99

System architecture has always been -- and will always be -- a type of art. People make systems for people, and esthetics of system design is a vital factor for motivation, both during design and utilisation of the outcoming system. The other way round, an esthetic design is somewhat as a proof that the designer has understood what we produced. By no means, however, these arguments must lead us to forget that architecture has also to become a science, viz. the conceptual phase of system design has to be based on well-established methods. Some ideas of those methods, and how they are to be applied are given by examples in the following.

> Architecture is a PROCESS - not a result.
> The result of the process is a conceptual MODEL.

When systems become complex due to the involved technologies, or due to a manifold of functions, and when, above all, interactions with their environment become essential, then architecture can no longer be a straightforward process. It must necessarily become a loop process such as sketched above. Dry runs will be followed by simulation runs, small-scale systems for lab experiments, pilot systems then, and real-world systems finally. It is the essential purpose of this loop process to bridge the gap between the product development engineer and the final user of the end product.

> The architect TRANSLATES
> ANTICIPATED user requirements
> into a FROZEN frame for design.

Designing real-world systems takes us into a controversial situation. The user of those systems is part of a DYNAMIC world. The engineer, however, needs a STATIC basis for system development. The problem is essentially to establish a basis in terms of detailed product specifications which keep unchanged during the whole product development process, while the user may not even be able to express his future requirements and desires. Therefore, as a primary task the systems architect has to involve himself into the user's world in order to foresee how the user will behave with the envisaged new system.

> Technology stimulates user desires,
> user requirements stimulate technology.

This is, by no means, a simple and straightforward task, because two evolutionary processes are stimulating each other. Caused by the introduction of new technologies an evolution of the user's requirements result in a growing complexity of the user's situation. Simultaneously an evolution of engineering concepts is the answer to a growing complexity of technologies and user-driven systems. As a consequence we need an evolution of systems architecture so that the systems architect can really act as a bracket around the two controversial worlds.

99 Evolution: from the primitive via the complex to the simple 99

About a decade ago the starting point was -- as it seems now -- rather primitive. The issue was that the archiving of diagnostic images -- predominantly X-rays at that time -- should be performed by means of digital technology rather than archiving the photographic film -- particularly since optical mass storage technology increased the chances of its realization. Technically, the digital picture archive was regarded as a giant data store with film input and hardcopy or softcopy output -- a straightforward concept. Then, as more and more digital imaging machines were implemented in hospitals, mass storage media for archiving the digital pictures were looked upon as peripheral devices to the picture-generating machines -- still a straightforward concept.

Meanwhile, it was becoming obvious that digital picture archives would be far more than isolated devices or extensions to existing isolated devices. Instead, they would become the kernel of networks that would connect a variety of imaging machines, interact with hospital databases and administration systems, and, last but not least, facilitate communication of pictorial and other information between medical personnel. This perspective added tremendous complexity to the concept.

It seems that we have reached the "complex" phase of the evolution of PACS concepts. How can we get more insight into the subject in order to disentangle the host of problems into a simple concept? (A. Einstein: "Make it as simple as possible -- but not too simple." We have to be aware that the "simple" concepts could be again primitive!) Yet there is a ray of hope that the problem itself will guide us to some novel architecural procedures. Let us try to visualize the reasons of the complexity of the problem. There are mainly three:

● An image is itself a complex type of information.

● The vast amount of data requires complex technologies.

● The diversity of user functions leads to complex system topologies.

Among the various types of documents -- let a "document" be the representation of information about a piece of reality -- an image is the representation with the least amount of abstraction. It makes excessive use of the capability of our most powerful information input organ -- the visual system. By means of images complex reality can be represented without loosing information by cutting it down to formal descriptions, because an image can be perceived as an entity. The price for that is that an image needs much more data for its digital coding than any other document. The price is, furthermore, that we do not understand in full detail how an image is perceived, and what has to be done to make handling and perception convenient and efficient (see also WENDLER, this issue |1|).

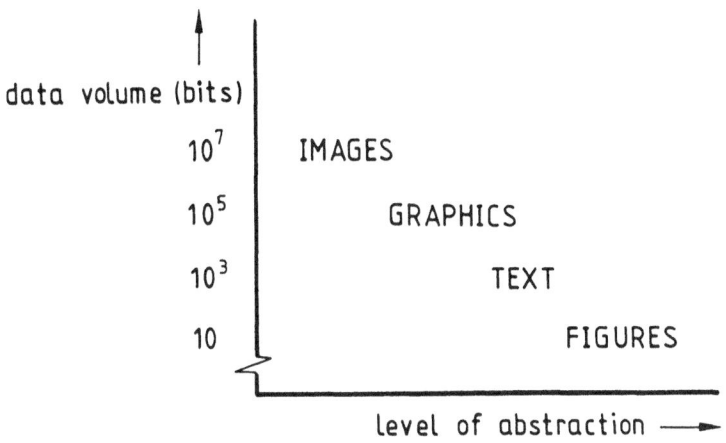

The second reason for complexity is a direct consequence of the above. So far data processing technologies have been developed for coded documents such as sets of figures or text, sometimes graphics. Systems are in wide-spread use which can handle millions of such documents of limited data volume, say some thousands of bits each. The same technologies are, with some restrictions, applied for the processing of images with a data volume of up to several million bit each. They are, however, limited to the handling of small volumes of images. Examples are digital picture generation machines like CT or picture analysis systems built for specific purposes. The shaded areas in the sketch below indicate the performance of today's digital systems. For a PACS, however, the system performance has to be boosted by about three orders of magnetitude: A PACS has to handle millions of documents with millions of bits each! Novel technologies have already been developed which can potentially solve the mass data problem. Experience with their application, however, is still lacking.

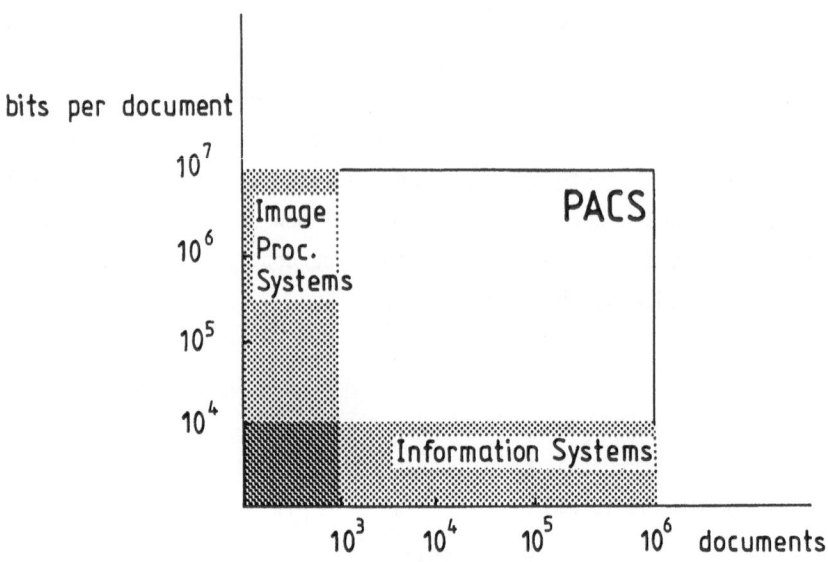

The third reason for complexity is the diversity of how medical disciplines make use of imaging and image processing, and how a PACS will supposedly have to interact with the diagnostic process. The range of imaging and image processing methods vs. the range of diagnostic disciplines each of which having its specific requirements on the methods to be apllied have often been described as a matrix. The diagnostic process, however, is a dynamic process. The steps along this process are necessarily distributed temporally and locally. A PACS must follow this process, and should support it during each step. For a PACS architecture it is, therefore, important to take into account this additional dimension, i.e. to extend the matrix description into a three-dimensional description as sketched below.

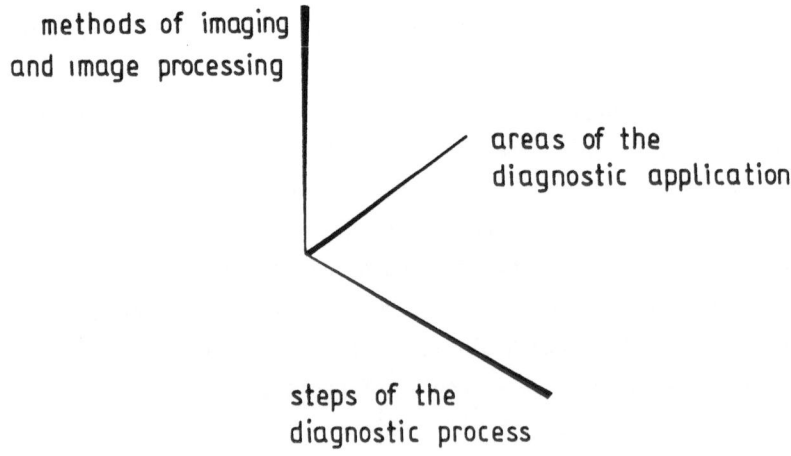

In view of this multitude of aspects the question is, how can the architect translate this into one consistent PACS concept? The crucial point is that we have to understand the various aspects as being perspectives from diffe-rent angles of observation of the same object. Restricting ourselves to one perspective only will, by no means, lead to a recognition of the kernel of the problem. Even worse, observers who are focussing on different perspec-tives each can not even communicate with each other, and thus benefit from a mutual understanding of their different views. In a more complex environ-ment we have to learn more complex understanding like kids learn to under-stand their three-dimensional world, and to express it by their two-dimen-sional drawings:

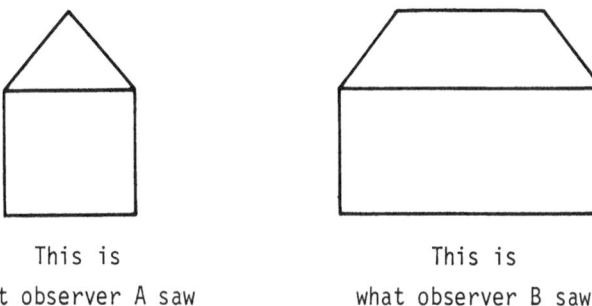

This is
what observer A saw

This is
what observer B saw

Two objects?

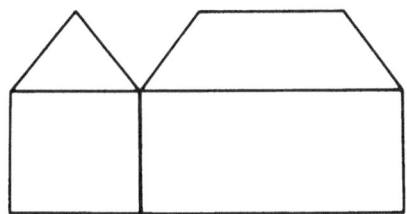

They communicated and recognized that they observed the same object.
But it doesn't fit together nicely ....

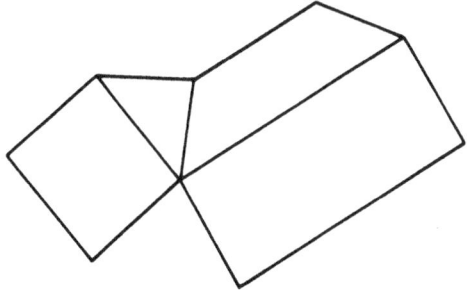

They try again. Now the roof fits. But the body?
This seems to be the status of PACS understanding now.

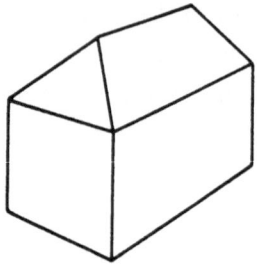

Aha! This is where we have to arrive to:
a model which takes care of the dimensions.

Again we have to stick to the statement that architecture is no straightforward process. Rather it is a loop process. Loopwise the architect has to get
himself involved into the subject, several times entering the loop from several directions. "Aha-effects" of the above type will suddenly bring structure into the subject -- structure which is the precondition for modeling.

> A MODEL is a formal description
> of the behaviour of a real process
> usually idealized and simplified.
>
> The TYPE of the model description
> depends on the purpose of the model.
>
> The STRUCTURE of the model depends
> on the observer's perspective.

A PACS is a typical example of a system with multiple dimensions, including
user, network, system management, data processing, and other dimensions. The
system architect's primary mission is to define the system's dimensions such
that conceptual models from the different perspectives become orthogonal projections of the same system. Orthogonality in this respect means that conflicting descriptions of the same matter can be avoided by decorrelating the
functional descriptions (models). A conceptual process which leads to a model
must be top down, but where is the top and where is the bottom? This depends
on the oberserver's perspective.

When starting the conceptual process the two most important -- and most controversial -- perspectives are the user's perspective and the engineer's perspective. Conceptual models are primarily needed for two main purposes: (1)
to describe the interaction between a PACS and its users, that is, the
user's PACS model; and (2) to establish a basis for a PACS development,

namely the technical PACS model. The type of model description will depend on the purposes of the two models, although they have to describe the same reality, and there must be rules for transforming one model into the other.

If our objective is a PACS technology that exactly meets the user's requirements instead of pressing him into the template of a technology-driven PACS, we cannot apply the usual method of top-down modeling. Instead, we must start with a user's individual situation, and after analyzing his requirements, set up the specific user's PACS model. At this point, a crucial question arises: How does the need to define an individual user's PACS model for every application relate to the engineer's need for a technical model that gives him solid guidelines for his product development? The proposed answer is that the technical PACS model must be a generic structure of well-defined functional modules with a set of well-defined rules for their interconnections. A corollary is that instead of fixing the user's model, we must fix a modeling procedure. This procedure in turn leads to a model description that can be translated one to one into the basic functional and structural elements of the technical model description. Thus we arrive at an overall modeling and implementation strategy as sketched below.

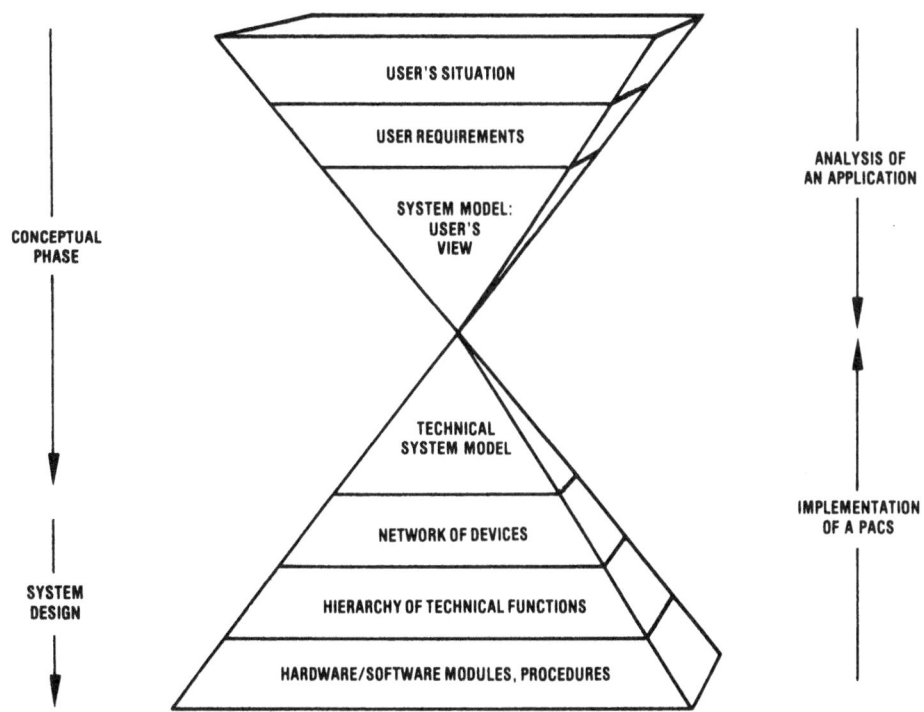

Following this conceptual process, we will describe the procedures for setting up the user's model, a decomposition strategy for arriving at the complementary modular technical model, and, finally, concepts for the implementation of functions that will fit into the proposed technical model. Consistency is required throughout the conceptual process and the subsequent
system design to make the two "pyramids" meet at the same point, i.e. to
make the final PACS product meet the user's requirements and desires.

Before we can go to specify the user's anticipated requirements and desires
we have to analyze the actual conditions of the specific user's work situation . The organization of a well-established working group -- e.g. an
X-ray department -- is, however, the result of long years of cooperation
during which an acceptable distribution of work elements has been evolved.
Usually this distribution is to quite a degree non-formal, and its analysis is impeded due to different levels of distribution:

- The WORK itself is distributed due to physical reasons (technical
  devices, floorplan, etc.).

- Also the KNOWLEDGE required for the various work procedures is
  distributed.

- Even the UNDERSTANDING of the organization (procedures of cooperation,
  etc.) is distributed amongst the members of the working group.

In this situation well-established TOP-DOWN modeling strategies do not apply.
Rather there is no other way than BOTTOM-UP modeling. But remember: The
user's model which has necessarily to be individual has to become CONSISTENT
with the engineer's model which is aimed at to be generic. This can only be
achieved by defining a formal description of the modeling PROCEDURE. Such a
procedure -- which, of course, has been derived also from an understanding
of basic technical system considerations (the loop!) -- is being described
in the following. It starts with a decomposition of the user's situation.

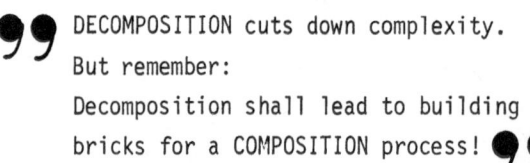

DECOMPOSITION cuts down complexity.
But remember:
Decomposition shall lead to building
bricks for a COMPOSITION process!

The proposed bottom-up procedure to set up the model description of the
user's situation -- the organization and work routines in a hospital department -- requires, first of all, rules for a decomposition into less complex
segments.

A natural breakdown would be patients, medical professionals, and technical
devices (imaging machines). Patients propagate through the diagnostic department following distinct temporal and geographical rules. Physicians and
assisting personnel perform activities requiring cooperation, or at least
communication. Imaging machines and support devices add another dimension
because they are also communicating with each other. Thus, three separate
networks can be defined.

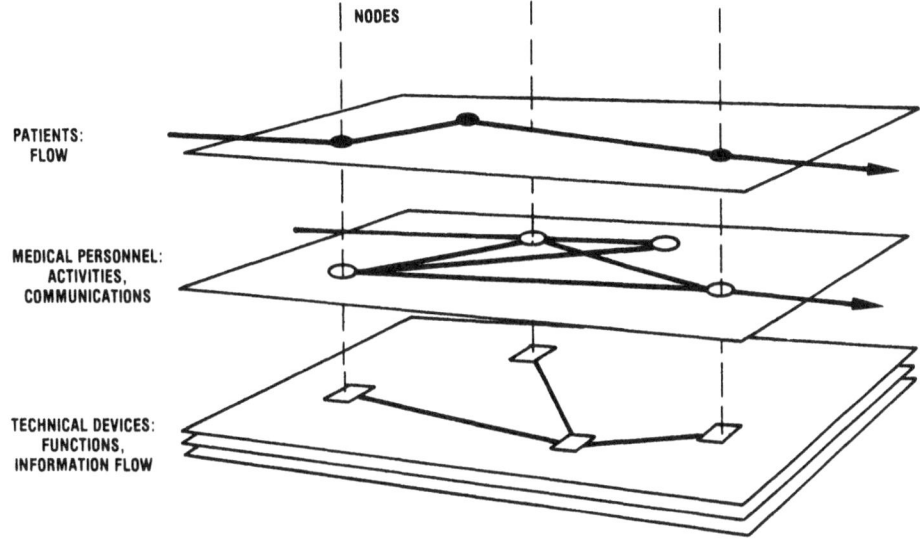

Common nodes can be recognized when the networks are superimposed. These
common nodes represent functions in the user sense. Analyzing user functions
leads to defining technical functions of the devices installed at the individual nodes. Analyzing the connections between the nodes leads to defining
the communication structure of the network of devices - the PACS. To the
user, the PACS appears to be a geographically decomposed system defined in
terms of user functions (nodes) and communication tasks (links between nodes).

With the description of nodes and links, the user's model has reached its
highest level of abstraction. Now it must be translated into the technical
model; user functions must be translated into hardware or software modules,
and communication tasks must be translated into transmission links. Not
least for the sake of economy, the number of elements of the technical model
must be kept at a minimum, and their networking should follow a single concept. Therefore, considering the variety of desired user functions and the
host of individual communication structures, strategies are needed to decom-
pose user functions and communication tasks into classes of basic technical

functions and procedures.

The pivot of the model translation process is being found by a comparison of the user's vs. the engineer's understanding of a PACS.

● The user understands

the local subsystem (imaging device, image workstation, etc.) to be his private tool,

the locally distributed devices of a PACS to supply his private tool with knowledge and to take care of information logistics.

● The engineer understands

a PACS as a network of distributed but more or less identical functions,

the functions of a PACS being separated into layers due to the different nature of information, or the different type of processing.

Thus the LOCAL decomposition of the user's model has to be translated into a FUNCTIONAL decomposition of the engineers model. The following functional decomposition approach is based on the consideration that in a PACS-like information system, categories of functions can be designated by the nature of information, the type of processing, and the typical size of data packets. Along these lines, we can define three categories of functions and associated communications.

The first is PICTURES, a special type of information represented by uncommonly large data packets. Their processing is unique, and the applied technologies will be unique as well. A strict separation of picture data functions presupposes the definition of the second and superior category: CONTROL of picture operations. This category includes the control of transmission, filing, and retrieval actions. It also covers the vital field of man-machine interaction. All control, however, is necessarily based on knowledge, which will be provided by the next category -- a separate, superior category of functions: MANAGEMENT of information. This category includes the supervision of the PACS operations and the department, and all support functions providing convenient user access to the current knowledge actually required.

Looking at the three categories of functions again as a model of separated system layers will help to modularize the technical model according to three principal objectives:

● FLEXIBILITY. Unified modules should be used to build autonomous devices
such as imaging systems; modular hardware/software will simplify adaption to specific user requirments. The resulting devices will be employed to configure a specific PACS.

● COMPATIBILITY. Interfaces between modules must be standardized. Separation of functional layers simplifies the optimization and reduces the
complexity of communication protocols. Thus, connecting devices that are
different in nature or have different manufacturers ceases to be an in-
surmountable problem.

● SPEED. Negligible response time is a must for routine applications. Dis-
entangling the three functional layers will prevent speed-reducing data
flow conflicts and suboptimal hardware/software compromises. On the picture layer in particular, where the massive data problem occurs, optimized hardware concepts can be introduced without affecting other layers.

A typical example of a modular PACS based on these principles is shown below.
Imaging stations and diagnostic workstations are situated at the user sites.
Central facilities consist of the picture base - a giant data store - and
the supervising data management system with patient database and support
functions for the departmental organization and logistics. These central
facilities will also provide connections to the outside world (hospital information network, remote users, etc.). A mix of communication networks can
be employed on the separated layers. One result is that management information and control commands can be connected to an existing LAN, while optical fibers build up an image highway with an optimized topology.

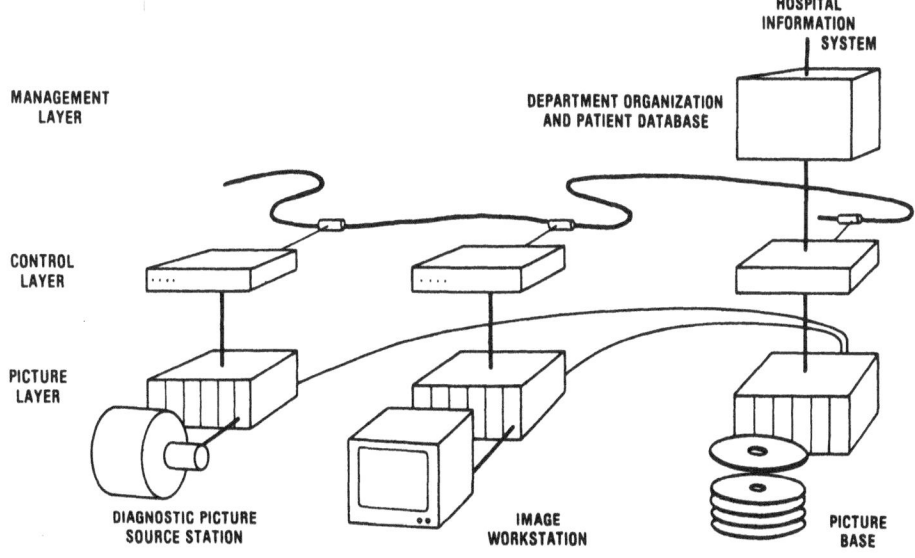

We were forced to generalize while describing the modeling process and the
resulting model structure, but now we must become specific about a number
of problem areas that are unique to a PACS. After establishing the proposed
separation into functional layers, we can easily employ existing mini- or
microcomputers for all management and control tasks. Existing database systems can be modified to manage a picture base as well. On the picture layer,
however, we must attack the massive data problem with new technologies,
techniques, and procedures.

The basic concepts of a PACS are storage, transmission, and processing. The
route of all information in an information processing system is a chain of
paths alternately in the space domain (transmission) and in the time domain
(storage), which are connected by processing steps (where "processing" also
includes very basic dataset transforms):

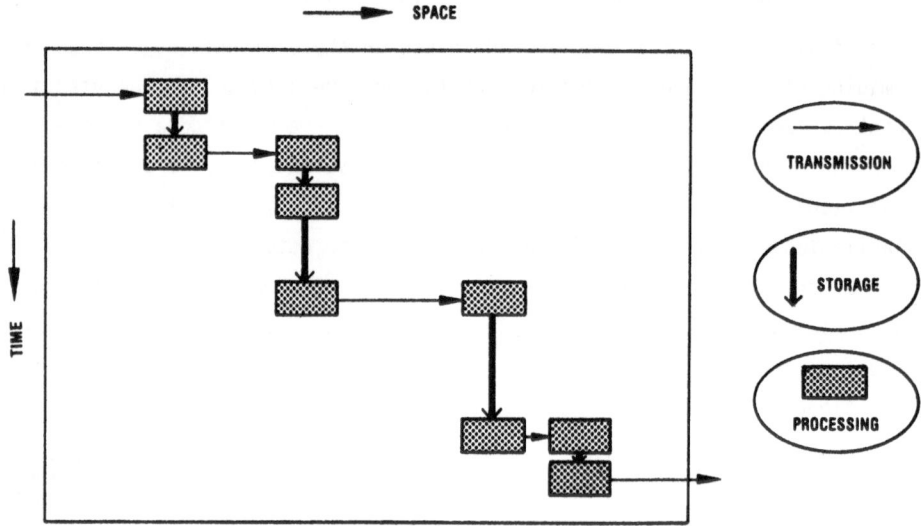

Architectural concepts for those three key areas will be described in the
following:

- STORAGE of millions of megabyte-size pictures requires specific strategies, particularly when new storage technologies such as digital-opti-
  cal recording are applied. A concept for a unifying strategy on all
  storage levels will be proposed.

- TRANSMISSION of several megabytes for each picture access requires optimized data structures, which, for the sake of compatibility,
  should be generic with respect to the size of the picture matrix. We

will propose a coding scheme that combines data compression with access-optimizing structuring and matrix size conversion features.

● PROCESSING of megabyte-size matrices at user-friendly times of one second or less requires a throughput rate of some 10M picture element operations per second, which, by definition, include pixel transfers and address generation. We will propose a modular picture processor architecture that could lead to unified hardware on the picture layer throughout all imaging devices.

*

The global decomposition process has allowed us to focus ourselves on the global problems. We will demonstrate now how we can further benefit from decomposition strategies on each of the above mentioned areas. We will start with the most challenging area PICTURE STORAGE.

Working with pictures necessitates the use of picture stores, that meet a large range of requirements regarding the storage period, access time, and the number of stored pictures. Extremely fast and flexible access is required within the user environment (workstation) while pictures are beeing interactively processed and displayed. But this requirement only applies to a small number of pictures that are quickly interchanged. While well-established storage technologies such as solid-state RAM and magnetic disks can be applied here, high-density optical storage will be a must for all central archival functions. One optical disk of 2G-byte storage capacity |2| will store 1000 to 10.000 pictures with a reasonable access time. In 10 years a medium-size hospital produces about five million pictures, most of them high-resolution X-rays |3|. Therefore, mechanical disk exchanges ("jukeboxes") for several hundred disks, or automatic magazines for several thousand, will have to be applied, significantly increasing the access time. Retrieval statistics indicate that a trade-off between access time and age of pictures would be acceptable. This trade-off requires structuring the archive into a small "actual," a larger "active," and a very large "permanent" section |4,5|.

These parameters have all been incorporated in a hierarchical scheme that identifies six levels of storage through which the pictures propagate -- downward during the filing process, upward during a retrieval process. The typical function of each level leads to the definition of performance parameters. In accordance with the logical storage hierarchy, the decivise parame-

ters -- access time, storage period, storage capacity, and the storage economy factor -- are found to be proportional functions of the level depth. The
location of islands in the parameter space allows us to assign the storage
technology best fitted to each level.

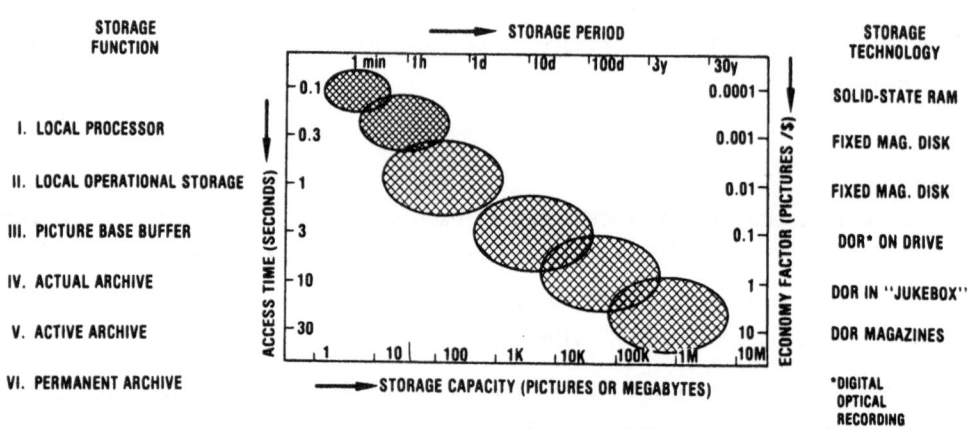

The concept of a storage hierarchy helps in managing the storage problem
physically, but it undermines efforts toward its logical management. To get
back to a simple and straightforward store management model, we introduced
a unifying virtual addressing concept. The storage spaces on the different
storage levels are all projected onto a single linear address space. During
filing, the address of a picture is shifted step by step to the right of the
address space. When a picture has been copied into a deeper storage level,
it also remains physically on the previous level as long as the occupied
storage space is not needed for something else. This space is then logically
transferred with the physical transfer of the picture. During retrieval,
when a picture propagates to higher storage levels, the occupied pieces of
storage space on the higher levels will always be cut out of their original
address range and will get the address of the deepest location the picture
has reached (see scheme on next page).

Every picture is now found only at the farthest right address it has reached.
A mapping procedure, which is performed individually on each storage level,
ensures that during a retrieval process the called picture is always accessed
on the highest available level. For example, a picture that has just been

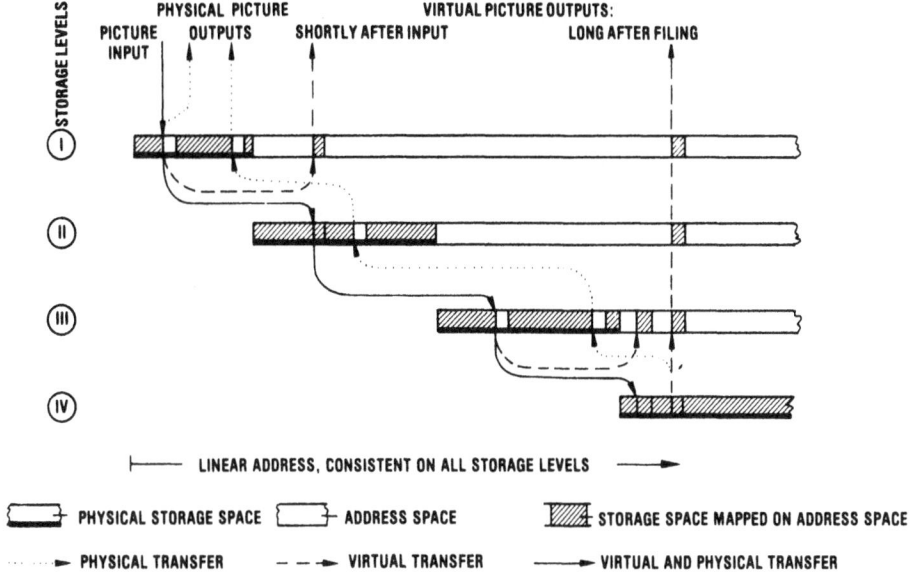

retrieved may still exist in the operational store of the workstation (level 2).
Calling this picture again does not require a new access to the picture base.
Also, when sequences of logically connected pictures have to be retrieved
(e.g., during the case study of a repeatedly treated patient), extended
access times caused by the scattering of the desired pictures over several
optical disks can be shortened. The pictures most likely to be retrieved next
will then be transferred in advance to a higher level (e.g., picture base
buffer store) by the local store manager.

The operation of the different storage media is quite sophisticated, especially that of optical disks, which cannot be erased. In spite of this, and
the complex partitioning of the overall storage space, the logical management of the picture base has become simple and straightforward. Futhermore,
it is consistent with the management of the workstations' decentralized
storage facilities. Any given database system (e.g., hospital or departmental information system) can perform this task by simply adding picture references to the patient records; these will point to the final addresses of
the related pictures. The simplicity and transparency thus achieved clearly
result from the hierarchical system concept, which decouples the management
layer from the physical picture stores by inserting appropriate address mapping and local transaction management on the control layer.

Complexity of DBMSystem

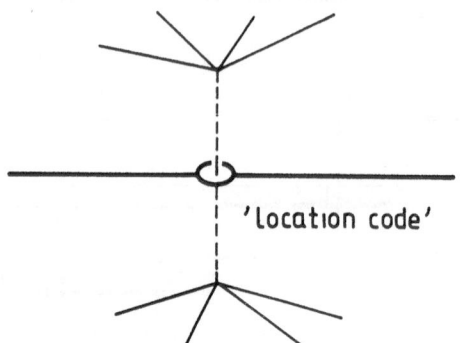

'Location code'

Complexity of Picture Storage System

*

The second essential area is PICTURE TRANSMISSION. Let us assume that appropriate technologies for high-speed data transmission are being developed for general applications. Using those technologies for a PACS, however, we have to take into account the specific nature of pictures. Therefore, we shall concentrate on the problem of PICTURE CODING.

Codes for the transmitted information are prerequisites for any communications. They must be efficient, flexible, and consistent throughout a system. With respect to these objectives, data formats based on a pixel-by-pixel chain are insufficient. The result would be as many codes as types of pictures in a PACS, and every introduction of a new imaging device would presumably add a new code. The application of individual data compression schemes would cause further complications. Finally, in view of the inevitable speed bottlenecks on the picture's route from the storage medium to the display, there would be an annoying wait for the last pixel to be transmitted before anything worthwhile could be done with the picture.

We therefore advocate adopting a picture matrix transformation scheme that transforms a given matrix into a coarse matrix (base matrix) and a set of detail information matrices |6|. A suitable algorithm is given by the S-transform coding scheme |7|, a derivative of the Hadamard transform. The basic transform step, which is successively applied several times to a given picture matrix, partitions the input matrix into subpictures two pixels

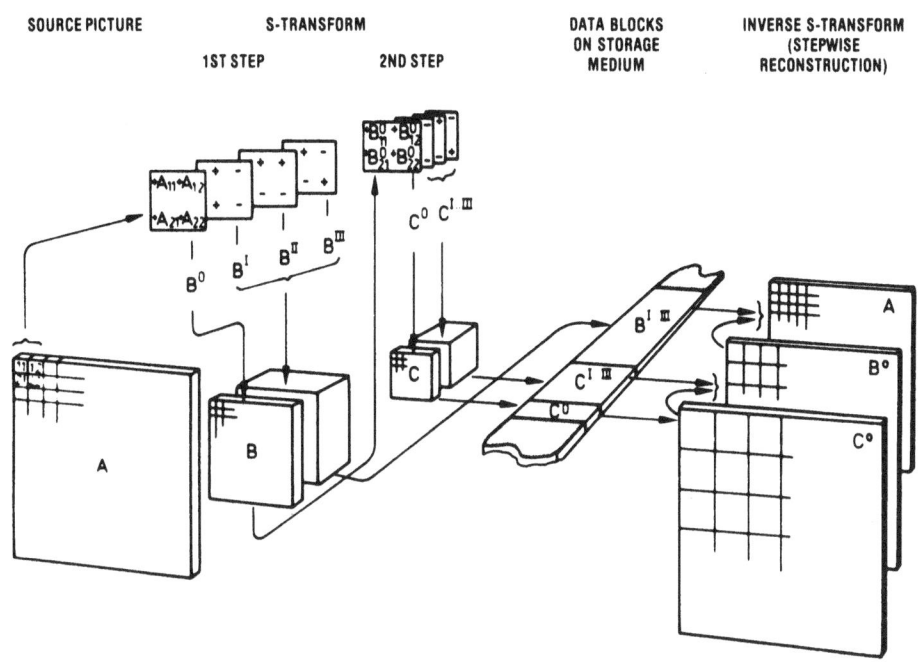

square. It then computes the sum of all four pixels and the differences between the two horizontal pixel pairs, the two vertical pixel pairs, and the two diagonal pixel pairs. The resulting matrix of pixel sums represents the input picture, but with half the original spatial resolution, while the three matrices of pixel differences contain the missing details.

The basic S-transform step must be repeated by taking in each subsequent step the sum matrix of the previous step until an original picture of arbitrary resolution has been reduced to a base matrix of unified size (e.g., 128x128 or 256x256 pixels). After each step, the difference matrices are separated and stored away. At this point, well-known nonlinear and/or adaptive quantization schemes can be applied to compress the data sets, because the pixel differences are likely to be close to zero. Because of the algorithm, similar compression factors have been obtained as with the well-investigated Hadamard-transform. Current research allows us to predict that with reasonable computing effort, a compression factor in the range of two to ten can be achieved.

Picture retrieval is accomplished by first accessing the final base matrix. For example, assuming that three transform steps have been applied to a

2048x2048-picture matrix, the first access step will present a 256x256 matrix. Even at moderate transfer rates (e.g., 2M bps) this near TV-quality survey picture will be accessed almost instananeously, since a data volume of only 1/64 of the original picture has to be transferred for this first access step. The blocks of detail information are transferred in subsequent steps as required by the user. In a continuous decoding process, akin to the adaption process of the human eye, the resolution of the picture is gradually increased.

From the system point of view, the proposed coding concept is an essential step toward meeting the diverging interests on the different conceptual levels. The concept combines several important features:

- ● ECONOMY. Compression algorithms can be optimized to specific types of pictures. "Empty" detail matrices will be truncated at the source already. Because of the simple add/subtract algorithm, the processing effort for the basic transform scarcely exceeds the effort which is necessary for data formatting anyway.

- ● EXTENDABILITY. Faced with increasing image quality requirements and progressing imaging technologies, existing systems can be extended without conflicts by implementing additional transform steps.

- ● COMPATIBILITY. Every picture will be transformed down to a standard base matrix by its source. During retrieval, the adaptation of high-resolution pictures to low-resolution retrieval stations needs only the truncation of detail data blocks; the inverse, the display of low-resolution pictures on high-resolution displays, is inherent in the decoding process. Thus, standard format sets for picture storage and standard transmission protocols can be defined that will be applicable throughout the system.

- ● USER FRIENDLINESS. The stepwise reconstruction of the final picture offers meaningful display information promptly, while the sharpness is continuously increased during the ongoing retrieval process. Retrieval can be aborted on demand. Quick browsing through stacks of pictures is supported.

*

The third area is PICTURE PROCESSING. All remote picture source subsystems and picture retrieval subsystems of a PACS need picture processing functions. The basic difficulty is that the megabytes representing the picture must be processed in times conveniently short for the user, i.e., approximately one

second. This results in a throughput rate of some 10M picture element operations per second, and this rate of processing cannot be achieved by conventional minicomputers. Although the processing functions are quite different
in the diverse subsystems, we can recognize a number of identical subfunc-
tions: display, display manipulation, input/output, coding/decoding, spatial
filtering, etc. Existing devices always use dedicated hardware for the speed-
critical functions; this hardware is connected to conventional mini- or
microprocessors, which perform the noncritical functions by software. A
breakthrough in imaging devices with network compatibility and high-power
imaging workstations, however, requires a universal approach that is more
economical and more flexible. We will describe an architectural concept for
picture processors which allows us to move on the shaded plane within the
flexibility vs. speed vs. cost space (see fig. below) in order to establish
the optimal trade-off for any type of dedicated system instead of confining
us to the three indicated "islands":

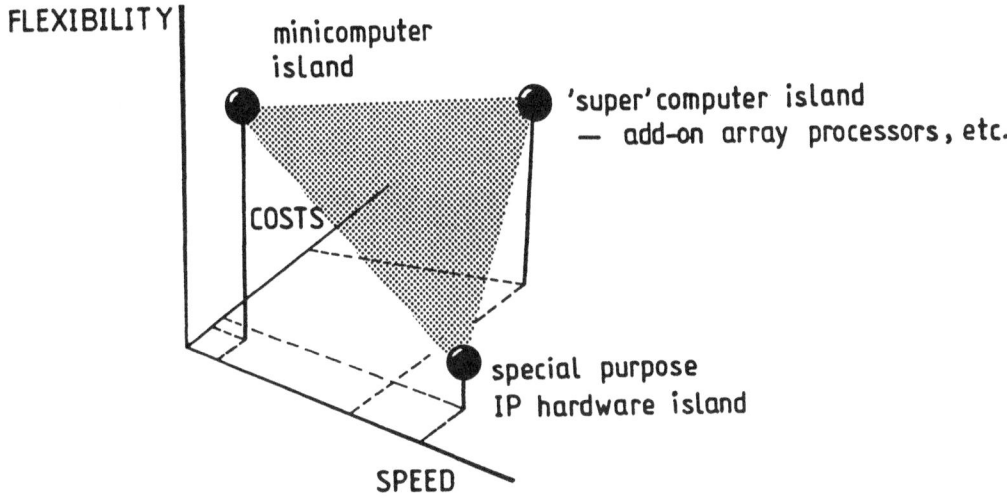

The layer structure of the overall concept features a strict separation of
picture processing functions from all supporting control and management tasks.
This enables us to think exclusively in terms of pictures when developing
a more general approach to a picture processor architecture. With the following approach, we tried to combine high speed with some essential advantages
of conventional computer systems, such as unified hardware structure, modu-
larity, and some degree of programmability |8,9|.

> Optimisation means specialisation.
> But don't diverge from proven
> concepts further than necessary!

The proven concept of data processing is to put everything necessery to control the process, manage the data, and run the process into one processing unit.

operating system
user interface
picture managem't
PIXEL managem't
PIXEL processing

In picture processing, however, there is obviously a tremendous unbalance between sub-tasks which are dedicated to the data elements of a picture -- the PIXELS -- and all remaining tasks. Therefore, let's cut off those tasks to keep the remaining efficient though beeing processed in the "old" style. PIXEL management and PIXEL processing -- now isolated -- appear as "brother and sister": two tasks of mostly equivalent weight which need each other but are of a basically different nature. Therefore, why don't separate both for the sake of individual optimisation -- but don't forget that they have to be carried out in a cooperative way!

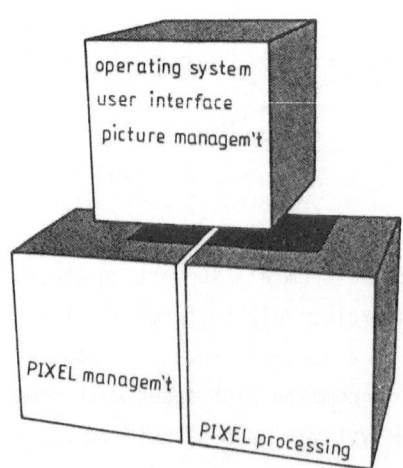

operating system
user interface
picture managem't

PIXEL managem't

PIXEL processing

Let's focus now on the PIXEL level. In picture processing, a meaningful entity of data is always a complete picture matrix. Generally, picture-to-picture processes can also be understood as kind of a matrix of the elements of the process, mostly homogeneous. However, only in very specialized cases source, process, and destination matrices fit element by element. Generally they don't.

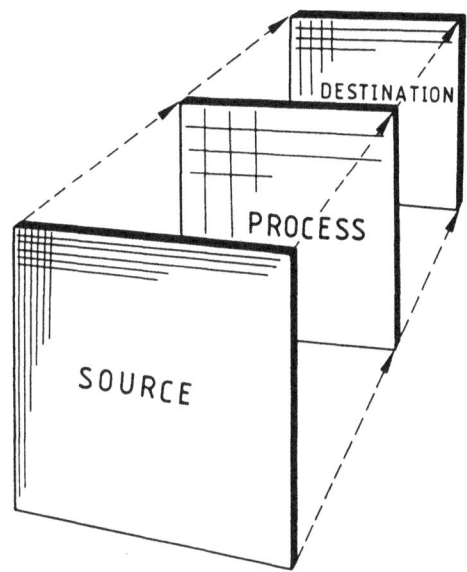

Numerous structures for array conversion have been proposed, all being necessarily limited concerning their range of application. For the sake of generality we adhered to a structure which, although being primitive, features utmost flexibility: Convert the spatial representation of a picture (set of storage elements, set of processor elements, set of spatial elements in the optical domain, etc.) into a temporal stream of elements (SPACE-to-TIME conversion), and convert them back into a spatial distribution (TIME-to-SPACE conversion) while assembling them in the desired way (see figure on next page).

The very basic scheme of information processing -- the alternating sequence of PROCESSING and TRANSPORT defined above -- still applies on this level. Picture sources, processor assemblies, displays, etc. are the processes in this scheme. Transport in time is done by a storage array, a giant RAM called PICTURE REGISTER, connected by means of a "pipe" to the processes for the transport in space.

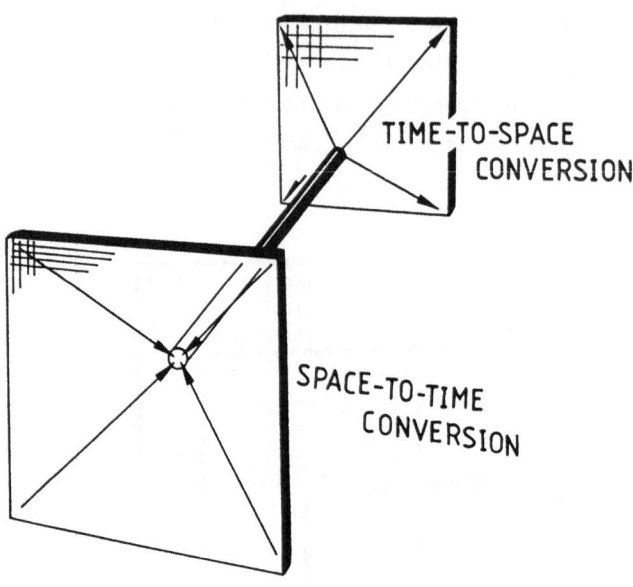

TIME-TO-SPACE
CONVERSION

SPACE-TO-TIME
CONVERSION

In a general-purpose system the "pipe" has to be "switched" to a variety of processes. A bus is suited to do this. It is called the PICTURE BUS. Because this bus system will be the physical backbone of the picture processor and the conceptual backbone of a unification attempt, it must be strictly defined and transparent in operation, and its performance should cover a reasonably large application range.

For an experimental system, a bus system was designed with a pixel data bus, a pixel adress bus, and a set of handshake and control lines. For simplicity and transparency, the bus is strictly synchronous, operating at 4oM pixels per second. The synchronous transfer of a pixel and its picture register address is necessary because pixel management is not provided in the picture register. Rather, throughout the processor, pixel management and pixel processing are defined as concurrent processes that are tightly related and must always be synchronized. Consequently, both processes are always combined in the processing elements. The clear definition of the processing unit and the management unit facilitates the design of a processing element's hardware and/or software optimized to a specific processing function. Furthermore, all supervisory and control tasks are physically separated from the picture data layer. A conventional mini- or microcomputer handles these chores by communicating with all processing elements via a slow control bus. Thus, the operation of the picture processor needs only simple macro programming.

175

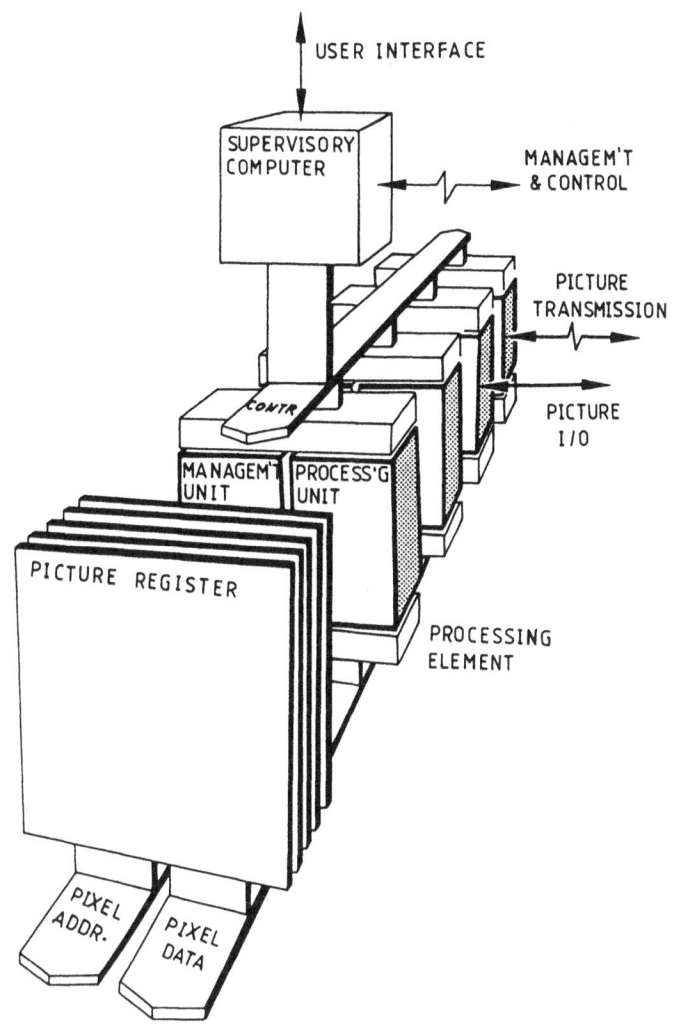

Again we profited by a thorough decomposition strategy. On the most speed-
critical lowest level -- the pixel level -- modularity was yielded by the
isolation of processing elements (LATERAL domain), freedom of PE-design was
yielded by hard-specified PE-to-bus interfaces (VERTICAL domain), and transparency and predictability of the behaviour of the bus was yielded by the
time-slot synchronisation and multiprocessor-type arbitration of the picture
bus (TEMPORAL domain) -- see sketch on next page.

A scenario involving the above concept may elucidate the benefits of well-
structured picture processing hardware. We will assume that the concept of
a separate picture layer has been accepted and that the specifications for
the high-speed synchronous bus - the backbone of the picture processor -
have been agreed on. Instead of developing hardware from scratch, designers
of imaging devices would then configure their customized designs from a set

Dimensions of Decomposition

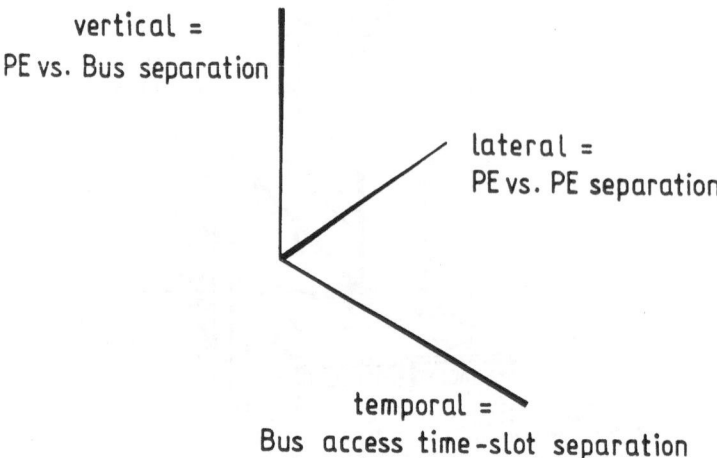

vertical =
PE vs. Bus separation

lateral =
PE vs. PE separation

temporal =
Bus access time-slot separation

of standard modules. These could include

● memory modules to configure the picture register to the required size;

● processing elements for picture I/O(displays, hardcopy units, image sensors, transmission lines, or operational picture stores such as magnetic disk drives);

● hardwired processing elements for the computation of frequently applied picture operations with provisions for pipelining;

● programmable processing elements based, for example, on bit-slice technologies, with an appropriate software backup, i.e., a problem-oriented, higher-level language to handle a microprogram library or to develop customized microcode;

● a bus arbiter to enable timeshared multiprocessing without involving the supervising computer;

● a generic structure for embedding arrays of microprocessors to create high-power processing elements; and

● dialog-oriented supervisory software to run one picture processor or even several cooperating processors.

Of course, this approach cannot provide high speed, unlimited flexibility, and low cost all the same time. Although we can assume a steady evolution of technologies over the next few years, today we must make tradeoffs involving these three attributes. The concept we have described will support and facilitate such decision making over a significant range of applications.

*

Observing the guidelines for an overall systems concept, we described approaches to open structures in the three essential areas: storage, transmission, and processing. The most important point is that only well-structured overall system architectures will make it possible to avoid conflicts. Naturally, conflicts will arise during the evolution of user requirements on the one hand and technologies on the other, and at some point the product designer must freeze specifications. Generic structures on all levels are prerequisites for modifications and extensions -- in effect, prerequisites for a living system.

We must also consider the extremely complex situation for the engineers involved in a PACS development. A successful systems architecture should have as its mirror image the organizational and managerial structure of development groups. Conversely, and often in actuality, ill-structured or nonexistent system concepts result in problems with the implemented systems that mirror the conflicts in communication and cooperation between development individuals or teams. On a reduced scale, our experience at bringing leading specialists from different fields into the conceptual process has been good. Instead of being limited to a specific system segment, their understanding of their individual task is top down. They understand the need to communicate and cooperate and are highly motivated to fit their part into the overall system.

So far, we have not mentioned the human-interface problem. Although we started with a user-oriented system model, and although quite a number of human interface problems have already been studied in laboratory or pilot project enviroments, we are quite sure that unforeseeable problems will arise after the introduction of PACS products. Again, open systems would support the implementation of ad hoc solutions and would free commercial interests from the economic conflicts that arise when customers call for individual modifications.

ACKNOWLEDGEMENTS

Philosophies, concepts, and specific approaches described in this paper result from many years of research in PACS and related fields in a fruitful cooperation with T. WENDLER, Philips Research Labs Hamburg. Essential input came from numerous medical professionals who gave us insight into the workings of hospitals. We had valuable discussions with psychologists and sociologists, in particular H. SCHMALE and H: JANSEN at the University of Hamburg. Many ideas were clarified with the help of colleagues of the Philips Medical Systems Division, represented by G. ARINK in Eindhoven, The Netherlands; R. HEU in Hamburg; and R. HINDEL in Shelton, Connecticut. The Philips Digital Optical Recording Laboratory at Eindhoven also provided valuable assistance.

The project owes much of its success to the practical experience gained from prototype hardware and software, and theoretical investigations in areas such as picture coding which has been carried out at the Philips Research Labs, Hamburg. All this work was performed by the members of our research team, D. Boehring, R. Grewer, K.-J. Moennich, J. Schmidt, H. Svensson, and, during the initial phase, P. Lux.

Finally, we want to mention the financial support by the German Ministry of Research and Technology. The underlying research was funded under grants 01 ZS 04/2-ZK/NT 02 and DV 4906-081 2074 A. The author solely is responsible for the content of this article.

*

The author and the whole research team he is working in hope that the results of their work will not be misused for military purposes. To the best of our knowledge we will not participate in any activities concerning military applications.

REFERENCES

1   TH. WENDLER, et al.:
    "Design Considerations for Multi-Modality Medical Image Workstations"
    This issue.

2   "INTRODUCTION TO DOR."
    Product brochure, Philips Data Systems, Nederland, B.V., Den Haag, The
    Netherlands

3   D. MEYER-EBRECHT, et al.:
    "Medical Picture Base Systems", Digital Image Processing in Medicine,
    Lecture Notes in Medical Informatics, K.-H. Hoehne, ed., Springer, Ber-
    lin, 1981, pp. 133-148

4   D. MEYER-EBRECHT, R. GREWER, K.-J. MÖNNICH:
    "Die digitale Bildbank - Grundstein für medizinische Bildinformations-
    und Bildverarbeitungssysteme", Proc. GMDS 27. Jahrestagung, Hamburg,
    Sept. 1982

5   Th. WENDLER, K.-J. MÖNNICH:
    "Strategies for Storage and Distribution of Medical Images and their
    Realisation in a Picture Base Prototype"
    Proc. 1984 I.J.A.S., Innsbruck, Feb. 1984

6   TH. WENDLER, D. MEYER-EBRECHT:
    "Proposed Standard for Variable Format Picture Processing and a Codec
    Approach to Match Diverse Imaging Devices,"
    SPIE-Proc. 318 PACS I, 1982, pp. 298-3o5

7   P. LUX:
    "Redundancy Reduction in Radiographic Pictures,"
    Optica Acta, Vol 24, 1977, pp 349-366

8   D. MEYER-EBRECHT:
    "The Management and Processing of Medical Pictures - An Architecture
    for Systems and Processing Devices,"
    Proc. IEEE Workshop PDDM, 1980, pp. 202-206

9   TH. WENDLER et al.,:
    "A Modular Multiprocessor Picture Computer Architecture for Distributed
    Picture Information Systems,"
    SPIE-Proc. 318 PACS I, 1982, pp. 125-132

THREE–DIMENSIONAL COMPUTER GRAPHIC DISPLAY IN MEDICINE: THE MIPG PERSPECTIVE

Gabor T. Herman
MIPG (Medical Image Processing Group)
Department of Radiology
Hospital of the University of Pennsylvania
3400 Spruce Street/G1
Philadelphia, Pennsylvania 19104

ABSTRACT

We concentrate on the problem of producing three-dimensional displays of organ surfaces from computerized tomograms. The theory of such surfaces is discussed. This theory leads to efficient methods of surface detection, visible surface determination, and rendering. The methods have been implemented in a software package especially designed for producing high quality displays inexpensively and efficiently for the medical end-user.

INTRODUCTION

In many three-dimensional (3D) imaging applications the 3D scene is represented by a 3D array of volume elements (or voxels for short). An example is computerized tomography (CT) which provides us with values assigned to abutting parallelepipeds filling a portion of 3D space occupied by a human body. Organs can be distinguished from their immediate surrounding if the densities of voxels just inside the organ are different from those of adjacent voxels just outside the organ.

In this article we discuss methods for defining, detecting, and displaying the surfaces of objects in such

NATO ASI Series, Vol. F19
Pictorial Information Systems in Medicine
Edited by K. H. Höhne
© Springer-Verlag Berlin Heidelberg 1986

discrete 3D scenes. We concentrate on the medical application. Other application areas are not explicitly mentioned, but the problems and approaches that are presented are not specific to organ surfaces based on CT.

The Medical Image Processing Group (MIPG, originally at the State University of New York at Buffalo, now at the Hospital of the University of Pennsylvania) has developed over the last nine years a comprehensive package of computer programs for the detection and display of surfaces in three-dimensional scenes. This package has now been perfected to the level where it can be routinely used on a mini-computer to obtain displays of complex organs or malformations in the human body based on standard series of computed tomograms. The package has been in regular use by physicians located in various parts of the world [1, 11, 13].

The most recent implementation of this package is a program called 3D83, which has been designed to run on a standard CT scanner: the General Electric 8800. This program allows the user to produce on the computer of the scanner (an Eclipse S/140) high quality 3D images from CT data with a minimum of fuss. We discuss the use of 3D83 in some detail.

In Figures 1 and 2 we illustrate the type of output that is produced by 3D83. For further examples, see [5, 6, 11, 13].

The images produced by 3D83 (and by similar programs) present medically significant information in a pictorial form. Since 3D display is a potentially very useful medical display modality [1, 7, 9, 10, 11, 13, 19, 22, 30], a program capable of producing such displays should be part of any pictorial information system that is designed for medical use.

Figure 1A

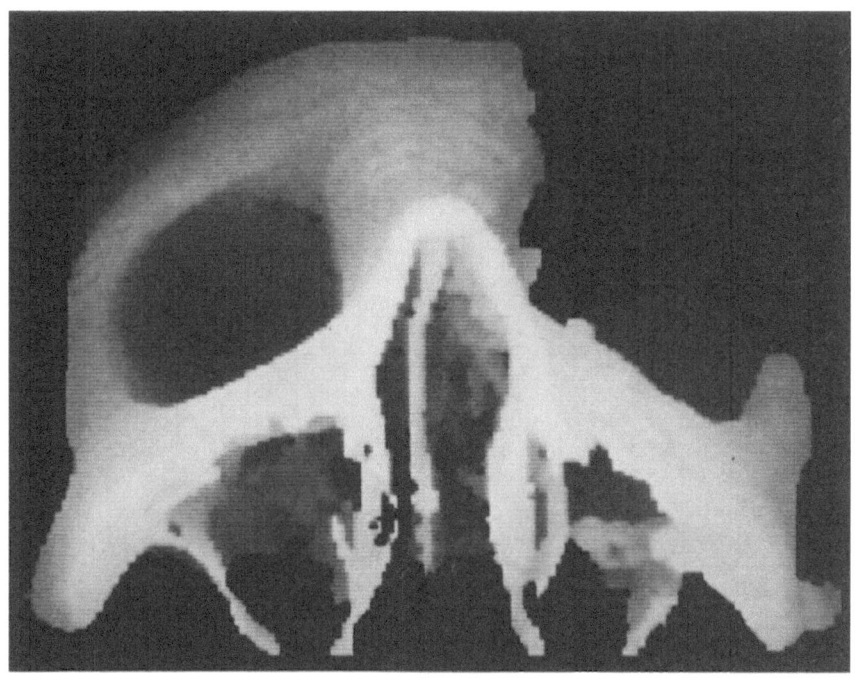

Figure 1B

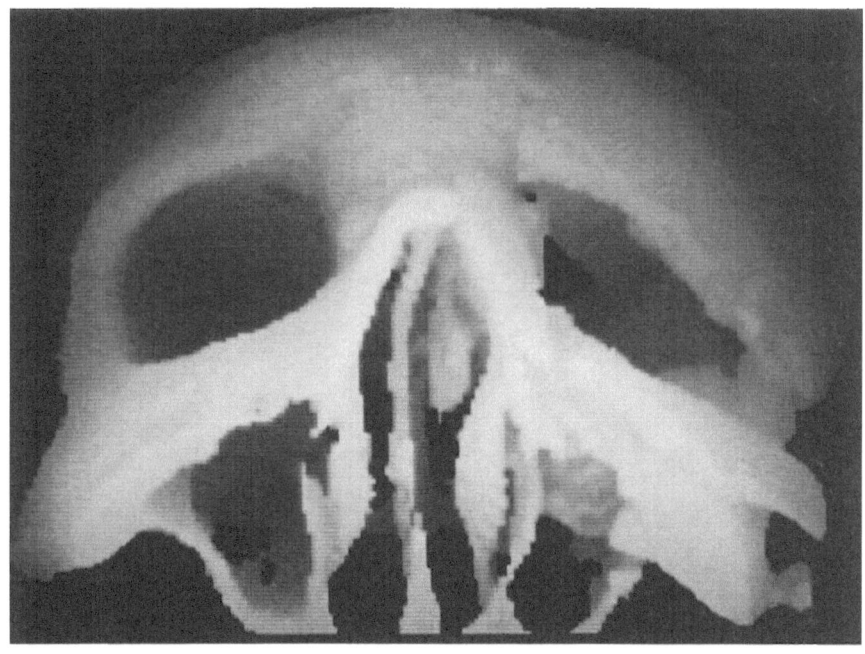

Images produced from pre- and postoperative CT scans
of the skull of a motorcycle accident victim.

Figure 2A

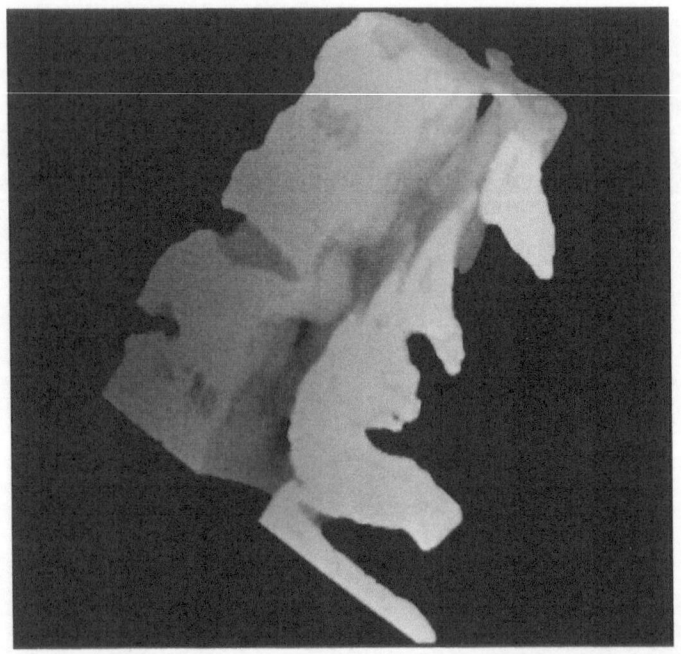

Figure 2B

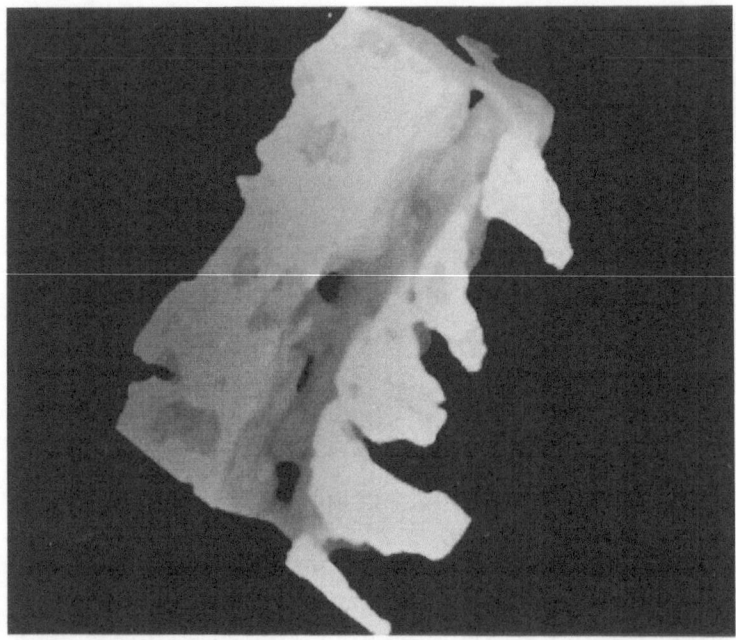

Images produced from pre- and postoperative CT
scans of the right half of a spine.

DISCRETE 3D SCENES

In this article a <u>discrete 3D scene</u> (or <u>scene</u> for short) is considered to consist of a rectangular parallelepiped (referred to as the <u>region</u> of the scene) which is subdivided by three sets of parallel planes into smaller identical parallelepipeds (referred to as <u>voxels</u>) each one of which has a value assigned to it (referred to as the <u>density</u> of the voxel). In application areas the scene is <u>digitized</u> in the sense that the density must be one of a finite set {L, L+1,...,U-1, U} of integers. If L = 0 and U = 1, then these are the only possible values, and we say that the scene is a <u>binary scene</u>.

We assume that together with the scene a rectangular coordinate system has been determined which assigns to each voxel a triple (i,j,k) of integers, where $1 \le i \le I$, $1 \le j \le J$, and $1 \le k \le K$. We refer to such a scene as an IxJxK scene. The IxJx1 scene formed by all voxels (i,j,k) for a fixed k is called the k'th <u>slice</u> of the scene.

An application area in which digitized 3D scenes arise naturally is CT [12], a medical technique for reconstructing the internal structure of the human body from x-ray projections. In CT the 3D scene is usually built up slice-by-slice. To obtain the densities for the k'th slice, the apparatus (called a <u>CT scanner</u>) sends many x-rays through the cross-section of the body which resides in the k'th slice, and estimates the densities based on the attenuation of these x-rays.

It is important to realize that, even though typically the computer of a CT scanner is a minicomputer with only 32K words of simultaneously addressable memory, the scenes produced by the scanner are large, quite typically 320x320x40. Hence, only about a third of a slice (less than one hundredth of the scene) can fit into the simultaneously addressable memory of the computer. Nevertheless, the demands of patient care and economics have resulted in fast and efficient programs for processing the data on the minicomputer of the CT scanner itself.

Scene processing is the 3D analog of picture processing [26]. In this article we mention three scene processing operations: subregioning, interpolation, and thresholding. Each of these operations produces a new scene from another scene.

A scene produced by subregioning a set of voxels is a subset of the voxels in the original scene, with densities unchanged. Taking a single slice out of a scene is an example of subregioning. Typically, subregioning is performed as an early step of scene processing: by isolating a region of interest the size of the data set that need to be handled by further processing is reduced.

Interpolation produces a scene in which the voxel size is different from that in the original scene. The densities assigned to the new voxels are estimated based on the densities of the original voxels. For example, for reasons to be explained below, we desire to have a cubic scene, i.e., a scene in which each voxel is a cube. On the other hand, the voxels associated with a CT scanner are typically not cubic; e.g., 0.5 mm x 0.5 mm x 3 mm. In such a case, we use linear interpolation to estimate the densities assigned to 0.5 mm x 0.5 mm x 0.5 mm voxels. If the original scene is an IxJxK scene, then the interpolated scene is an IxJx[6(K-1)+1] scene.

Thresholding produces a binary scene. The original and the resulting scenes are of the same size and are subdivided into voxels in the same way. A voxel in the binary scene has density 1, if the density in the original scene is above a predetermined (threshold) value. For the illustrations given earlier (Figures 1 and 2) we were interested in the bony structures of our patient, we therefore chose the threshold so that it is below the densities of the voxels containing mostly bone, but it is above the densities of the other voxels. Thresholding is the simplest mode of producing from an arbitrary scene a binary scene of the same size. Other techniques are often required to deal with more complicated situations (these often arise when we wish to see soft tissue surfaces or when the CT scans contain artifacts such as those due to fillings in teeth or breathing motion), but they will not be discussed in this paper, see [5, 28].

OBJECTS AND THEIR SURFACES

Even though human observers identify objects in 3D scenes without any apparent difficulty, the mathematically precise definitions of an "object" and its "surface" are far from trivial, and equally reasonable definitions can be (and have been) proposed giving rise to different types of entities being identified as objects. In this section we survey some of the approaches.

First, one has to recognize that in most application areas the discrete 3D scene is considered only to be an approximation to the underlying domain of interest. Usually, we assume that there is a function $f(x,y,z)$ of three real variables, and the density assigned to a voxel is an estimate of the average (or central) value of $f(x,y,z)$ in the voxel. For example, in computerized tomography $f(x,y,z)$ is the x-ray attenuation coefficient of the human body at the point (x,y,z). The object of our interest in this underlying continuous space will occupy some voxels totally, will miss some others, and will partially intersect a third set of voxels. In the process of discretization some information regarding this "true" object is irrevocably lost; nevertheless recovery of its surface (or a good approximation to it) may be our heuristic aim. In Figure 1A, for example, we show the skull of a motorcycle accident victim who injured one side of his head; our aim is to see the surface of the broken bones (and to compare it with the surface of the unbroken bones on the other side) to help with the planning of corrective surgery [13].

Although we have not yet defined "object" and "surface", the following can be understood just using the intuitive meanings of these words. Given a discrete 3D scene, an object in that scene corresponds to a real object in continuous space. Can we (or even if we can, should we) try to estimate the surface of the real object by a surface other than the surface of the discrete object? Such questions cannot be answered in general, and even in a particular application area they may be subjects of controversy.

We now indicate an approach to defining surfaces of objects.

We are dealing with ordinary three-dimensional Euclidean space, R3. We use topological terminology in the standard way (see, e.g., [24] or the Appendix). In particular we talk about closed, open, and connected sets in R3. The object that we are interested in is a subset of R3. The complement of the object can be divided into a number of components. (Components are defined, as usual, to be connected subsets, which are not contained in yet larger connected subsets, see [24].) The surfaces of the object are the frontiers of these components, one surface for each component. (Frontier of a set D is the set of points which are closure points both for D and its complement.)

Figure 3 gives a two-dimensional example. Consider a subset Q (we assume it to be closed) which is indicated by leaning lines "/". The connected component labelled by E may be an object under a suitable definition. The leaning lines "\" indicate the complement CE of E. This open set CE has two components: the infinite one labelled D and the finite one enclosed by E. (These are separated, since Q and hence E was assumed to be closed.) The surface of E associated with D is the frontier indicated by the "hairy" line around E. The object E has one other surface, not indicated in the figure, associated with the other component of CE (the one which is enclosed by E).

We have now described a definition of a surface which is usable with any definition of an object. This of course would lead to some very peculiar surfaces, unless we put some further restrictions on the nature of objects. Note

189

Figure 3

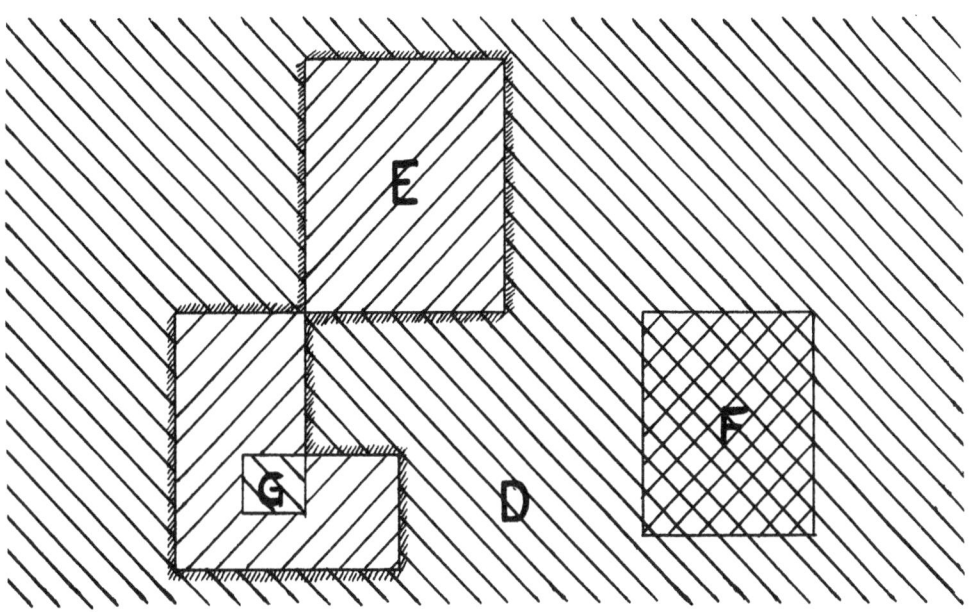

A two-dimensional illustration of our definition of
a surface. The closed set Q is marked by lines "/".
It has· two components, one is E and the other is F.
The complement of E is the open set CE, marked by
lines "\". Since F is a subset of both Q and CE, it
appears cross-hatched. CE has two components, the
open sets D and G. (The common closure point of D
and G is in neither of them; it is in fact in Q.)
The "exterior surface" of E is the frontier of D,
marked by the "hairy" line around E. E also has an
"interior surface", namely the frontier of G.
(Reproduced from G.T. Herman, "Three Dimensional
Imaging from Tomograms", Digital Image Processing in
Medicine, K.H. Höhne (ed.), Springer-Verlag, Berlin,
Germany, pp. 93-118, 1981.)

that in the example given above, E is closed, bounded and connected, and that the surface we obtained corresponds to what one intuitively would describe as the "exterior surface" of E.

We now give a precise definition of an object in a discrete scene. We assume that a region is given and is already subdivided into voxels. In what follows we use the word voxel to mean a closed set of points in R3. In other words, a voxel includes all its faces. We put the following restrictions on the notion of an object in the scene.

Property 1. An object is the union of voxels.

Property 2. An object is connected.

The boundedness of the region of the scene and Property 1 imply that an object is a bounded closed subset of R3. Another consequence of Property 1 and our definition of a surface is that a surface of an object in the scene is the union of a set of faces of voxels (closed sets including their edges), where each face is contained in exactly one voxel which is inside the object. Property 2 and standard results of topology [24] imply that a surface of an object in a scene is a connected set. (Recall that as a result of our definition of a surface, one object may have several surfaces.)

These properties, which make no reference to the densities of voxels, can be used to exclude certain subsets of the region as objects in the scene, but tell us nothing about how to find subsets which are objects. Many different approaches can be, and have been, taken.

For simplicity, let us restrict our attention (for now) to binary scenes. A possible definition is the following. Let Q denote the union of all voxels with density 1. We may then define an object as a connected component of Q. Such a definition has been used, for example, by Rhodes [25]. We refer to objects defined this way as Q-objects.

In our own work [3, 4, 18] we have found it more convenient to use the following definition. A 1-object is a 0-object which cannot be described as the union of two 0-objects U and V whose common points are isolated. (Geometrically this means that we do not allow 1-objects to be connected by vertices only, they must be connected at least by edges, which are 1-dimensional entities; hence the name 1-object.)

One can similarly define 2-objects (not allowing edge connectivity). For a discussion of such notions see [21, 29]. Similar distinctions exist in 2-dimensional pictures; see, e.g., [26].

Given an arbitrary scene, one way of defining objects in it is to use thresholding and then to apply the definitions discussed above to the resulting binary scene. This is the approach of 3D83 and will be further discussed below.

The approach of thresholding and then defining objects in the resulting binary scene is appropriate if there is a single threshold which clearly separates the object from its surrounding. If this is not the case, then we have to work with the original scene. Sometimes it is possible to reverse the approach described above; we may be able to detect the (external) surface first and (intrinsically) define the object as the set of voxels "inside" the surface. Such an approach has been taken, for example, by Liu [20], who used local gradient properties to put together surfaces consisting of voxel faces. For generalizations of the approach see [14, 23, 31]. Such approaches are of more general applicability, but are likely to be considerably slower, than surface detection for binary scenes. We give an example of the latter in the next section.

SURFACE DETECTION OF 1-OBJECTS IN BINARY SCENES

It can be shown [18] that, in a binary scene, for any
face F which belongs to exactly one voxel of density 1,
there is a unique 1-object O and a unique surface S of O
such that S contains F. In this section we discuss an
algorithm which, given F, finds S.

There is a complex definition [18] (details of which
are not repeated here) which with any face F in S associates
exactly two other faces F1 and F2 in S to which F is said to
be P-adjacent. The important thing about P-adjacency is
that, given F, it is computationally inexpensive to find F1
and F2.

The following algorithm has been first proposed by
Artzy [4]. Its input is a binary scene and a face F0 in a
surface S. Its output is a list L of all the faces in S.
It makes use of two auxiliary data structures, a queue X and
a list M. We use the verb "mark" to mean "put in the list
M" and the verb "queue" to mean "put in X".

BEGIN

1. Output and queue F0. Put two copies of F0 in M.

2. While X is not empty,

    a. remove a face F from X;

    b. determine the two faces F1, F2 in S to which F is
       P-adjacent;

    c. for i = 1,2,

            if Fi is in M,

                    then remove it from M;
                    else queue, mark, and output Fi;

            end if,

        end for,

    end while.

END

It is proved in [18] that the above algorithm terminates, and that at termination the output list L contains precisely one copy of each face in S.

For a detailed discussion of the rather ingenious nature of this algorithm see [3]. The important aspect of it is in step 2c, where membership of Fi in M is checked and Fi is removed from M if it is found there. This way the size of M is kept small, typically less then 2% of the output list L. The importance of this is easily appreciated when we consider that the size of L in typical medical applications is over 150,000 [3, 6, 17, 18] and can be greater than 500,000.

DISPLAY OF OBJECT SURFACES

Property 1 of objects in a scene results in the surface being a union of voxel faces. Such a surface is unrealistic, "real" objects are not unions of parallelepipeds. One can in principle make the voxels arbitrarily small, using interpolation, but that can increase computer costs associated with surface detection and display to an unacceptable level in a medical environment where the patient has to pay for the services rendered.

Two basically different approaches have been taken to resolve this difficulty. We now state what they are and then critically compare them.

One of the approaches has been illustrated in this article so far. It has been referred to as the "cuberille" approach in [6, 15] where detailed discussions of its nature are given. The underlying idea of this approach is to approximate surfaces by a large number of very simple surface elements. Surfaces of objects in discrete scenes, especially in cubic scenes, are automatically of this type. This approach is useful due to the fact that the simple nature of the surface elements allows us to prove powerful mathematical theorems. Such theorems lead to efficient computer procedures to handle the large amount of data. An example of such a theorem, leading to efficient boundary detection, has been given in the last section. Another example (related to surface display) is a theorem of Herman and Liu [15] which states that in a cubic scene the distances from an observer of two points A1 and A2 on the surface are ordered in the same way as the distances from the observer of the centers of the faces which contain A1 and A2, respectively. Such a theorem allows rapid determination of the visible part of the surface.

The interpolation which resulted in the cubic scene in our medical examples has been used partly so that we can apply the rapid computer graphic display procedures which can only be applied to cubic scenes. The other reason is that we believe that it results in more accurate approximation to the "true" surface than what we would have obtained using the alternative of thresholding the original scene and then displaying the resulting surface. The surface produced by this alternative method would have a different "resolution" in one direction from that in the other two. The use of original densities for interpolation and the thresholding of the interpolated scene is likely to result in a more accurate approximation to the true organ surface than the alternative. This is a statement of personal belief which is not shared by all workers in this field. Since it is easy to create single objects both for supporting and for refuting this claim, its validity can only be resolved by long practical experience in the field of application.

An alternative to the cuberille approach is to determine a relatively few points on the surface (as compared to the number of faces used in the cuberille approach) and put together a surface by locally fitting surface patches to small collections of such points. In this paper we shall refer to this as the "patching" approach, to distinguish it from the cuberille approach. (An alternative name that has been used is "tiling".)

An example of the patching approach is the following. Suppose for now that the surface in the scene is sufficiently simple so that for every slice the intersection of the surface with the central plane of the slice is a simple closed curve. We identify a number of points on this curve so that the straight line segments connecting these points give a reasonable approximation to the curve. Then we fill in the surface between curves in two consecutive slices by triangles which have two points on one of the curves and one point on the other. Such triangles can be chosen according to some optimization criteria; for example, Fuchs, et al. [8] produced an algorithm which selects the triangular patches so that the resulting surface has minimal area for the selected points. Triangular patches for displaying the surfaces of organs have been reported in the literature since 1970 [10]; for works in this direction involving CT see [7, 9]. Examples of other publications using the patching approach are [19, 22, 27], the last of which uses cardinal splines to fit the data. We now compare these two approaches from various points of view.

(i) The intersection of a complex object with a single slice of the scene usually contains many disconnected components. Looking at a single slice it may be impossible to tell which voxels belong to the object. Even if this task is solved, the slice-to-slice connection in a complex situation (such as in the presence of bifurcation) may be very difficult using the patching approach. Such problems simply do not occur in the cuberille approach.

(ii) In the patching approach an orientational bias may be visible due to the way the surface is put together. Whether this accurately reflects the state of knowledge due to the noncubic nature of the original voxels or hides information which exists in the actual densities assigned to the voxels is a matter open to dispute. This author believes the latter to be the case.

(iii) In spite of the much larger number of surface elements, the reported computer times indicate that the cuberille approach is at least an order of magnitude faster

in the medical application area.   This is due to the fact
that powerful mathematical results lead to special purpose
programs which can deal with cuberille surfaces very
rapidly.   This advantage is lost if general purpose display
software (applicable to arbitrary patches) is applied to
cuberille surfaces.

(iv)   In general, a well designed patching program
will produce prettier images than what can be obtained by
the straightforward cuberille approach.   It is the belief of
this author that, at least in the medical area, this
prettiness is usually at the expense of accuracy of
representation of the "true" surface.

RECENT ADVANCES IN THE CUBERILLE APPROACH

We first briefly reiterate the cuberille model and
discuss the usual conventions for rendering cuberille-based
surfaces.

A cuberille is a dissection of the 3D space into equal
cubes (called voxels) by three orthogonal sets of equally
spaced parallel planes.   In the intended use of the model, a
finite subset of voxels is to be identified as forming the
object of interest, and the surface to be displayed is a
collection of faces of these cubes.   In our discussion we
deal only with orthographic projections, i.e., those in
which a point P on the surface of the object is mapped onto
the point Q on the display in such a way that QP is always
orthogonal to the display screen.   We have found that for
the medical applications of our interest (where the
distances between the eye and points on the surface of a
relatively small organ do not vary much for natural viewing
positions), the orthograhic projection produces acceptable
images.   For similar reasons, we can assume that the
illumination source is emitting parallel rays of light.   For
the sake of computational simplicity (e.g., avoidance of
shadows), it has traditionally been assumed in the cuberille

approach that the light rays are also orthogonal to the display screen. We also adopt this lighting model.

Accurate representation of objects as collections of cubes and their surfaces as collections of faces demands that these cubes and faces be small. Thus the number of square shaped surface elements that represent a particular surface in the cuberille approach is likely to be much larger than the number of surface elements demanded by other approaches [7-10, 19, 22, 27]. Nevertheless, total computational demand is not necessarily increased, and in some cases drastically decreased. This is because the simplicity of the basic units of the cuberille approach allows us to develop powerful mathematical theorems about them, which in turn can be used to significantly simplify surface detection, visible surface determination, and rendering. We have already discussed surface detection in some detail. For recent developments on speeding up visible surface determination in the cuberille environment see [16].

At the early stages of our work [14, 15, 20] the quality of the displays based on the cuberille approach was not quite as good as we desired, especially if they were displayed in a dynamic sequence. A number of techniques have been suggested to deal with this [12]. Recently, we have proposed [6, 17] an approach which improves the quality of the displays, but retains those properties of the cuberille approach which make it computationally efficient. In this new method, visibility of a point on the surface is determined in the same fashion as in the previously published cuberille approaches [2, 15]. It is the shading of the visible points which is to be done differently. The convention of assigning a single shading value to all points of a face is retained, however the value depends on the direction cosines of all the faces having an edge in common with the given face. We call this <u>contextual shading</u>. Additional computation for contextual shading is minimal, since all necessary information regarding the faces sharing an edge with the given face can be gathered while the boundary surface is detected and can be stored as attributes of the face in question. Striking improvements in the quality of displays due to contextual shading have been illustrated [6, 17].

Another recent development is a set of interactive techniques which permits isolation of arbitrary subregions of the 3D scenes [5, 28]. These can be used to supplement thresholding in producing binary scenes from arbitrary 3D scenes. This can be useful if (i) the discrete representation of a real world object of interest is not adequately separable from its background by using thresholding alone, or if (ii) we are interested in observing what an object would look like after an alteration to it (e.g., what is the shape of a complex bony structure after surgical removal of a part of it?).

An alternative recent approach is one by Vannier et al. [30], in which the thresholded 3D scene is displayed directly (i.e., without surface detection). This has the advantage of saving the initial time overhead of boundary detection before the first image is produced. On the other hand, once the surface is detected, the techniques described in this paper produce images more rapidly. The break-even point seems to be at around 20 images; if fewer images per patient are required then the method of Vannier et al. [30] will work somewhat faster. On the other hand, the flexibility in determining what is to be imaged and the appearance of the images appears to be superior in the methods described in this paper. Whether these differences have any effect on the clinical efficacy of the procedures remains to be seen.

THE COMPUTER PROGRAM 3D83

We explain what we are doing with special reference to a program 3D83 which was developed by members of the Medical Image Processing Group (in particular, L.S. Chen, G.T. Herman, C.R. Meyer, R.A. Reynolds, and J.K. Udupa) for the purpose of three-dimensional display. Such a program can be implemented on the computer of any CT scanner; ours happens to be a GE8800. All the work reported in this article and in [1, 11, 13] was produced by using a standard GE8800 without any additional equipment whatsoever.

The input to 3D83 is a CT file identified by a run number. The CT file contains all the reconstructed images for a sequence of patient sections; it is produced by the scanner as part of its standard operation. (We mention here by the way that the spacing of these patient sections does not have to be uniform; in particular, 3D83 can handle automatically a file in which some scans are missing.) The output of 3D83 is a movie file, also identified by a run number. The movie file contains a sequence of images of an organ surface, viewed from a number of different directions. These images are similar to what would be obtained if the organ was surgically removed, possibly cut open, and then rotated in front of the viewer. The way a movie file is stored in the CT scanner is exactly the same as the way a CT file is stored, and so images in a movie file can be viewed and filmed using the standard procedures, just as if they were images in a CT file.

When planning a 3D83 run, we use a flow sheet (see Figure 4) to organize the answers needed by the 3D83 program. We now briefly describe the how and why of completing the flow sheet.

In 3D83 it is possible to create a whole sequence of movies of (possibly) different organ surfaces rotating in (possibly) different directions. In making one movie there are two distinct processes involved: the detection of the surface to be displayed and the rotation of the surface as the images of the movie are created. If a movie is desired which contains a different rotation of the same surface which has been used in the most recently created movie, then there is no need to go through the surface detection process again. Question 0 allows the user to specify whether or not this is the case. Questions 1-8 are used to specify the surface to be detected. In order to have the answers ready

200

Figure 4

FLOW SHEET

(0) DO YOU WANT ANOTHER MOVIE OF THE MOST
      RECENTLY DETECTED SURFACE? (Y OR N)
      If you have answered Y, go to Question 9.   _____

(1) ENTER THE RUN NUMBER OF THE INPUT FILE: _____

(2) ENTER # 1 FRAME LOCATION:
      (SLICE, ELEMENT, LINE NUMBER)   ____ ____ ____

(3) ENTER # 1 FRAME SIZE:
      (WIDTH AND HEIGHT, IN PIXELS)   ____ ____

(4) ENTER # 2 FRAME LOCATION:
      (SLICE, ELEMENT, LINE NUMBER)   ____ ____ ____

(5) ENTER # 2 FRAME SIZE:
      (WIDTH AND HEIGHT, IN PIXELS)   ____ ____

(6) ENTER THE MODE - IDENTIFY (I) OR MEASURE (M) - TO SPECIFY THE
      RANGE OF CT NUMBERS FOR DETECTING THE OBJECT OF INTEREST: ____

(7I) If you have answered I to Question 6
      ENTER VALUES OF LEVEL AND WINDOW: ____ ____

(7M) If you have answered M to Question 6
      ENTER THE VALUE OF LEVEL: ____

(8) SPECIFY STARTING POINT FOR SURFACE DETECTION: ____ ____ ____
      (SLICE, ELEMENT, LINE NUMBER)

(9) SPECIFY AXIS OF ROTATION (X/Y/Z/T):   ____
      (WARNING: T WILL RESULT IN SLOWER PROCESSING.)

(9T) If you have answered T to Question 9
      ENTER DEGREE OF TILT: ____

(10) ENTER NUMBER OF IMAGES IN THE MOVIE: ____

(11) ENTER RUN NUMBER OF MOVIE FILE: _____

(12) DO YOU WANT ANOTHER MOVIE OUTPUT?: (Y OR N)   ____
      If you have answered Y, you have to complete another flow sheet.

Patient's Name: _____   Patient's ID: _____
Physician's Name: _____   Radiologist's Name: _____
Date of Scan: _____   Date of 3D Imaging: _____
Comments: _____
                                          Page _____ of _____

Flow sheet for the program 3D83.

to these questions we use a program DYTEST, which is part of the standard software provided by the General Electric Company.

The answer to Question 1 is simply the run number of the CT file which is to be used as input.

3D83 displays an object (or part of an object) which is within a "box"; the size and location of this box is provided to 3D83 by answering Questions 2-5.

First we discuss a special simple case, but one which is used very frequently. In this case the box is rectangular (see Figure 5). The box intersects each slice in a rectangle which is determined by using a frame. Figure 6 shows an image from a CT file with a frame over it. The program DYTEST reports to the user the location and size of the frame (answers to Questions 2-5). In the case of a rectangular box the answers to Questions 3 and 5 are the same. Choosing a rectangular box is a form of the subregioning procedure discussed earlier.

While a rectangular box is appropriate for many purposes, it sometimes becomes necessary to use a general box. An example of this is when we wish to demonstrate the size of the spinal foramina by displaying an anterior portion of the spine as if the posterior processes were removed. Since the spinal column is usually not parallel to the direction of the table motion, the requirements of keeping the anterior portion of the spine within the box and the posterior processes outside the box mandates a general box (see Figure 7). The subregion of interest is then the smallest scene containing the box.

The next step is the specification of the object of interest within the box. This is done using either the IDENTIFY or the MEASURE mode on the system. The selection of which mode to use depends on the object to be displayed. Thresholding corresponds to the MEASURE mode; therefore this is the only mode that we discuss. On the CT scanner the selection of the threshold is done as follows.

We press the so-called MEASURE button and adjust the LEVEL (threshold) control so that those voxels which belong

Figure 5

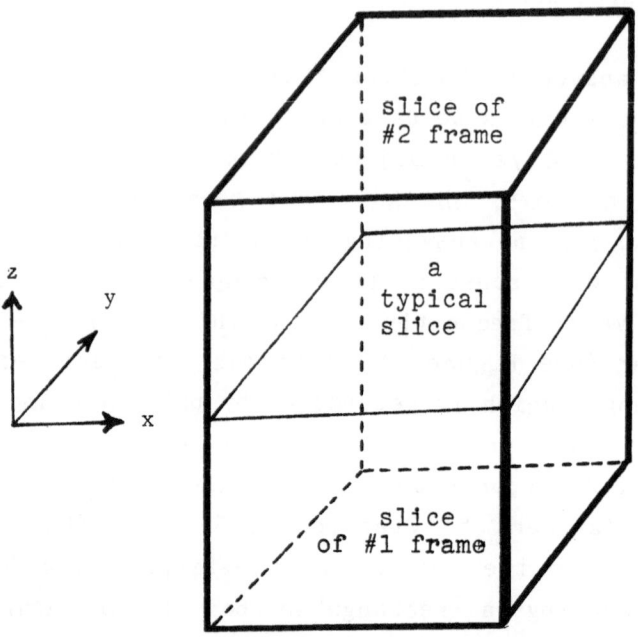

A "rectangular" box. The size and location within
the slice of the #1 frame and the #2 frame are the
same.

Figure 6

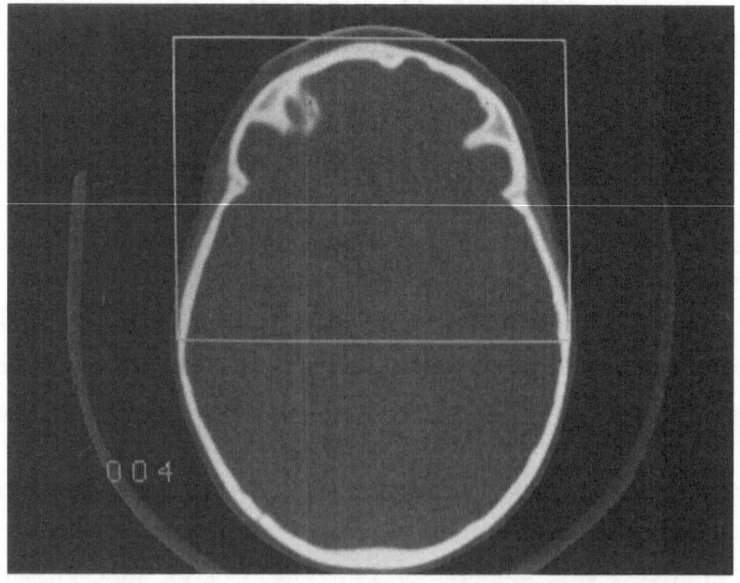

A CT slice with a frame marked on it.

Figure 7

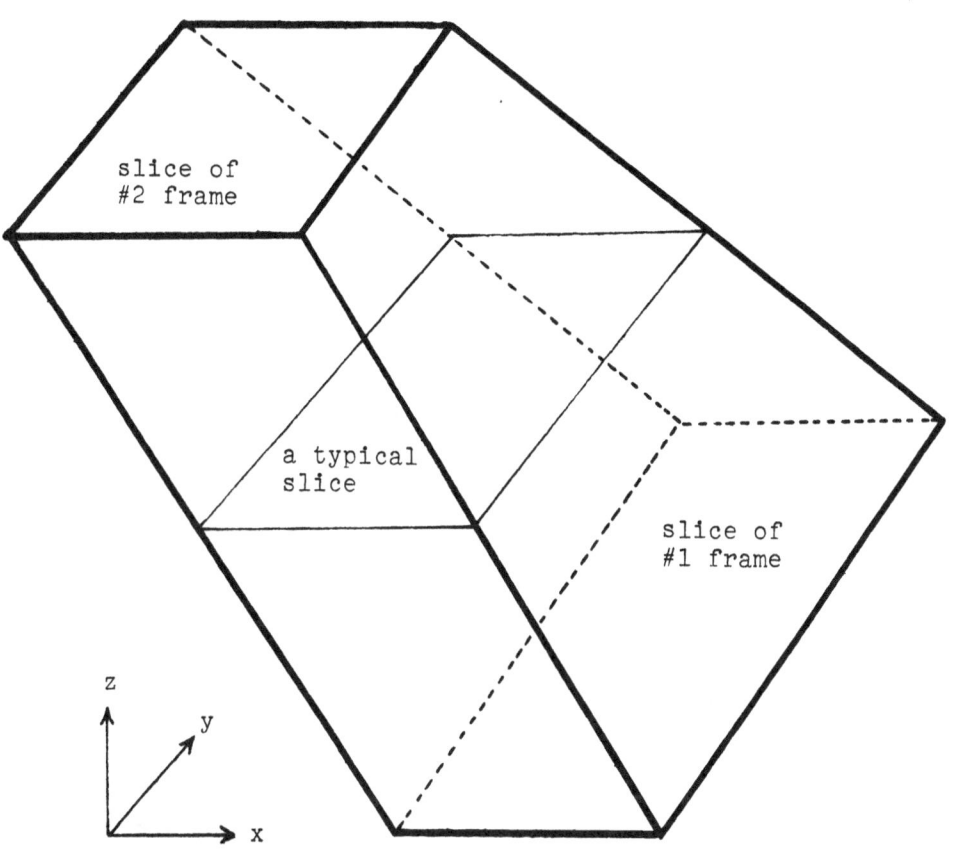

slice of
#2 frame

a typical
slice

slice of
#1 frame

z

y

x

A "general" box. The size and location within the
slice of the #1 frame and the #2 frame may be
different.

to the object that we wish to display appear bright white.
When we find a LEVEL which accurately specifies the object
of our interest, we can read off its value from the display
screen.

At this stage we have specified a box, and some voxels
within the box (those whose CT value is at or above the
level given in answer to Question 7M). The surface to be
displayed is the surface of a 1-object which is entirely
made up from such voxels. There may be more than one such
object, but the surface of only one will be displayed. The
choice of the surface to be displayed is indicated to 3D83
by the answer to Question 8.

Questions 9 and 10 define how the surface will be
displayed. In 3D83 we can rotate the surface around any one
of three mutually perpendicular axes, referred to as X, Y
and Z. The X-axis is horizontal in the plane of the screen.
Rotation of an object (based on the usual transverse slices)
around the X-axis will result in a "head-over-heels" type of
display sequence of the object (illustrated in Figures 1A
and 1B). The Y-axis is vertical in the plane of the screen.
Rotation around the Y-axis will result in a display sequence
in which the object is seen as if we walked around the
scanner to get our different views. The Z-axis is
perpendicular to the screen. Rotation around the Z-axis
will result in a display sequence in which the object is
seen as if we first made the patient stand up and then
walked around him to get our different views.

Option T is a generalization of the Y and Z options.
In all of these options the different views correspond to
walking around the patient. The T option allows any tilted
position in between the assumed positions of the Y and Z
options (illustrated in Figures 2A and 2B). Thus choosing
the T option and specifying "0" as the degree of tilt in
Question 9T results in the same effect as the Y option,
while specifying "90" as the degree of tilt results in the
same effect as the Z option. Specifying "45" as the degree
of tilt will result in a display in which the patient's
position is exactly half way between those depicted by the Y

and Z options. (Special hidden surface removal procedures make the X, Y, and Z options less expensive than the T option [16].)

The number of images that are created is determined by the answer to Question 10. The different views of the object are computed such that the object rotates 360 degrees in equal incremental steps.

The answer to Question 11 can be any run number which is not already in use by the system.

Question 12 allows us to do a whole sequence of 3D83 runs, all with separate run numbers for the movie files. The beauty of this is that all the information for all the 3D83 runs can be entered to the computer in a single seating, and then the computer will produce all the movie files without any further interaction with the user. Thus a sequence of 3D83 runs can be initiated last thing at night and the movie files will be ready for viewing and filming by the morning.

Typical timings for medical applications for the 3D83 program running on the Eclipse S/140 computer of a CT scanner are as follows.

Production of a binary scene from the CT file (i.e., subregioning, interpolation, and thresholding)	3-6 mins
Surface detection of the specified 1-object	3-12 mins
Production of an image in the movie file using X, Y, or Z-axis rotation	1-2 mins

In a recent test, the total time for making three movies of twelve images in each, two of which were of the same surface (a skull) and the third of a different surface (a spine) was 75 minutes.

An additional useful feature of 3D83 is its ability to make quantitative measurements. Once the movie files have been completed, the user may place a cursor over the surface in any image and the program will report the location of that point on the surface (in the coordinate system of the original CT scans). If two such points are specified (possibly using different images in the same movie file), then the program will report the distance between them in 3D.

A version of 3D83 which will operate on the GE9800 CT scanner is under current development. This version will also have a number of additional features, such as measuring angles in 3D and simultaneously displaying two surfaces with an option of making one of them transparent. (Using this feature, one can simultaneously display a skin surface and the underlying bony surface, or the bony surface and a globe surface.)

SUMMARY

We have discussed a theory of three-dimensional image processing and display which depends on the dissection of space into voxels, and especially on the special dissection called the cuberille. We have shown how objects and their surfaces can be defined in such an environment, and indicated how mathematical theorems about cuberilles lead to efficient computer algorithms for the processing and display of 3D objects. Finally we have illustrated how these ideas have been implemented in an easy-to-use, versatile, and computationally efficient program, which is especially designed to produce high quality 3D displays of organ surfaces from CT scans.

ACKNOWLEDGEMENTS

The work reported in this paper has been partially supported by NIH grant HL28438. The 3D83 program is based on an earlier program, called 3D82, which was developed in conjunction with L. Brewster, L. S. Chen, P. Dillard,

H. Levkowitz, C.R. Meyer, R.A. Reynolds, and J.K. Udupa. The author is grateful to M. A. Blue for typing and to S. Strommer for photography. Useful comments on an earlier version of this paper have been given by L. S. Chen, H. Levkowitz, C. R. Meyer, E. Levitan, A. Rosenfeld, S. S. Trivedi, J.K. Udupa, and M.W. Vannier.

APPENDIX

For the sake of completeness we give here precise definitions of some of the standard topological terminology used in the paper.

The three-dimensional Euclidean space R3 is the set of all points (x, y, z), where x, y and z are real numbers.

A point (x_0, y_0, z_0) is said to be a closure point of a set S of points in R3, if for every positive real number ε, there is a point (x, y, z) in S such that

$$(x-x_0)^2 + (y-y_0)^2 + (z-z_0)^2 < \varepsilon .$$

A set of points in R3 is said to be closed if it contains all its closure points.

The complement of a set S of points in R3 is the set CS of all points in R3 which are not in S.

A set S of points in R3 is said to be open if its complement is closed.

A set S of points in R3 is said to be connected, if there do not exist two open sets P and Q of points in R3 with all of the following properties:

(i)   P and Q have no points in common.
(ii) P and S have at least one point in common.
(iii) Q and S have at least one point in common.
(iv) All points in S are either in P or in Q.

REFERENCES

[1] Armstrong, E.A.; Smith, T.H.; Currarino, G.:   3D CT
      reconstruction in children.   Radiol. 149(P), 60
      (1983).

[2] Artzy, E.: Display of three-dimensional information
      in computed tomography. Comput. Graphics Image
      Process. 9, 196-198 (1979).

[3] Artzy, E.; Frieder, G.; Herman, G.T.: The theory,
      design, implementation and evaluation of a three-
      dimensional surface detection algorithm. Comput.
      Graphics Image Process. 15, 1-24 (1981).

[4] Artzy, E.; Herman, G.T.: Boundary detection in
      3-dimensions with a medical application. Comput.
      Graphics 15, 92-123 (1981).

[5] Brewster, L.J.; Trivedi, S.S.; Tuy, H.K.; Udupa, J.K.:
      Interactive surgical planning.   IEEE Comput. Graphics
      and Appl. 4(3) 31-49 (1984).

[6] Chen, L.S.; Herman, G.T.; Reynolds, R.A.; Udupa, J.K.:
      Surface Rendering in the Cuberille Environment.
      Technical Report MIPG87, Medical Image Processing
      Group, Department of Radiology, University of
      Pennsylvania (January 1984).

[7] Dwyer III, S.J.; Cook, L.T.; Fritz, S.L.; Lee, K.R.;
      Preston, D.F.; Batnitzky, S.; DeSmet, A.A.: Medical
      image processing in diagnostic radiology. IEEE Trans.
      Nucl. Sci. NS-27, 1047-1055 (1980).

[8] Fuchs, H.; Kedem, Z.M.; Uselton, S.P.: Optimal
      surface reconstruction from planar contours. Comm.
      ACM 20, 693-702 (1977).

[9] Glenn, Jr., W.V.; Rhodes, M.L.; Altschuler, E.M.;
      Wiltse, L.L.; Kostanek, C.; Kuo, Y.M.: Multiplanar
      display computerized body tomography applications in
      the lumbar spine.   Spine 4, 282-352 (1979).

[10] Greenleaf, J.F.; Tu, J.S.; Wood, E.H.: Computer
      generated three-dimensional oscilloscopic images and
      associated techniques for display and study of the
      spatial distribution of pulmonary blood flow.   IEEE
      Trans. Nucl. Sci. NS-17, 353-359 (1970).

[11] Hemmy, D.C.; David, D.J.; Herman, G.T.: Three-
      dimensional reconstruction of craniofacial deformity
      using computed tomography. Neurosurgery 13, 534-541
      (1983).

[12] Herman, G.T.: Image Reconstruction from Projections:
      the Fundamentals of Computerized Tomography. Academic
      Press, New York, New York (1980).

[13] Herman, G.T.: Applications of three-dimensional
      computer graphics to surgical planning and evaluation.
      NCGA '84 Technical Sessions Proceedings 2, 66-74,
      Anaheim, California (1984).

[14] Herman, G.T.; Liu, H.K.: Dynamic boundary surface
      detection. Comput. Graphics Image Process. 7, 130-138
      (1978).

[15] Herman, G.T.; Liu, H.K.: Three-dimensional display of
      human organs from computed tomograms. Comput.
      Graphics Image Process. 9, 1-21 (1979).

[16] Herman, G.T.; Reynolds, R.A.; Udupa, J.K.: Computer
      techniques for the representation of three-dimensional
      data on a two-dimensional display. Proc. SPIE 367,
      3-14 (1982).

[17] Herman, G.T.; Udupa, J.K.: Display of three-
      dimensional discrete surfaces.   Proc. SPIE 283, 90-97
      (1981).

[18] Herman, G.T.; Webster, D.: A topological proof of a
      surface tracking algorithm. Comput. Vision, Graphics,
      and Image Process. 23, 162-177 (1983).

[19] Ledley, R.S.; Chan, M.; Ray, D.R.: Application of the
      ACTA-scanner to visualization of the spine. Comput.
      Tomogr. 3, 57-69 (1979).

[20] Liu, H.K.: Two- and three-dimensional boundary
      detection. Comput. Graphics Image Process. 6, 123-134
      (1977).

[21] Lobregt, S.; Verbeek, P.W.; Groen, F.C.A.: Three-
      dimensional skeletonization: principle and algorithm.
      IEEE Trans. Pattern Anal. and Mach. Intell. PAMI-2,
      75-77 (1980).

[22] Mazziotta, J.C.; Huang, K.C.: THREAD (three-
      dimensional reconstruction and display) with
      biomedical applications in neuron ultra-structure and
      computerized tomography. Proc. Natl. Comput. Conf.
      241-250, New York, New York (1976).

[23] Morgenthaler, D.; Rosenfeld, A.: Multidimensional
      edge detection by hypersurface fitting. IEEE Trans.
      Pattern Anal. and Mach. Intell. PAMI-3, 482-486,
      (1981).

[24] Newman, M.H.A.: Elements of the Topology of Plane
      Sets of Points, (second edition). Cambridge
      University Press, Cambridge, England (1951).

[25] Rhodes, M.L.: An algorithmic approach to controlling
      search in three-dimensional image data. SIGGRAPH '79
      Proc. 134-142, Chicago, Illinois (1979).

[26] Rosenfeld, A.; Kak, A.C.: Digital Picture Processing,
      Academic Press, New York, New York (1976).

[27] Sunguroff, A.; Greenberg, D.: Computer generated
      images for medical applications. SIGGRAPH '78 Proc.
      196-202, Atlanta, Georgia (1978).

[28] Udupa, J.K.: Interactive segmentation and boundary
      surface formation for 3-D digital images. Comput.
      Graphics and Image Process. 18, 213-235 (1982).

[29] Udupa, J.K.; Srihari, S.N.; Herman, G.T.: Boundary
      detection in multidimensions. IEEE Trans. Pattern
      Anal. and Mach. Intell. PAMI-4, 41-50 (1982).

[30] Vannier, M.W.; Marsh, J.L.; Warren, J.O.: Three
      dimensional CT reconstruction images for craniofacial
      surgical planning and evaluation. Radiol. 150,
      179-184 (1984).

[31] Zucker, S.W.; Hummel, R.A.: A three-dimensional edge
      operator. IEEE Trans. Pattern Anal. and Mach. Intell.
      PAMI-3, 324-331 (1981).

Psychovisual Issues in the Display of Medical Images

Stephen M. Pizer, Ph.D.
Departments of Computer Science and Radiology
University of North Carolina
Chapel Hill, North Carolina, U.S.A.

Abstract

Given a recorded image as a continuous or discrete array of measured or computed intensities, display is the process by which that image is presented to the human viewer as a light image. PACS imposes certain requirements on display such as the use of digital, electronic display devices, the provision of wide-ranging interactions, and the presentation of multiple images simultaneously so that they can be compared. At the same time the digital displays give one considerable flexibility in specifying the display process, providing options that importantly affect the information transmitted from the recorded image to the observer. Essentially one must match the display process to the needs of the observer and capabilities of the display devices. In this paper the parameters of display will be set forth, relevant properties of the human visual system and of display devices will be surveyed, and display processes to provide the required match will be described. In particular, matters related to the size of the display, the number of display pixels, interpolation, the display scale, and the assignment of recorded intensity levels to the display scale will be covered.

1. Introduction

For the purposes of this paper the term display refers to the process by which images recorded in a computer memory are made visible using electronic displays. The concepts presented herein will also be applicable to hard-copy and non-computer based displays. When starting with a computer, the image is originally represented as an array of numbers representing intensity in a picture element (*pixel*). We call these numbers the *recorded intensities* of the image. Examples of recorded intensities are CT numbers, digitized light levels from a radiographic film, and numbers of scintigraphic events in a pixel. In display these recorded intensities are possibly resampled and are transformed to displayed intensities.

The objective of the display process is not to correct the distortions that exist in the recorded image but to transmit most effectively the relevant information in

NATO ASI Series, Vol. F19
Pictorial Information Systems in Medicine
Edited by K. H. Höhne
© Springer-Verlag Berlin Heidelberg 1986

it. Thus noise, blurring, spatial sampling, spatial distortion, intensity distortion, and intensity nonuniformities over space that come from the way the image was measured or processed are not to be corrected at this stage but rather considered part of the information that may be displayed.

Psychovisual issues related to the display of medical images may usefully be divided into those related to spatial parameters and those related to intensity. In space the major parameter of concern is the size of the display pixels. In intensity the major concern is the means by which each recorded intensity is made to correspond with a displayed intensity, commonly an amount of light. It is useful to think of specifying this correspondence in two parts: first one selects a display scale to be used, and second one chooses a scheme of assigning recorded intensities onto the displayed intensities on this scale. In this paper first spatial issues will be discussed, then intensity issues. Finally a short discussion of issues of perception of motion and of the third dimension will be given.

2. Spatial Issues of Display

In the space domain we must be concerned with resolution, distortion, and sampling. These properties of the display process must be distinguished from properties of the recorded image with the same name. As has been indicated above, the latter will be assumed as given, while the former are to be determined. Thus we will assume that the resolution of the imaging device is known. Further, we will assume that the spatial sampling in the recorded image is adequate to either that resolution or a lower resolution suitable for the use to which the image is to be put. For display, resolution has to do with the accuracy with which pixels can be placed on the display screen. Normally this is much higher than the recorded image or the visual system requires, so we will treat display resolution no further. Nor will spatial distortion of display be covered further, as it is normally small and the visual system, sensitive to local context, is quite forgiving in this regard.

2.1. Spatial Sampling

The spatial sampling of display is a crucial matter. The principal objective here is to avoid the observer seeing the individual display pixels while allowing him or her to make all the spatial distinctions that are allowed by the resolution of the recorded image. Due to humans' great sensitivity to texture, visible pixels decrease sensitivity to contrast in the recorded image (this is called the *pixel artifact*). Since acuity increases with luminance in the video range, our constraint is that even at high video intensities such as 30-70 footlamberts, depending on the device, the display pixels should not be visible, while even at relatively low video intensities all resolution provided in the recorded image should be accessible.

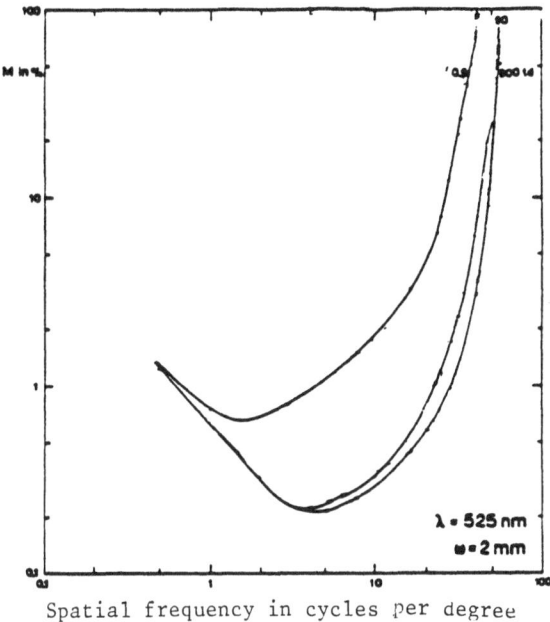

Spatial frequency in cycles per degree

Figure 1. Threshold-modulation curves for green light $\lambda = 525mm$, at three retinal illuminance levels (0.9, 90, and 900 trolands) and a pupil diameter of 2 *mm*. From [van Nes, 1968].

Curves giving contrast threshold as a function of spatial frequency are given in figure 1 for luminances of 44 footlamberts (high video – about 90 trolands for a 3mm diameter pupil), 440 footlamberts (about half of a high luminance from a light box – about 900 trolands for a 3mm diameter pupil), and 0.44 footlamberts (low video – about 0.9 trolands for a 3mm diameter pupil) for a typical observer [van Nes, 1968]. It must be realized that these curves are dependent on the spatial structure of the pattern being perceived. Nevertheless, from these curves we can conclude roughly that pixels ought to be less than 1' of arc in diameter and that the highest resolution to be seen in the image, measured in line pairs, should correspond to no less than 6' for video, or 3' for a light box. Since the sensitivity to pixels is masked by image noise, pixels of well over 1' of arc can be tolerated with noisy images, such as scintigrams [Sharp, 1981].

We conclude that for low-noise images displayed on video we should use approximately 6 display pixels to represent a resolution distance in the recorded image. On the other hand, the sampling theorem together with the fact that summation across a pixel causes some blurring leads to the well known conclusion that to represent the imaged scene without significant loss of information, the recorded image should have approximately 3 pixels per resolution distance. We conclude that normal display sampling should be approximately twice the storage sampling for each spatial dimension. Of course, it is possible to increase the storage sampling, but the gain is only in removing the need to resample at display time, not in image information. Methods of spatial resampling (interpolation) for display will be discussed in section 2.2.

What are the effects of screen size and viewing distance on spatial display requirements? The viewing distance determines the size that corresponds to the numbers of minutes of arc specified above. At a normal viewing distance of 50-60cm 6' corresponds to about 1mm. For a lifesize 35cm x-ray displayed in video, the observer can see only 1 lp/mm resolution, only a fifth (or less) of that available in the measured image (for an easily achieved resolution of 5 lp/mm). However, it must be emphasized that only the greater illumination of a light box improves this for hard copy display viewed at the same distance, and then only by about a factor of two. The advantage of hard copy display is that one can move closer if one wants to achieve the full resolution available from the recorded image.

With the requirement of 1 pixel per minute of arc the viewing distance of 50-60cm implies a display pixel size of 0.15mm or less. Once this pixel size is established, moving towards the screen provides little information increase because of the contrast losses caused by the increasing pixel artifact as you move. Similarly, decreasing the number of display pixels without incurring the pixel artifact is possible only with a smaller displayed image.

In practice it appears [Burgess, 1985] that observers choose a viewing position that varies by task over about a factor of two. Also, performance has a quite flat optimum as a function of viewing distance, falling off by only 5% a factor of two from the optimal distance. However, observers' comfort falls by a considerable amount when they are a factor of two from the distance at which they are most comfortable.

With a display pixel size of 0.15mm an image width of 35cm requires about 2000 display pixels. One can use these pixels to achieve a perceived resolution of approximately 1/350 of the image size if the recorded image is sampled to about 1000 pixels. Thus a CT scan with 1 lp/mm resolution and an (abnormal) sampling of 1000 × 1000 to support this resolution would require a screen of approximately 35cm with 2000 display pixels. No smaller screen would do unless the viewing distance were decreased and the pixel size changed to accommodate the same number of pixels. And to take advantage of the full resolution of a radiograph (at least 5 lp/mm) while maintaining the 50-60cm viewing distance, one would require a display screen of over 1.75 meters with about 10,000 pixels in each dimension. Since this is clearly undesirable, both economically and ergonomically, an ability to zoom and roam is desired. This would allow part of a recorded image to be viewed at full resolution on a reasonable size screen.

In summary, digital display differs from analog display in the fact that the pixel artifact determines the viewing distance and thus the perceivable resolution in a displayed image. Thus the perceivable resolution can be increased significantly only by zooming (if the recorded image is sampled finely enough to support the increase). The numbers given above can be modified by a small factor depending on the conditions of the image and viewing environment, but the basic limitation of perceivable resolution by the pixel artifact remains.

2.2. Interpolation

The comparison of two images from different imaging modalities is most conveniently done when the displayed images are of the same size. However, the discussion above implies that the displayed size of the image should be in inverse proportion to the resolution of the recorded image, and information losses will be obtained if the image is simply enlarged without resampling. Thus, image comparison across imaging modalities normally implies interpolation. Similarly, the convenience of displaying an image at life size can only be effectively achieved via interpolation. Finally, we have shown above that the benefits of minimizing storage requirements imply that interpolation be part of the display process, at least for relatively noise-free images. Therefore, in this section the methods of interpolating finer sampling from a recorded image are discussed.

Parker et al [1983] give an excellent survey of interpolation concepts and methods. The following is paraphrased from their article.

Generally interpolation is accomplished using a dimensionally separable weighting function $f(x)$ to produce the interpolated image $i'(x', y')$ from the recorded image $i(x, y)$ by the convolution

$$i'(x', y') = \sum_{x,y} f(x' - x)\, f(y' - y)\, i(x, y).$$

Probably the two best methods are bilinear interpolation in which

$$f(u) = 1 - |u| \text{ for } |u| < 1 \text{ and zero otherwise;}$$

and high-resolution cubic spline interpolation in which

$$f(u) = u^3 - 2u^2 + 1, \text{ for } |u| < 1,$$

$$f(u) = -u^3 + 5u^2 - 8u + 4, \text{ for } 1 < |u| < 2,$$

and $f(u) = 0$ otherwise.

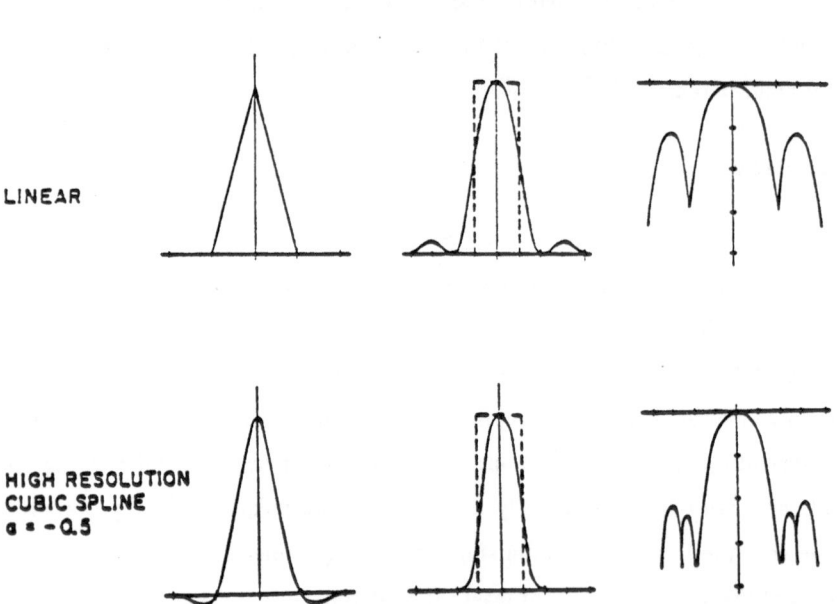

Figure 2. Two interpolating functions. From [Parker, 1983].

For both of these $i'(x,y) = i(x,y)$ at the original sample points, i.e. the result is a true interpolation leaving the image unchanged except for filling in new pixels, a desirable situation. Generally, Parker shows (see figures 2 and 3) that bilinear interpolation causes some resolution loss but is quite efficient, since each new pixel only involves a weighted sum of four original pixel values; whereas the high resolution cubic spline method better transmits the resolution in the recorded image, but it requires sixteen original pixel values per new value and thus is about four times slower.

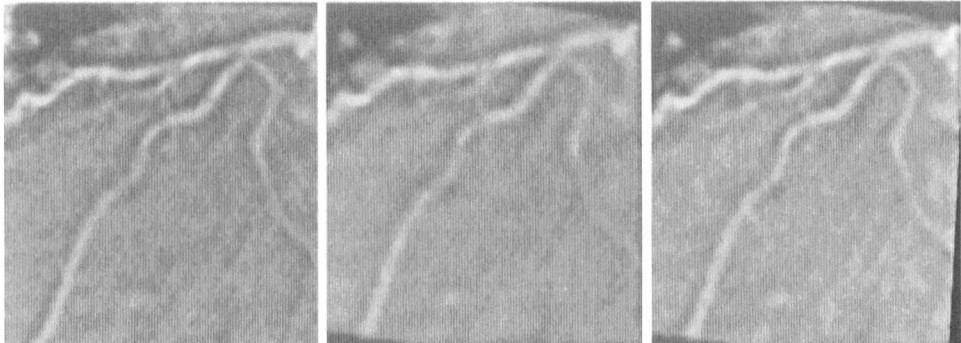

Figure 3. Image resampling. (a) Initial image of a coronary angiogram. The primary data is 64 × 64 with a display dimension of 128 × 128. (b) Resampling using the bilinear interpolating algorithm. Notice the loss of sharpness at the edges of the vessels. (c) Resampling using the high- resolution cubic spline. From [Parker, 1983].

3. Intensity Issues of Display

For display, at each pixel the recorded intensity must be transformed into a displayed intensity. The displayed intensity is normally in the form of luminance or color, but it may sometimes involve some other parameter such as apparent height or motion. This section focuses on matters related to choosing this transformation.

More completely, the transformation that we are concerned with is from recorded intensity to perceived intensity (see figure 4), since it is in the perceived image that the information needs to be optimally available. Thus, we can distinguish three different intensities: the *recorded intensity* that is input to the process, the *displayed intensity*

218

that is produced by the display device, and the *perceived intensity* (i.e. brightness) that can be thought to be generated in the observer's visual system. This perceived intensity will be more carefully defined in section 3.1.

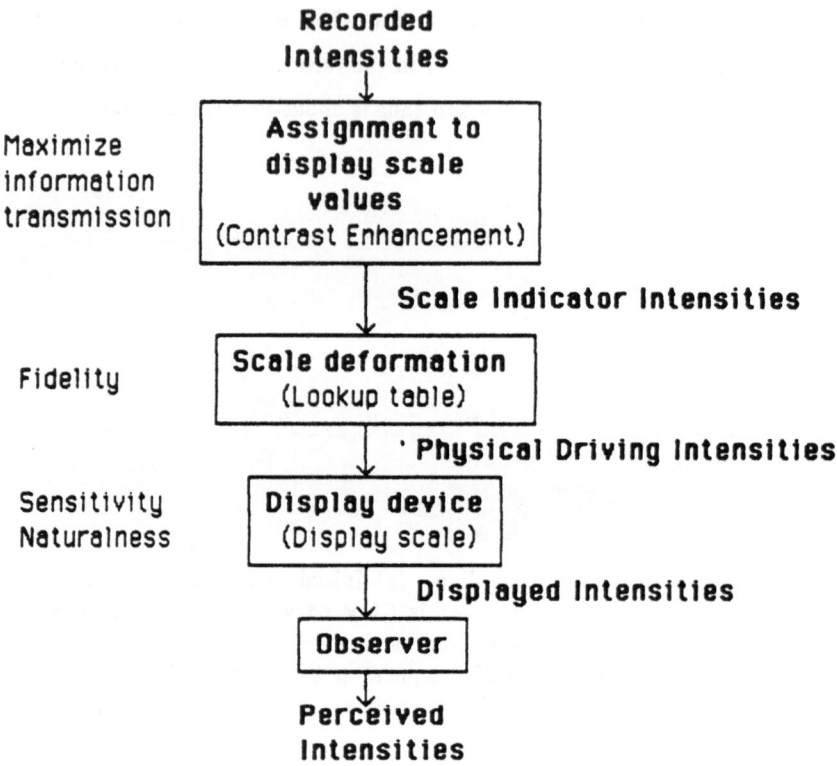

Figure 4. Intensity types and their transformations.

We can control only the transformation from recorded to displayed intensities. It is useful to divide this transformation into three components (see figure 4) in order to separate the concerns of

(1) maximizing information transmission (dependent on the image and the viewing task),

(2) choosing a display scale, i.e. a path through color space or some other space, that provides satisfactory intensity distinctions (dependent on the observer), and

(3) controlling the rate at which we move along the display scale path (dependent on the display device).

In the first component one attempts to achieve the property that intensity differences increase with the importance of seeing the difference and thus information is optimally presented. This step is sometimes called *contrast enhancement*. The result of this step is used to select an intensity on whatever display scale is used; it is the *scale indicator intensity*. The display scale, a path through color or other space, is considered at this point to be controlled in regard to the rate at which the path is traversed. This control over the local stretching or compressing of the display scale is achieved by interposing a transformation between the scale indicator intensity and the *physical driving intensity* (e.g. one or more voltages), which together with the basic display device (e.g. CRT) determine the displayed intensity.

In order to choose the above-mentioned component transformations, we must specify what is meant by optimal information transmission, display scale path, and rate along this path. These issues are treated in reverse order in the next sections.

3.1. Display Scale Linearization

To make it possible to design contrast enhancement so as to maximize information transmission, the succeeding transformations of display scale position assignment, display scale deformation for distortion correction, and perception (see figure 4) must not distort the contrast relationship. That is, they must faithfully transmit intensity ratios: perceived intensity must be linear with the display scale indicator intensity that is the output of the contrast enhancement. Achieving this linearization will have the secondary benefit of standardization across display devices: Any image will appear the same on any display device except for differences due to variations in overall display scale sensitivity. That is, the relative values of perceived contrasts will be independent of the display device.

A method for linearization requires first a knowledge of the perceived intensity corresponding to any physical driving intensity, and this in turn requires a definition of perceived intensity. These can be defined in terms of absolute intensity judgements of observers or of judgements of intensity differences. Since the latter is more relevant to pattern perception in medical images, the definition in terms of intensity differences is more appropriate here. The natural units for this perceived intensity are those in which the observer perceives intensity differences, that is *just noticeable differences (jnd's)*. It is useful to define the perceived intensity of zero as the intensity which is at the bottom of the display scale. Then other scale locations will produce a perceived intensity that is a specified number of jnd's above the minimum intensity.

The jnd is defined as that change in physical driving intensity that results in a just perceivably different displayed intensity, where we must carefully define what we mean by just perceivably different. The jnd is in units of driving intensity, and it is a function of the reference value of driving intensity in which a change is being perceived. This function we call the *jnd curve*, giving jnd vs. reference intensity (see figure 5).

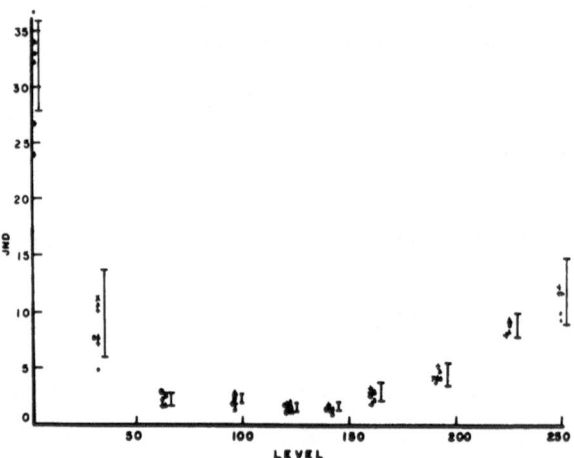

Figure 5. Measured jnd curve for grey scale on a Tektronix 690SR monitor.

A definition for "just noticeably different" must include a full specification of the target, background, and viewing environment together with a probability of correctness (true positive probability) defining the detection of a change and a false positive probability specifying the conservatism of the observer. Many definitions of jnd have failed to include this last factor, the probability that the observer will see a difference when none exists, and since observers can easily vary their conservatism, the resulting jnd can vary significantly in some situations. Pizer and Chan [1980] have suggested using 50% as the true positive probability defining detection and 5% as the false positive probability defining the level of conservatism. Determining the jnd involves in principle carrying out an ROC experiment [Green & Swets, 1974] to determine the change in driving intensity producing an ROC curve passing through this (0.05,0.50) criterion point. Since the two displayed stimuli being distinguished are so close in intensity, it is reasonable to expect, and Chan [1982] showed, that the variances of the two decision variable distributions corresponding to change and no change are the same, that is the ROC curve can be described by the single variable conventionally called d' describing the standard-deviation-normalized difference between the means of the decision variable distributions. The criterion ROC curve can be shown to have a value of d' equal to 1.645, or equivalently correspond to a two-alternative forced choice experiment with a fraction correct of 87.8%.

Johnston [1985], Pizer [1980, 1982, 1983], et al have measured jnd curves of various display scales with ROC rating and two-alternative forced choice experiments using a target consisting of two nearby but separated squares in a background with intensity chosen so that an 8 degree region viewed at 60cm centered at the target has a fixed average adapting intensity. Given such a jnd curve, $jnd(i)$, over a range of reference intensities i_{min} and i_{max} defining the full range of driving intensities for a particular display device, Pizer [1982] showed that perceived intensity, $P(i)$, is given by

$$P(i) = \int_{i_{min}}^{i} \frac{jnd'(i)}{jnd(i) \, \log \, (1 + jnd'(i))} di.$$

Thought of as a function of i, we will call this the perceived intensity function.

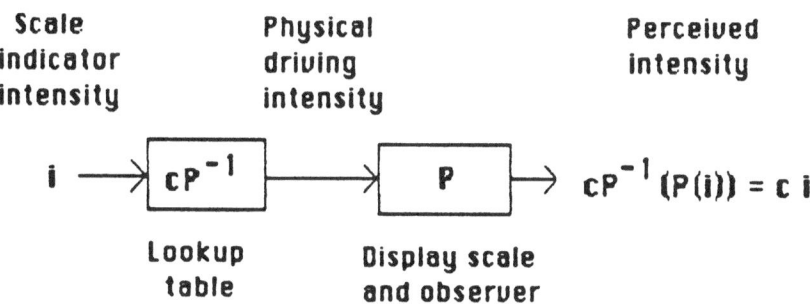

Figure 6. Linearization as the inverse of P(i).

By definition a linearizing function transforms scale indicator intensity to physical driving intensity so that the relation between scale indicator intensity and perceived intensity is linear. Thus (see figure 6) the linearizing function required by a particular device and observer is proportional to the inverse of the perceived intensity function, normalized so that the range of physical driving intensity matches that of the display device. This linearizing function is frequently implemented by insertion in a lookup table of a digital display system. Experience shows that using the physical display scale directly (an identity lookup table) usually results in a strongly different perceived image from one produced with linearization (see figure 7). In particular, the grey scales of video displays normally have an abnormally large jnd (low sensitivity) for small physical driving intensities (see figure 5), and linearization increases sensitivity to contrast in the lower levels of recorded intensity. We will assume in the following sections that all scales being considered are linearized.

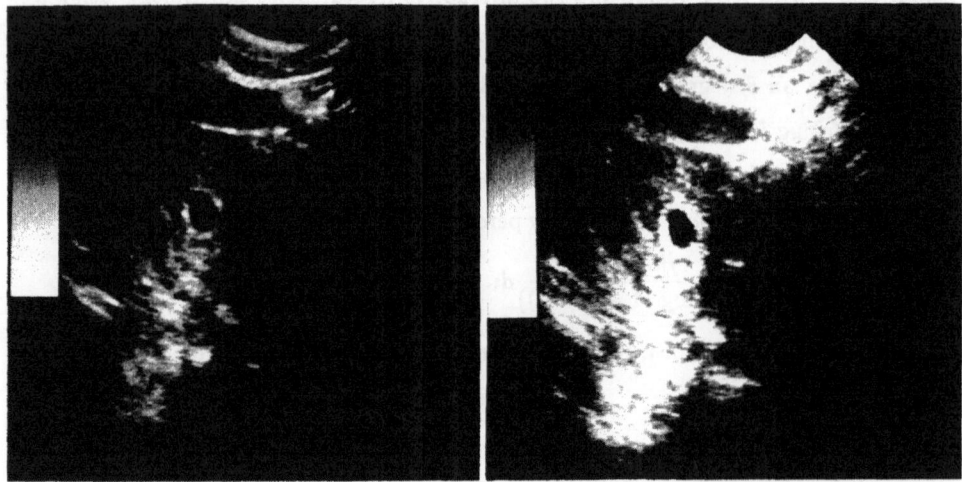

Figure 7. Ultrasound image of the intrauterine sac shown on an original unlinearized (left) and linearized (right) video display.

Since the jnd curve is in principle a function of observer, viewing environment, and image structure, we need to know how sensitive the linearizing function is to these factors. That is, do we need to change linearizing functions as the observer, viewing environment, or image structure changes? Johnston [1985] has demonstrated that inter-observer variations in jnd curve are comparable to intra-observer variations and thus a single linearizing function is satisfactory across observers. Johnston and Pizer respectively have pilot results indicating that although the jnd curve does change with viewing environment and image structure, the linearizing function, determined by the ratio between values along the jnd curve, does not change significantly with these factors. Thus it appears that a single linearizing function suffices for each display device. But more research is necessary to determine the correctness of this result.

The jnd curve is determined by noise in both the observer and the display device. To the extent that the observer noise is the dominant factor, the linearizing function can be determined by measuring photometrically the relation between physical driving intensity and displayed intensity and then using the relation between displayed intensity and jnd derivable from earlier observer experiments. We have developed a program to accomplish this determination of the linearizing function based on photometric measurements for grey scales, and we are happy to distribute copies of this program.

3.2. Display Scale Choice

Assuming linearization, i.e. fidelity of the transformation between scale indicator and perceived intensities, the display scale should be chosen on the basis of overall sensitivity, naturalness, and edge production. Sensitivity indicates the lowest contrast in scale indicator intensity that can be perceived to a criterion degree. Naturalness specifies the ease with which an observer can determine the the relative difference between pairs of scale indicator intensities. Edge production determines informativeness in that false edges limit the comprehension of the recorded image and true edges are the major factor in producing comprehension.

3.2.1. Sensitivity

The overall sensitivity of a display scale is measured by the total number of jnd's across the scale. We call this number the *perceived dynamic range (PDR)* of the scale. Remember that these jnd's were measured in terms of some fixed target, and thus the absolute value of the PDR is target-dependent. But the relative values for different display scales are very informative, and the absolute values reported below are in terms of jnd's that match reasonably well our sense of what is "just noticeable".

The PDR is the same as the perceived intensity corresponding to the top of the display scale:

$$PDR = \int_{i_{min}}^{i_{max}} \frac{jnd'(i)}{jnd(i) \log\left(1 + jnd'(i)\right)} di.$$

Its value for various scales, using a target of separated squares on a uniform background (see section 3.1), is indicated in Table 1. It can be seen that various pseudocolor scales have a greater PDR than the grey scale. But other factors must be taken into account before we can conclude which is the best scale.

Scale	PDR (jnds)
grey	90
heated-object	120
magenta	113
rainbow	200

Table 1. Perceived dynamic range for various display scales. From [Pizer, 1982].

Even for a natural, continuous scale such as the grey scale, it is not clear whether informativeness increases monotonically with the PDR. It is possible that we can be too sensitive to contrast in the recorded image. Increasing the sensitivity, for example by choosing a display scale with a higher PDR, eventually causes the observer to see image differences that are dominated by noise or distracting image structure. Does such an increase of the contrast of these noise differences decrease informativeness even though the contrast of differences due to signal are increased proportionally?

The present scientific evidence [Burgess, 1982], based on artificial images, indicates that informativeness does not fall as sensitivity is increased to show the noise better and better. However, the behavior of radiologists in limiting contrast so that noise is not well seen when reading medical images and one's intuition suggest the opposite. It appears that further study is needed with realistic images before this question will be settled.

3.2.2. Edge Presentation and Artifacts; Number of Discrete Intensity Levels

The informativeness provided by a sensitive scale is also affected by the way in which it causes edges to appear. The human visual system is very sensitive to edges. It both carries out edge enhancing transformations and seems to encode information in terms of edges. Edges determine not only image objects [Hubel & Weisel, 1974], but the way in which object surrounds affect object detection [van der Wildt, 1985]. Therefore, isointensity contours that appear as edges in the display but do not represent edges in the recorded image (the *contour artifact*) must be avoided because they limit what is seen. On the other hand edges should be used to present important information.

Edges may be seen when adjacent groups of pixels differ by a large fraction of a jnd or more (with jnd's defined by the separated-square target mentioned above). Thus the difference in the displayed intensity across real edges should always be more than one jnd. To avoid the contour artifact when the objective is comprehension of image patterns (*qualitative display*), one must assure that adjacent digital display scale intensities differ by well less than a jnd. On the other hand, if one wants to present quantitative information such as absolute recorded intensity values or differences, edges are an important means, and adjacent digital display scale intensities should differ by at least 1.5-2 jnd's. Thus display scales for qualitative and quantitative display should be different.

It follows from the above discussion of false edge avoidance that the number of discrete intensities in linearized digital display scales for qualitative display must be in proportion to the total number of jnd's across the scale, i.e. the PDR. The PDR values given in Table 1 are in jnd's determined for separated target regions. But difference detection of adjacent regions is more sensitive than for separated regions, due to the Mach effect of enhancing contrast at edges. Experience indicates that to make every discrete display level significantly less than one jnd from its neighboring levels, in order to avoid the contour artifact in qualitative display, we should arrange that the number of digital levels in the scale is 1.5-2 times the PDR. Thus, 256 is an appropriate number of levels to avoid false edge artifacts with grey or heated object scales.

3.2.3. Naturalness; Pseudocolor Scales

Qualitative display is not just used to compare intensities that are nearby on the display scale. Scales that have no natural order, such as the rainbow scale with the large PDR recorded in Table 1, confuse the observer by distorting the relation between scale indicator intensity and perceived intensity for considerably differing intensity values. Although the matter is poorly understood, it appears important for the observer to comprehend immediately which of two intensities corresponds to more, and in fact to have some feeling for how much more. There is considerable literature on the perceived difference between perceivably distinct, possibly colored intensities, that is, on the distance function $d(v_1, v_2)$ between two displayed intensities v_1 and v_2 that an observer effectively imposes on the space of displayed intensities (e.g. colors). We say that a display scale is *natural* if perceived differences do not contradict the differences given by integrating differences along the scale. That is, if i_1, i_2, and i_3 are any three scale indicator intensities such that $i_3 > i_2 > i_1$, and v_1, v_2, and v_3 are the corresponding displayed intensities, then $d(v_1, v_3) > d(v_1, v_2)$; an intensity on the display scale does not appear closer to another intensity on the scale than it does to a third that is closer to it on the scale.

It has long been said that pseudocolor scales can increase overall sensitivity, and Table 1 confirms this fact. PDR results for three linearized color scales all monotonically increasing in brightness are given: a heated object scale going from red through orange and yellow to white, a magenta scale going from red through magenta to white, and a rainbow scale approximately going through the hues of the rainbow (while decreasing in saturation so that monotonic increase in brightness can be achieved). But these four scales (fig.8, p.496) are not equally natural; observers informally report [Pizer and Zimmerman, 1983] that the heated object and magenta scales have a natural order, but the rainbow scale does not (we have not yet applied the above process for testing naturalness to these scales). And despite its considerably greater PDR, the rainbow scale appears to give far less information and is preferred by no observers over the grey scale. In contrast, the more natural pseudocolor scales, and especially the heated object scale with the slightly greater PDR seems to give more information than the grey scale. Even this result is disputed by various experiments faulted by the use of nonlinearized scales [Todd-Pokropek, 1983; Burgess, 1985]. Moreover, the apparent advantage of even natural pseudocolor scales is also brought into question by the fact that the visual system has low spatial acuity in distinguishing chromanence changes, as changes in nearby pixels may not be more sensitively perceived as a result of chromanence changes.

It is interesting to compare the curves giving red physical driving intensity, green

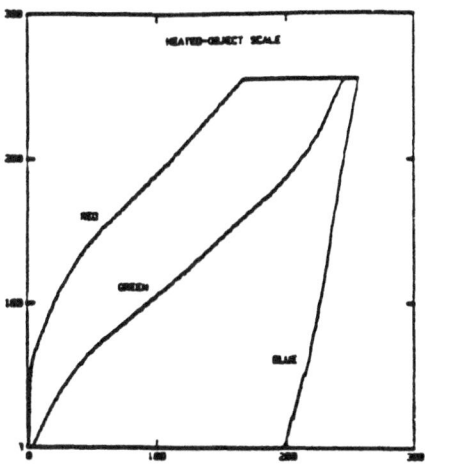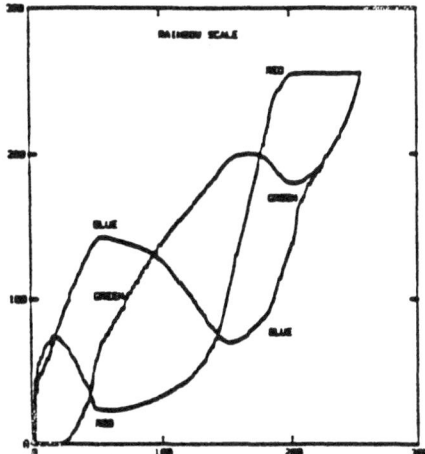

Figure 9. Mappings from scale indicator intensity to color gun-driving intensity for two linearized display scales.

physical driving intensity, and blue physical driving intensity vs. scale indicator intensity for the rainbow and heated object scales (see figure 9). Pizer and Zimmerman report that scales for which these three curves do not cross, as with the heated object and magenta scales, always appear to be natural, while those for which the three curves do cross appear to be unnatural, even when the scales are monotonic in brightness. The heated object scale has in our experience the greatest PDR of the natural scales with the aforementioned non-crossing property. It is thus possible that pseudocolor scales can be used to give increased informativeness over the grey scale, but even if so, it can apparently provide an increase in the PDR of less than 50%.

3.3. Intensity Inhomogeneities

Finally in the area of properties of the display device and the observer, we must face the fact that displayed intensity as a function of physical driving signal is often quite nonuniform across the display area. Cathode ray tubes *(CRT's)* frequently vary by as much as 20% in luminance across the screen for the same driving signal (voltage). But these variations are normally smooth and result in only proportional changes in the brightness scale at each pixel. Therefore they cause quite low frequency variations in the image. The human visual system is principally sensitive to changes in local context; it is quite insensitive to low frequency changes (see figure 1). Therefore correction for intensity inhomogeneity is normally unnecessary even for the rather large inhomogeneities commonly encountered.

3.4. Assignment of Display Scale Levels to Recorded Intensities

The most common method for assigning display scale indicator levels to recorded intensities is windowing. In this method the user interactively selects a range over which recorded intensities are mapped linearly to the full range of the display scale. While this method has the advantage of giving control to the user, it limits him to a particular type of assignment function and requires that this function be the same everywhere in the image. The following discussion suggests that neither of these choices is close to optimal.

Cormack [1980] has suggested that the assignment function be chosen to minimize information loss in the transformation from recorded to scale indicator intensities (and thus to perceived intensities if the display scale has been linearized). Here information is used in the sense of information theory. While it can be argued that information should be defined in a more task-related way, we will accept Cormack's suggestion as a good starting point.

Cormack goes on to suggest that information be measured on an average per pixel basis, assuming independent pixels. This is clearly a poor approximation to reality, since pixel intensities are heavily correlated, but it does simplify the mathematics. Given this approximation, it can be shown that if the probability distribution describing the noise at a pixel is independent of the intensity of the pixel, the method called histogram equalization [e.g. Castleman, 1979] provides the assignment function. In histogram equalization the assignment function is the cumulative recorded intensity histogram, normalized so that the full range of the display scale is used. With this assignment function if a pixel has an intensity at the pth percentile of recorded image intensities, it is displayed at the pth percentile along the display scale. Thus the display is sensitive to changes in popular ranges of recorded intensities, while few displayed intensity levels are wasted in recorded intensity ranges in which there are few pixels with intensities to distinguish.

If we drop Cormack's approximations but accept his approach of minimizing information loss in the information theoretic sense, we will obtain some assignment function, perhaps approximately that provided by histogram equalization. But in any case this will be a single assignment function designed to optimize information transfer across the whole image. But the human visual system is not equipped to receive information in one part of the image relative to the whole image. It preprocesses the image by recording contrast relative to local context [Cornsweet, 1970]. It is simply unable to make accurate comparisons in either brightness or color between distant locations. Therefore the context in terms of which an information-loss-minimizing assignment should be computed should be quite local; the assignment function should change across the image. It is said that the assignment should be *adaptive*.

If histogram equalization is the basis of this adaptive approach, the method called *adaptive histogram equalization (ahe)* [Pizer, 1984] is obtained. The method attempts to optimize contrast enhancement everywhere in the image relative to local context, and as a result it provides a single displayed image in which contrasts in widely varying recorded intensities in different image regions, and thus organs, can be easily perceived. A comparison to windowing and to global (nonadaptive) histogram equalization is given in figure 10. Zimmerman [1985] has compared its results to many other methods of adaptive display scale assignment and found it distinctly superior in both producing high contrast and avoiding artifacts. Research is now proceeding to evaluate its effectiveness in a controlled experiment on simulated clinical images (inserted lesions in clinical normals). If it is shown as effective as present anecdotal experience indicates, it is a candidate for a standard assignment method that will avoid the need for interactive contrast enhancement with the large majority of images. We have therefore been facing questions of developing software and hardware that will make its implementation fast. But in this paper we address none of these matters of making the method speedy.

Besides its clinical effectiveness, the major questions with regard to *ahe* are the following.

(1) What should the contextual region size be? Our work with the method to date indicates that when the method operates as described above with each pixel having its own contextual region, a square region that is between 1/4 and 1/8 of the image width on a side is appropriate, with little sensitivity of the result to changes of region size between these extremes.

(2) Should the contextual region be related to the boundaries of nearby objects? The visual system most likely works in just this way [van der Wildt, 1983], but it is hoped and present experience suggests that pattern recognition will not be a necessary part of *ahe* .

(3) Should nearby points in the contextual region be given more effect by some kind of weighting scheme? Locality plays an important role in the operation of the visual

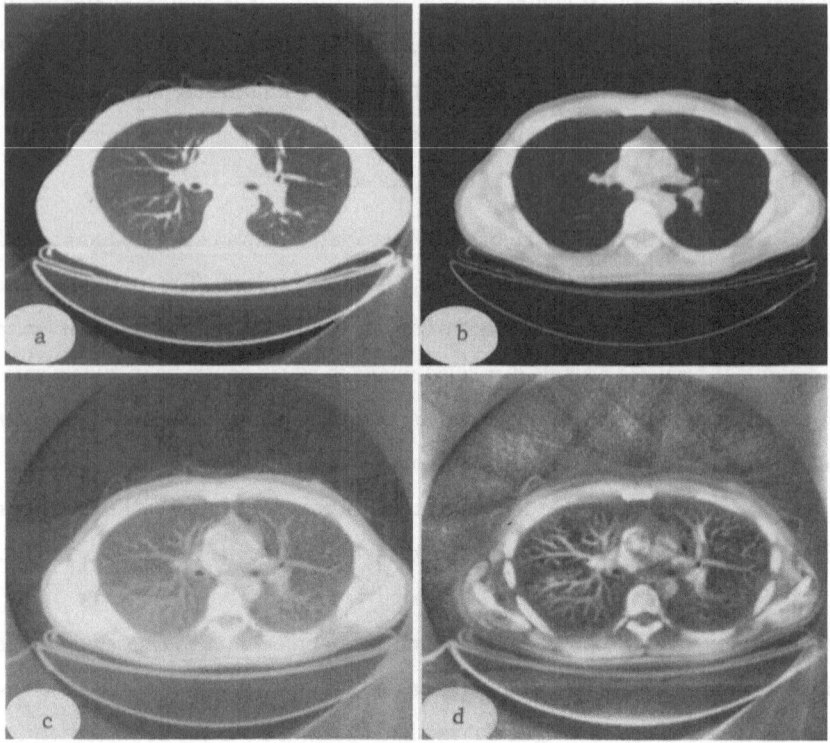

Figure 10. The effect of (a,b) windowing, (c) global histogram equalization and (d) adaptive histogram equalization on a CT scan of the
chest.

system.

(4) Should the degree of contrast enhancement be limited, e.g. by limiting the height
of the histogram? The method sometimes shows image noise disagreeably well,
while also showing signal contrast. This returns us to the previously mentioned
issue of whether sensitivity to contrast in the recorded image can be too high. A
positive answer to that question would indicate that we must investigate ways of
limiting contrast enhancement when it varies locally.

4. Other Visual Dimensions: Motion and 3D

It is now not uncommon for medical image display to show objects in motion by
cinematic approaches or to show the third dimension. It also seems worth considering
the use of motion and the third dimension to enhance the display of one parameter
varying in two dimensions or allow the simultaneous display of additional parameters.

A few comments on visual aspects in motion and 3D are thus in order.

The visual properties of motion detection are well surveyed by Nakayama [1985]. Humans have low sensitivity to low velocities. For moderate velocities the sensitivity to change in velocity is described by a Weber's law: we are equally sensitive to fractional changes in velocity, with a threshold near 5% of the present velocity. It is interesting to note (see figure 11) that the spatial frequency to which we are maximally sensitive decreases with velocity. Our resolution for matching directions of motion is about 1 degree.

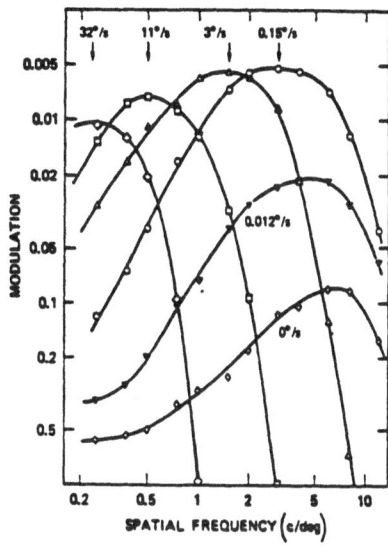

Figure 11. Stabilized contrast-sensitivity curves measured at constant velocity. Data are shown for six different velocities, ranging from 32 deg/s down to zero. From [Kelly, 1979].

In the depth dimension visual resolution in terms of visual angle is about the same as for the other two spatial dimensions. However, when this translates trigonometrically into absolute spatial units, for normal viewing distances such as 1 meter our resolution is on the order of a few mm, far poorer than that in the other two dimensions. The cues to the third dimension are many, including stereopsis (horizontal disparity of the images seen by the two eyes), vergence (the relative angle at which the two eyes are facing, operating up to about 6 meters), accommodation (the force of the muscles focusing the lens, important only for near objects), linear perspective (only important when there are long, straight, parallel edges), head motion parallax, interposition, surface texture, and surface shading. While head motion parallax and interposition appear to be among the most important cues, the stereoscopic effect has received by far the most research attention, and there has been little research in the combination of all of these cues into a single percept. Until such research is done, 3D display will have to be intuitively based.

5. Summary

As can be seen from the preceding, studies of visual perception have much to tell us about the important medical image display issues of spatial sampling, display scale choice and linearization, and assignment of display scale levels to recorded intensities, as well as the use of cinematic and three-dimensional displays. But much research remains to be done, both in basic studies of visual perception and applied studies in regard to the display of medical images. Especially, more understanding of the effect of the display task and of image structure in images of clinical complexity is required.

6. Acknowledgements

I thank Diane Rogers, Arthur Burgess, R. Eugene Johnston, John Zimmerman, and Jan Koenderink for useful discussions, J. Anthony Parker for providing figures, and Richard Bader, Scott Fields, Ching-Man Kwan, and Edward Staab for providing digitized images. I am indebted to Joan Savrock for manuscript preparation and to Bo Strain and Karen Curran for photography. This paper was prepared partially with the support of National Institutes of Health grant number 1-R01-CA39060.

7. References

A. E. Burgess, R. F. Wagner, R. J. Jennings, "Human Signal Detection Performance for Noisy, Medical Images", *Proceedings of the International Workshop on Physics and Engineering in Medical Imaging*, IEEE Computer Society No. 82CH1751–7:99-105, 1982.

A. E. Burgess, Personal Communication, 1985.

F. H. Chan, *Evaluating the Perceived Dynamic Range of A Display Device Using Pseudocolor*, Ph.D. Dissertation, University of North Carolina, Curriculum in Biomathematics and Medical Engineering, 1982.

J. Cormack and B. F. Hutton, "Minimisation of Data Transfer Losses in the Display of Digitised Scintigraphic Images", *Phys. Med. Biol.*, *25*:271-82, 1980.

T. N. Cornsweet, *Visual Perception*, 1970.

D. M. Green and J. A. Swets, *Signal Detection Theory and Psychophysics*, 1974.

D. H. Hubel and T. N. Weisel, "Sequence Regularity and Geometry of Orientation Columns in the Monkey Striate Cortex", *J. Comp. Neur., 158*(3):267-294, 1974.

R. E. Johnston, J. B. Zimmerman, D. C. Rogers, S. M. Pizer, "Perceptual Standardization", *SPIE Conference on PACS*, 1985.

D. H. Kelly, "Motion and Vision, II, Stabilized Spatio-temporal Threshold Surface", *Journal of Optical Society of America, 69*(10) : 1340-1349, 1979.

K. Nakayama, "Biological Image Motion Processing: A Review", to be published in *Vision Research*, 1985.

J. A. Parker, R. V. Kenyon, and D. E. Troxel, "Comparison of Interpolating Methods for Image Resampling", *IEEE Transactions on Medical Imaging, MI-2*, (1): 31-39, 1983.

S. M. Pizer and F. H. Chan, "Evaluation of the Number of Discernible Levels Produced by a Display", *Traitment Des Information en Imagerie Medicale* (Information Processing in Medical Imaging), R. DiPaola and E. Kahn, eds., Editions INSERM, Paris: 561-580, 1980.

S. M. Pizer and J. B. Zimmerman, "Color Display in Ultrasonography", *Ultrasound in Medicine and Biology*, 9 (4): 331-345, 1983.

S. M. Pizer, J. B. Zimmerman, and E. V. Staab, "Adaptive Grey Level Assignment in CT Scan Display," *J. Comp. Ass. Tomo.* 8 (2): 300-305, 1984.

S. M. Pizer, J. B. Zimmerman, and R. E. Johnston, "Contrast Transmission in Medical Image Display", *Proceedings of ISMIII '82* (International Symposium on Medical Imaging and Image Interpretation), IEEE Computer Society (IEEE Catalog No. 82CH1804-4): 2-9, 1982.

P. F. Sharp and R. B. Chesser, "The Influence of Data Interpolation on Image Quality", *Proc. VIIth Int. Conf. on Info. Proc. in Med. Imaging*, Division of Nuclear Medicine, Stanford University School of Medicine, Stanford, CA: 225-239, 1981.

A. E. Todd-Pokropek, "The Intercomparison of a Black and White and a Color Display: An Example of the Use of Receiver Operating Characteristic Curves", *IEEE Transactions on Medical Imaging, MI-2* (1): 19-23, 1983.

G. J. van der Wildt and R. G. Waarts, "Contrast Detection and Its Dependence on the Presence of Edges and Lines in the Stimulus Field", *Vision Res., 23* (8): 821-830, 1983.

F. L. van Nes, *Experimental Studies in Spatiotemporal Contrast Transfer by the Human Eye*, Ph.D. dissertation, Rijksuniversiteit Utrecht, 1968.

J. B. Zimmerman, *Effectiveness of Adaptive Contrast Enhancement*, Ph.D. dissertation, University of North Carolina, Department of Computer Science, to appear in 1985.

Systems for 3D Display in Medical Imaging

Stephen M. Pizer, Ph.D.
Departments of Computer Science and Radiology
University of North Carolina
Chapel Hill, North Carolina, U.S.A.

Abstract

Three-dimensional display can be accomplished by simulating either of the two means of three-dimensional presentation in nature: reflection from object surfaces and self-luminous objects. Both forms of display benefit significantly from real-time interaction.

Reflective displays are produced by shaded graphics, which require raster graphics hardware: a frame buffer, scan-out components, and often a processor. Hardware to speed display transformations is presently becoming available. The structures of both the standard and fast systems will be described. Also briefly presented will be the design of interactive head or hand trackers for moving and pointing within a 3D image.

Self-luminous display can be produced for surface representations made by dots or lines or for space-filling grey-scale distributions. In both cases one may calculate one or more projections which may be presented at various times or to different eyes to achieve the percept of depth. Alternatively one may place all of the individual points in 3-space by appropriate optical maneuvers. Display systems for both projective and optically based 3D display will be described. The former category includes vector, point, and raster graphics systems. The latter includes holography and systems with rotating or vibrating screens or mirrors, most interestingly the varifocal mirror and rotating LED-panel systems.

1. Introduction

There are two ways in which the three-dimensional world presents itself: by reflection and self-luminosity. Similarly, there are in essence only two forms of three-dimensional display: reflective and self-luminous. Shaded graphics is the method by which reflective display is normally presented (see figure 1). Self-luminous display is provided by many approaches, including holography, vibrating or rotating mirrors and screens, and vector graphics of wire frames (see figure 2) using, for example, the kinetic depth effect.

NATO ASI Series, Vol. F19
Pictorial Information Systems in Medicine
Edited by K. H. Höhne
© Springer-Verlag Berlin Heidelberg 1986

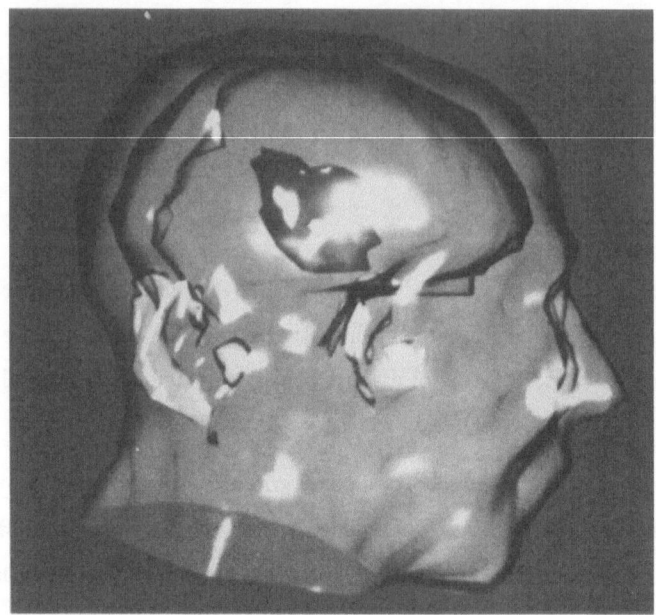

Figure 1. Shaded graphics display of a tumor in a brain in a head, from CT scans.

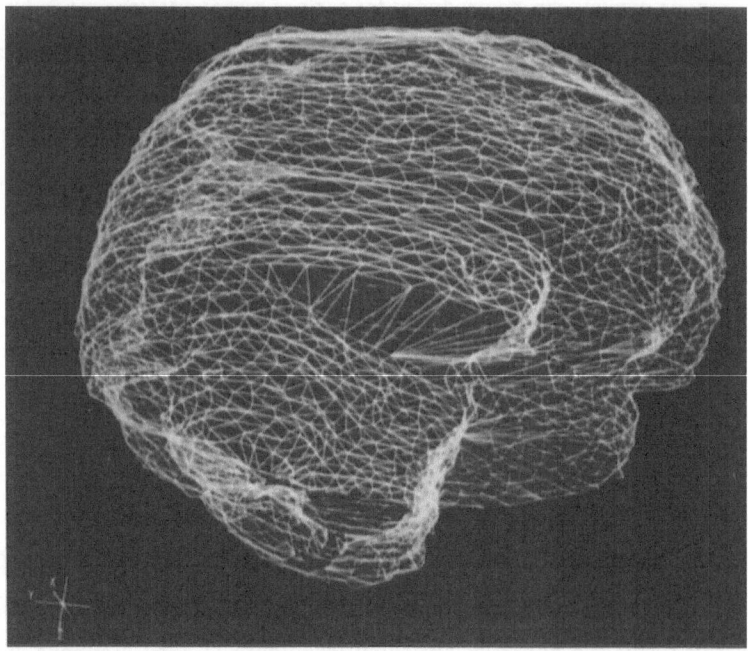

Figure 2. Wire-frame display of a brain, from CT scans.

It appears [Fuchs and Pizer, 1984] that reflective display is superior to self-luminous display for the presentation of known surfaces, e.g. the surface of an organ to a surgeon or radiotherapist who will use the result for planning. However, the need to know the surfaces before the display is produced makes reflective display weak for the explorational needs of a radiologist. In contrast, self-luminous display seems to have some real strengths for explorational purposes. Therefore, in the following first reflective and then self-luminous display systems will be covered.

2. Reflective Display Systems

The hardware currently supporting reflective display is the raster graphics system, the major component of which is the *frame buffer* (see figure 3). The frame buffer is a memory holding values for each of red, green, and blue intensities for each picture element (pixel) on the display screen. Alternatively, each pixel may hold an intensity value only, or values representing hue, saturation, and intensity. Provided with the

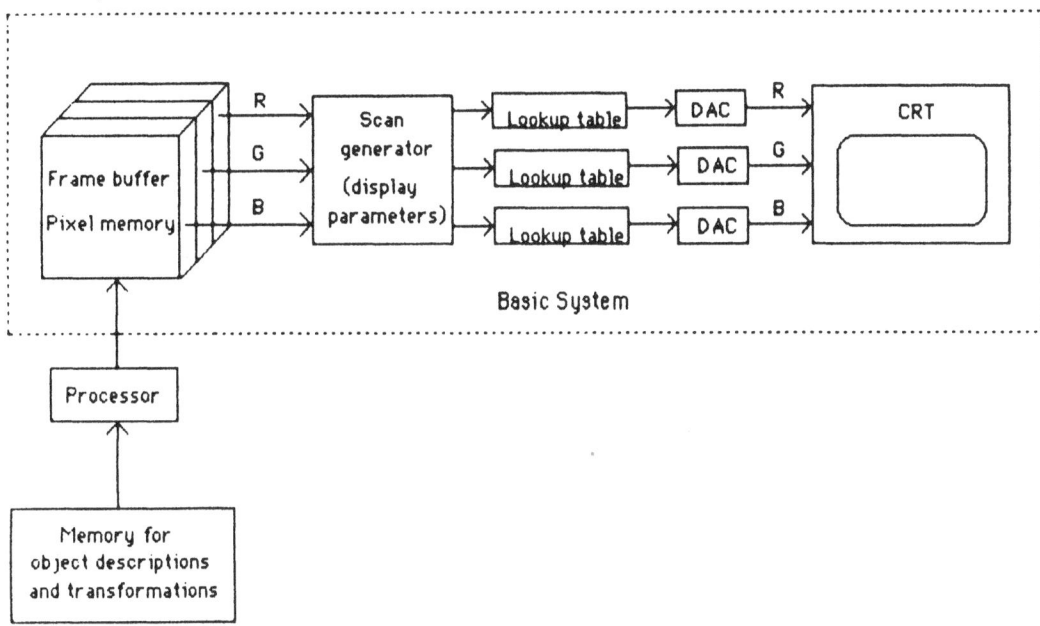

Figure 3. The structure of a raster graphics system.

frame buffer is a mechanism for scanning out these values to control the brightness of three color guns as the image is swept out onto the screen as a raster at 25-60 frames/sec. This mechanism, called the *scan generator,* includes parameters that cause the appropriate portion of memory to be read out, thus allowing pan and zoom. The three brightnesses can be modified, and thus the resulting color controlled, by passing the digital intensities through lookup tables before they are made analog.

Both effective display and image manipulation depend on interactive modification of the contents of the frame buffer and thus of the displayed image. For example, one may wish to rotate an object or move a cursor. This is accomplished by keeping descriptions of image objects in a separate memory and then filling the frame buffer by applying transformations on these objects [Newman and Sproull, 1979]. The object descriptions are normally in the form of lists of polygonal tiles. These tiles are frequently triangles fit between contours of the surface on successive slices [Fuchs, 1977] or rectangles representing the faces of object voxels (3D pixels) [Herman, 1985]. The transformations, represented by matrices and function subroutines, allow such operations as rotation, translation, scaling, perspective division, and lighting calculations. A processor, frequently a special one internal to the display system, applies the transformations, does *visibility* calculations to determine at each pixel which is the frontmost polygon at that position, and *shading* calculations to determine the light reflected from that polygon at the pixel in question, possibly smoothing the polygonal surface as it shades [Sutherland, 1973]. The result is loaded into the frame buffer.

The speed of calculation and loading of pixel values can be slow, since calculations are required at every pixel or for every tile. Speeding up this process is important in that it allows interactive control of the display parameters, and this improves the 3D perception as well as making the system convenient to use. Two hardware solutions, both involving parallel computation, are presently being developed to provide this speedup. In the first solution, processors are pipelined so that, for example, of a sequence of operations to be carried out on each tile, operation n is being carried out on tile k while operation n-1 is being carried out on tile k+1, etc. In the second solution, some processing capability is placed at each pixel so that in effect each pixel can do its own hiding and shading calculations. The latter approach is taken by the elegant design of Fuchs.

Fuchs's *Pixel-planes* structure is based on the realization that the calculations for determining 1) whether a pixel is in a particular tile, 2) which tile is hidden by which at each pixel, and 3) what shade each pixel should take, involves or can be well approximated by the calculation of a linear function $Ax+By+C$. Fuchs et al [Poulton, 1985] therefore have created a structure in VLSI (see figures 4 and 5) in which a tree connection of the pixels allows every pixel in the frame buffer (each with a position

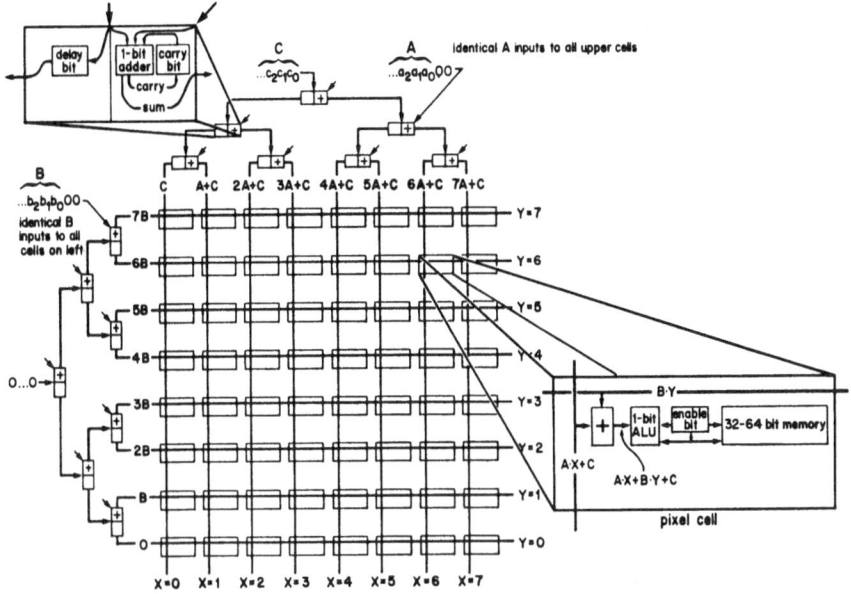

Conceptual design of an 8 x 8 pixel PIXEL-PLANES image-buffer memory chip.
Scan conversion, hidden-surface elimination and color-rendering commands are translated
outside the memory system into A,B,C coefficients and associated PIXEL-PLANES commands.

Figure 4. The structure of the Pixel-planes memory.

x,y) to calculate this linear function for its own value of x,y such that this operation is completed for all pixels in 30 clock cycles (presently under 3 microseconds). This machine, which on the basis of prototypes promises to take only about twice the chip space as an ordinary frame buffer, will be able to do the hiding and shading calculations for over a thousand tiles in the 1/30 second frame time of a video display. Thus the displayed image for a complex medical 3D object should be calculated in a fraction of a second, allowing interactive control of translation, orientation, lighting, transparency, etc.

2. 3D Trackers

Interaction in 3D, for example to control location and orientation in 3-space, requires special interactive devices. These may be hand-held or, even better, head-mounted [Sutherland, 1968] so that the display can be controlled by the viewing po-

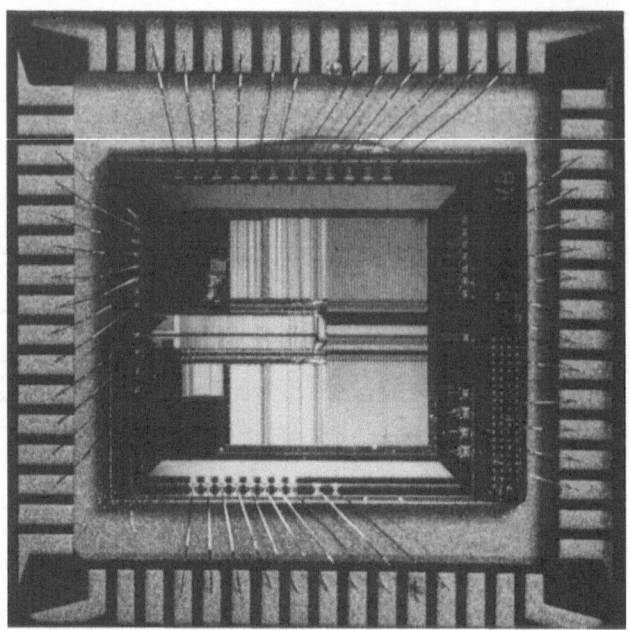

Figure 5. A prototype Pixel-planes chip (stage 3).

sition and direction of the viewer. With an adequately fast display system and a head-mounted display device, the viewer can feel that he is in the image space, such that ordinary movement of his body makes him feel that he is moving through the image (this has been called a *walkaround* display).

Among the 3D trackers available are mechanical devices such as trackballs and joysticks, pens that use the delays of sound to determine position, devices based on a magnetic field, those following special lights, either mounted on the observer and followed from special room locations or vice versa, and a newly developed prototype that follows natural features of the room. The 3D mechanical devices are in common use but are not very portable and allow the specification only of orientation, or only translation for some joysticks. Furthermore, 3D joysticks frequently do not provide continuous rotation or the specification of all possible angles, and 3D trackballs provide only relative orientation. Similarly, the sonic pen allows the specification only of location. The magnetic devices and those based on fixed or user-mounted lights overcome this restriction. However, the former have a somewhat limited range, and the latter provide somewhat limited accuracy, at least in a room-sized environment.

The "self-tracker" of Bishop [1984a,b] is designed to avoid all these difficulties without requiring special instrumentation of the viewing environment with lights or sensors, other than the necessary computing equipment.

Figure 6. A mockup of the Bishop self-tracker (photograph by Jerry Markatos).

The Bishop self-tracker (see figures 6 and 7) consists of many 1D sensors looking out in various directions from the viewer to the surrounding environment. It operates by correlating pictures at successive intervals of one or more milliseconds and calculating movement from the shifts required to achieve maximal correlation. The basic chip acts both as the 1D sensor and as a the correlating computer, and a central system combines the reports from the various chips into a single report of 3D location and orientation. Prototypes of this chip have been built and tested, and simulation based on their properties suggests that system development, now ongoing at Bell Laboratories, will produce a successful system by perhaps 1987.

Besides determining rotation and translation, 3D interactive devices are useful for selecting slices from 3D images for 2D display, for spatial clipping to limit obscuration or hiding problems, and for contrast enhancement in self-luminous displays. The devices mentioned above for determining rotation and translation can also be used for

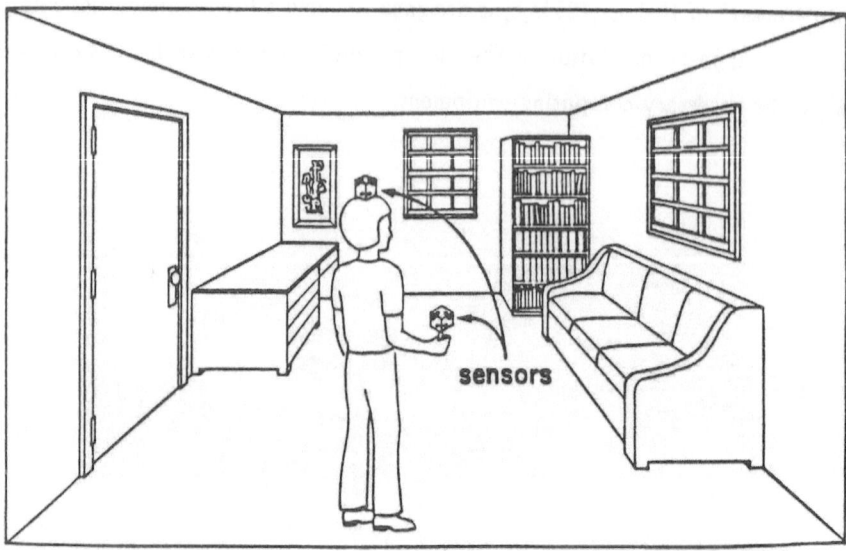

Figure 7. Use of the self-tracker in a "walkaround" display. From [Bishop, 1984a].

slice selection and spatial clipping. Contrast enhancement for self-luminous display is normally controlled by sliders, knobs, or mice.

3. Self-luminous Display Systems

Self-luminous display can be achieved via a directly viewed CRT or by doing the optical equivalent of putting points in 3-space by means such as holography or vibrating or rotating screens or mirrors. With any of these we may display 3D distributions of points, each with its own intensities, or of structures made from lines, such as wire frames. With space-filling self-luminous display, obscuration is such a problem that one never wants to display a full 3D raster of points, and interaction is frequently important to select the region to be displayed or the orientation of view.

3.1 3D Self-luminous Display by CRT

With CRT's the three-dimensional effect is given by providing stereopsis, by producing the kinetic depth effect, in which the object appears to move continuously in 3-space, or by using head-motion parallax via a head-mounted sensor as discussed above. Combining many of these cues is also possible and increases the strength of the 3D percept. All of these CRT-based methods require that the 3D image, consisting of a set of points or lines in 3-space (with co-ordinates x,y,z), be projected onto the 2D CRT (with co-ordinates x',y') from one or more orientations (along z').

Projection involves computing x', y', and z' from the recorded x, y, and z values for each point to be displayed, discarding z', and displaying at screen location x',y' a point with the recorded intensity of the point to be displayed. The transformation to the primed (*viewer*) co-ordinate system from the unprimed (*world*) co-ordinate system involves matrix multiplication, as well as division if perspective projection is used. This calculation must be done for every point.

For the display of lines the endpoints of each line must undergo a projection calculation as above, and one of the following three steps then must be undertaken with these endpoints together with any individual points to be displayed.

(1) Interpolate points intermediate to the line's endpoints, and put the result into a frame buffer, summing intensities in pixels that each picture element (pixel) on the display screen. Alternatively, each receive more than one result. One may possibly remove the stairstep artifact caused by discrete pixels by an approach called *anti-aliasing*, which involves a slight blurring of the line [e.g., Crow, 1981].

(2) Interpolate the points on the line as in method 1 and place the result on a point-plotting CRT, or deliver the endpoints of the line to a vector graphics display system, which will draw the line using analog means. Both point-plotting and vector graphics displays are given x,y,intensity information, rather than a 2D array of pixel intensities, as with a raster graphics system (frame buffer).

(3) Interpolate the points on the line as in method 1, and use a color frame buffer in the following nonstandard way [Fuchs, Pizer, et al, 1982] to make it behave like a point plotter (see figure 8). Load the locations in the frame buffer not with red, green, and blue values as is normally expected, but rather with x,y, and intensity values. Replace the video CRT at the output with a point-plotting CRT, with the three wires that are outputs of the frame buffer (labeled red, green, and blue) plugged into the x, y, and intensity terminals of that point-plotting CRT. With this scheme the intensity lookup tables normally provided for each color in a frame buffer system can be used to transform x, y, and intensity respectively, thus allowing, clipping, scale change, translation, and contrast enhancement.

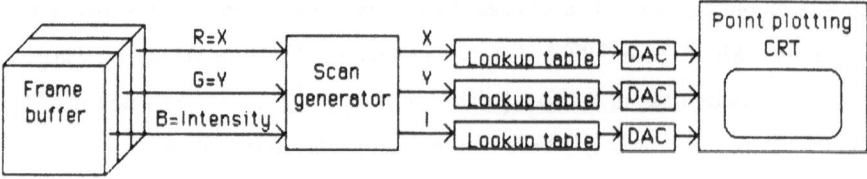

Figure 8. Use of a frame buffer to produce a point-plotting display.

Stereo display simply requires the calculation of two projections, and either the simultaneous display of both images or the placement of the results into two display refresh buffers and the alternation between these buffers. With simultaneous display, mirrors or lenses are necessary to direct the pictures separately to the respective eyes. If the two projections are alternated, one must have a means of alternately blocking the two eyes. Common means for this blocking are via a rotating mechanical device, via goggles with a birefringent material that can by electrical means be made to change between opaque and transparent, and by superimposing a polarizing plate on the screen that can under electrical control change polarization [Bos, 1983], while the viewer wears a differently polarized lens on each eye. The latter method allows many viewers to view the same display and for the goggles to be very inexpensive.

Display using the kinetic depth effect can be achieved by precomputing the projections corresponding to a rocking or rotating motion, but interactively controlled motion provides a more convincing 3D effect [Lipscomb, 1983]. Furthermore, such interactive control is useful for choosing viewing directions that avoid obscuring objects. But this requires calculation of the required projection(s) at on-line speeds. Even fast processors that operate serially on the points cannot provide adequate speed for more than on the order of 10,000 points. What is needed is a system that can transform many points in parallel, if necessary interpolate lines from endpoints in parallel, and if using a raster graphics system add the results in parallel into the frame buffer.

3.2 3D Self-luminous Display by Optical Means

Failing such new fast projection hardware, one can avoid the need to do projection calculations and add the (probably most) important depth cue of head-motion parallax by putting points in 3-space by optical means. The first means that comes to mind is holography, but it is orders of magnitude too slow in producing a new result to allow the critical step of interaction, and it is technically quite demanding. An attractive alternative is to write planar images while moving the writing plane cyclically, fast enough for the planes to fuse into a 3D image. Either screens or mirrors can be moved to provide the required motion of the writing plane.

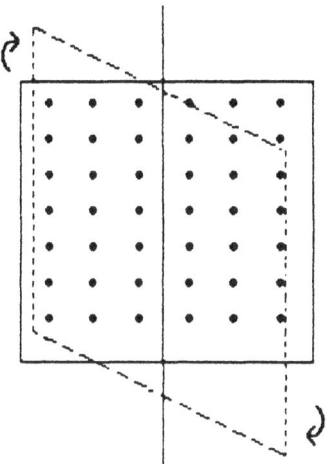

Figure 9. The rotating LED-panel display [Jansson, 1984].

(1) Among the many systems using rotating or translating screens, the most attractive seems to be the rotating LED panel [Jansson, 1984]. In this device (see figure 9) a closely packed array of LED's is rotated about an axis centered within the plane of the array, so that at any time the array forms a one-sided radial slice of a cylinder. The intensities within that slice are formed by controlling the fraction of the time that each LED is on while the array is near some angle. The strength of the device is that it allows all viewing angles and a viewing position close to the 3D display region. Its weaknesses are the need to store and transmit the zero intensities that must dominate a self-luminous display if obscuration is not to be debilitating, and limitations in the spacing of the LED's and the size of the panel for it to be rotated with adequate speed.

(2) Among the many systems using rotating or translating mirrors, the most attractive

is the varifocal mirror because of the small mechanical motion that is required to obtain adequate depth [Traub, 1967]. The principle of the device is shown in figure 10. The viewer looks at a screen reflecting in a mirror that can be made to vibrate between convex and concave, normally by placing a loudspeaker behind the mirror membrane or plate. Because the apparent depth of the reflected screen behind the mirror varies with the convexity or concavity of the mirror, the writing plane is swept over 20-30cm for just a few millimeters of motion of the center of the mirror. Thus, points written on the screen within a short time interval appear on a thin slab parallel to and behind the mirror face at a depth dependent on the time during the display cycle at which the points were written.

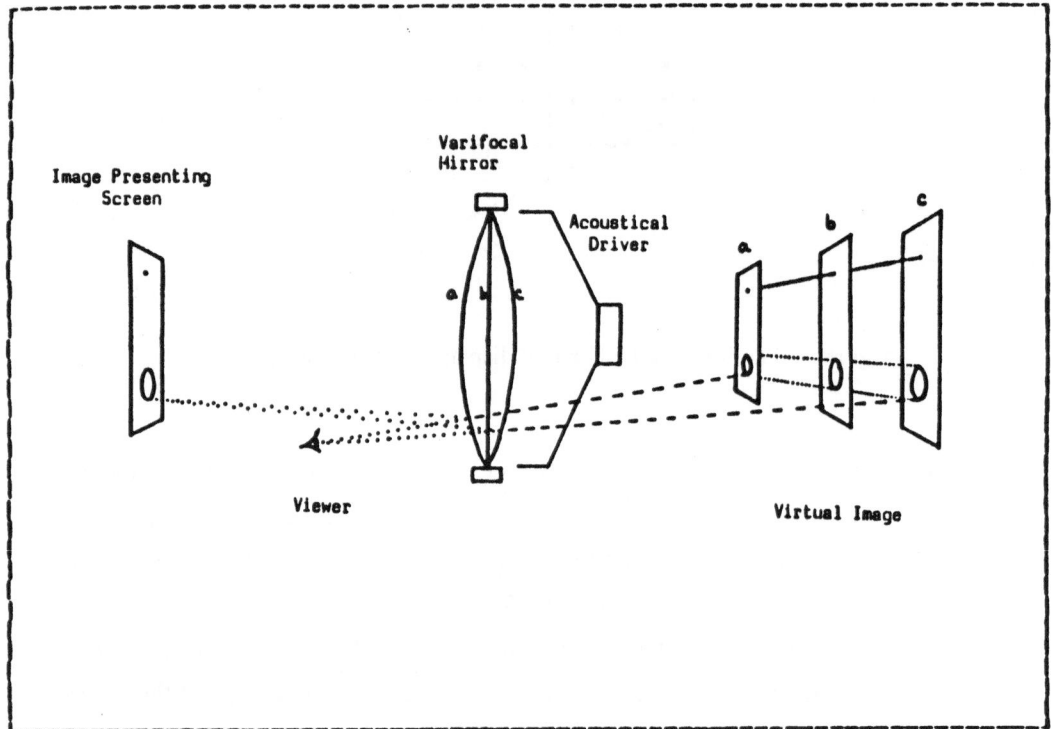

Figure 10. Principle of varifocal mirror display.

Only points with nonzero intensities need to be written onto the screen. These can be written in the form of a 3D raster or as 3D line segments [Baxter, 1980;

Sher, 1980]. Alternatively, if the screen is a serial point-plotting CRT, one can use the method given earlier of turning a raster graphics system into a point plotting system (see figure 8) with the additional restriction that the location in the frame buffer (time in the display cycle) at which a point's co-ordinates are stored must be related to the depth at which it is to appear [Fuchs, Pizer, et al, 1982]. The lookup tables can be used for clipping in the x and y dimensions and for contrast enhancement, and clipping in the z dimension, very important for avoiding obscuration, can be achieved using the registers normally used to specify the y-viewport (which region in y appears on the video screen) when the system is used for raster display.

The fact that the mirror's change from convex to concave causes a variable magnification with depth (so-called anomalous perspective) can be counteracted by varying the x and y scale of the image over time as the writing plane (the screen's reflection) moves in depth. Specifically, the x and y analog signals, taken relative to the screen center, need to be multiplied (gain controlled) by the speaker's sinusoidal input, offset by a constant.

Interactive modification of rotation, objects selected, blinking, scale, and other display parameters can be achieved by the standard graphics approach of having basic object descriptions as points in the world space and transformations from these to viewer space by a processor that frequently is part of the display system (see figure 3). If this processor is not fast enough to transform all of the points in one display cycle, as is commonly the case, a useful approach is that of *successive refinement*, in which a selected group of points are first transformed and displayed, forming a coarse display, with the remainder being transformed and added to the display in later cycles if no change in the interactive parameters controlling the transformation is made in the meanwhile. This approach appears to have some merit as well with reflective display.

4. Conclusion

The preceding has been a brief overview of 3D display devices. While 3D display has great promise in medical imaging, further work is needed in developing devices that are adequately speedy and provide adequate interactive control. New systems will almost certainly be invented, and developments and evaluation of applications in medical imaging are being carried out in large numbers.

5. Acknowledgements

I thank Henry Fuchs and Edward Chaney for useful discussions. I am indebted to Joan Savrock for manuscript preparation and to Bo Strain and Karen Curran for photography. The preparation of this paper was prepared partially with the support of the National Institutes of Health grant number 1-R01-CA39060.

6. References

B. Baxter, "3-D Viewing Device for Interpretation of Multiple Sections", *National Computer Conference, NCC-80*: 437-440, 1980.

G. Bishop, *Self-Tracker: A Smart Optical Sensor on Silicon*, Ph.D. dissertation, University of North Carolina, Department of Computer Science, 1984a.

G. Bishop and H. Fuchs, "The Self-Tracker", *Proc MIT VLSI Conf*, Artech House, Dedham, MA, 1984b.

P. J. Bos, P. A. Johnson, Jr., K. Rickey Koehler/Beran, Tektronix, Inc., "A Liquid-Crystal Optical-Switching Device (π cell)", *Proc. Soc. for Info. Disp.*: 30, 1983.

F. C. Crow, "The Aliasing Problem in Computer-generated Shaded Images", *Comm. ACM, 20* (11): 799, 1977.

H. Fuchs, Z. M. Kedem, and S. P. Uselton, "Optimal Surface Reconstruction from Planar Contours", *Comm. ACM, 20* (10): 693-702, 1977.

H. Fuchs, S. M. Pizer, E. R. Heinz, L. C. Tsai, and S. H. Bloomberg, "Adding a True 3D Display to a Raster Graphic System", *IEEE Computer Graphics and Applications, 2* (7): 73-78, 1982.

H. Fuchs and S. M. Pizer, "Systems for Three-Dimensional Display of Medical Images", *Proc Int Joint Alpine Symp on Med Comp Graphics and Image Commun and Clin Adv in Neuro CT/NMR*, IEEE Comp Soc [IEEE Catalog No 84CH2006-5]: 1-6, 1984.

G. T. Herman, "Computer Graphics in Medicine", in this volume.

D. G. Jansson and R. P. Kosowsky, "Display of Moving Volumetric Images", *Processing and Display of Three-Dimensional Data II, 507*: 82-92, 1984.

J. S. Lipscomb, "Motion Decomposition, Orthogonality, and Stereo Display in a Molecular Computer Graphics System", *ACM Transactions on Graphics*, 1983.

W. M. Newman and Robert F. Sproull, *Principles of Interactive Computer Graphics*, McGraw-Hill, New York, 1979.

J. Poulton, H. Fuchs, J. D.Austin, J. G. Eyles, J. H. Heinecke, C. Hsieh, J. Goldfeather, J. Hultquist, and S. Spach, "Implementation of a Full Scale Pixel-Planes System", to appear in *Proceedings of Conference on Advanced Research in VLSI*, UNC-Chapel Hill, 1985.

L. Sher, *The SpaceGraph Display: The Utility of One More Dimension*, internal paper, Bolt, Beranek, and Newman, 1980.

I. E. Sutherland, "A Head-Mounted Three Dimensional Display", *Proc. Fall Joint Comp. Conf.*: 757, 1968.

I. E. Sutherland, R. F. Sproull, and R. A. Schumacker, "A Characterization of Ten Hidden-Surface Algorithms", *Comput. Surv. 6*(1): 1, 1974.

A. C. Traub, "Stereoscopic Display Using Rapid Varifocal Mirror Oscillations", *Applied Optics, 8*(6): 1085-1087, 1967.

Issues in the design of human-computer interfaces

Jurg Nievergelt

Informatik, ETH, CH·8092 Zurich, Switzerland

Abstract

Interactive use of computers in many different applications by users with a wide variety of backgrounds is ever increasing. This proliferation has increased the importance of designing systematic human-computer interfaces to be operated by casual users. Novel styles of interfaces that exploit bitmap graphics have emerged after a long history of interactive command languages based on alphanumeric terminals. This development is repeating, with a 20-year delay, the history of programming languages: Large collections of unrelated operations are being replaced by systematically structured modes that satisfy general principles of consistency. This tutorial illustrates some *general* issues that designers of human-computer interfaces encounter, by means of examples chosen from one *particular* research project. We do not attempt to give a comprehensive or balanced survey of the many approaches to human-computer interface design.

A survey and classification of design errors common in today's command languages leads to concepts and design principles for avoiding such errors. But the very notion of *command language*, which emphasizes the input, is of doubtful utility for understanding the principles that govern the design of good human-computer interfaces. Observation of users shows that the most common difficulties experienced are expressed well by such questions as: *Where am I? What can I do here? How did I get here? Where else can I go and how do I get there?* Such questions reveal the fact that the fundamental problem of human-computer communication is *how to present to the user what he cannot see directly, namely the state of the system, in such a manner that he can comprehend it at a glance*. If the *output language* of a system, what the user sees or can get onto the screen easily, is rich and well structured, the command language proper, or *input language*, can be simple - ideally the minimum set of *selecting the active data, selecting the active operation, and of saying "do it"*, all by just pointing at the proper places on the screen. Instead of emphasizing the *command part* of the interface, designers should focus their attention on the *display part*.

The questions above also provide hints about how the display of an interactive system should be structured. *Where am I?* and *What can I do here?* refer to the state of the system. A good design answers them by displaying to the user on demand his current data environment (*site*) and his current command environment (*mode*). The questions *How did I get here? Where else can I go?* refer to the past and future dialog (*trail*). A good design presents to the user as much of the past dialog as can be stored: to be undone (in case of error), edited and reinvoked. And it gives advice about possible extrapolations of the dialog into the future (*help*). This *Sites, Modes, and Trails* model of interaction has been implemented in two experimental interactive system. They presents the state- and trail information by means of *universal commands* that are active at all times, in addition to the commands of an interactive application. This gives the user the impression that "all applications on this system talk the same language".

With the current spread of personal computers to be operated by casual users for short periods of time, the quest for systematic, standard man-machine interfaces must proceed beyond the design of "user-friendly" operating systems. It will be necessary to identify the most important dialog control commands, and to standardize these in a minimal piece of hardware that the user can carry with him and plug into any personal computer. We describe an experiment in this direction, which involves a *"mighty mouse"*.

NATO ASI Series, Vol. F19
Pictorial Information Systems in Medicine
Edited by K. H. Höhne
© Springer-Verlag Berlin Heidelberg 1986

Programming methodology has emphasized different criteria for judging the quality of software during the three decades of its existence. In the early days of scarce computing resources the most important aspect of a program was its *functionality* - what it can do, and how efficiently it works. In the second phase of *structured programming* the realization that programs have a long life and need permanent adaptation to changing demands led to the conclusion that functionality is not enough - a *program must be understandable*, so that its continued utility is not tied to its "inventor". Today, in the age of interactive computer usage and the growing number of casual users, we begin to realize that *functionality and good structure are not enough - good behavior towards the user is just as important.* The behavior of interactive programs will improve when the current generation of commercial software has outgrown its usefulness, and the next generation of programmers is free to make a fresh start.

Key words and phrases

Interactive systems, human-computer communication, system design, standardization

Contents

1. The computer-driven screen as a communications medium

A versatile new mass-communications medium has come into existence: *The computer-driven screen.* Considering the computer to be a communications medium, in competition with television or overhead projector, is perhaps an unusual idea to the users of traditional "data processing"; but it is a useful point of view for the designer and user of an information system.

The computer-driven screen is the only *two-way mass communications medium* we have. It allows *two-way communication between user and author of the dialog.* You don't talk back to newspapers, radio or television - or if you do, at least you don't expect them to react to your outburst - there is no feedback. But you *can* control a well-designed computer-generated dialog: *"Show me that picture again", "go slower", "skip these explanations for the time being", "what else do you have on this topic", "I want more detail on this figure", "remind me of this fact again tomorrow"* - these are some of the commands a human-computer dialog *could* accept - but usually doesn't.

A dialog between man and machine is not symmetric. On the contrary, it should exploit the different capabilities of the two participants: The humans's prowess at trial-and error exploration based on pictorial pattern recognition, and the machine's ability to call upon large collections of data and long computations extremely accurately, and to display results rapidly in graphic form. There is no point in trying to mold the human-computer interface on the model of human-human communication. In fact, *implementing the "commands" above requires no breakthrough in artificial intelligence - it simply requires a good systems design.* Designing the interface so as to exploit the computer's speed and accuracy, and the human's abilities in pattern recognition and intuitive selection, may well be the problem of greatest economic impact computer scientist face in this decade.

2. Survey and classification of errors

Until recently only a small minority of computer users had the opportunity to use interactive systems - mostly professionals who use the computer daily. They are concerned with *the inherent power of the computer for their application* and get so used to its idiosyncrasies that a mysterious or even illogical interface doesn't bother them any more. If they are system designers, they are also concerned with the *internal structure of their software*, but have often treated the behavior of their system towards the user as a superficial problem, addressed in the style *"if you don't like it, change it".* An example of this situation is provided by UNIX, widely acclaimed as a model of a well-designed operationg system. [No 81] reaches the conclusion "the system design is elegant but the user interface is not". The programmers' traditional attitude explains why today's interactive software is full of blemishes. Let us illustrate typical errors of dialog design, and try to explain them through a neglect of fundamental concepts of human-computer interaction. Two of the most fundamental concepts are the **state of the system** and the **dialog history.** Interface design is the art of presenting the state and the dialog history in a form the user can understand and manipulate.

The *state* is concerned with everything that influences the system's reaction to user inputs. The *interface* is concerned with interaction: what the user sees on the screen, what he hears, how he inputs commands on the key board, mouse or joy stick. Today's fad is to engineer fancy interfaces: fast animation in color, sound output and voice input. This is desirable for some applications, unnecessary for others, but in any case it does not attack the main problem of human-computer communication.

The fundamental problem of human-computer communication is how to present to the user what he cannot see directly, namely the state of the system, in such a manner that he can comprehend it at a glance.

So let us start with common errors of dialog design which hide vital state information from the user.

2.1 Insufficient state information

Imagine that you leave your terminal in the middle of an editing session because of an urgent phone call. When you return ten minutes later, the screen looks exactly as you left it. Even if the system state is unchanged, you may well be unable to resume work where you left off:

D: *"What file was I editing?"*, *"Is there anything useful in the text buffer?"*;

C: *"Am I now in search mode?"*, *"What is the syntax of the FIND command?"*;

T: *"Has this file been updated on disk?"*, *"Has it been compiled ?"*.

Such questions indicate that part of the state information necessary to operate this system has to be kept in the user's short term memory. When the latter is erased by a minor distraction, the user needs to query the system to determine its state. Hardly any system lets him do this systematically, at all times.

Today's systems provide state information sporadically, whenever the programmer happened to think about it. Examples abound. The file directory can be seen in the file server mode, but not in the editor; in order to inspect it you must exit from the editor, an operation that may have irreversible consequences. In order to see the content of the text buffer you may have to insert it into the main text, thus polluting your text. To find out whether you are in search mode you may have to press a few keys and observe the system's response - not always a harmless experiment.

A designer who observes the following principle avoids the problems above:

The user must be able AT ALL TIMES to conveniently determine the entire state of the system, WITHOUT CHANGING THIS STATE.

The notion of "entire state" is meant with respect to the user's model of the system, not at the bit or byte level of the implementation. Thus the details of the system state presented to the user will differ from system to system, but two of its components are mandatory. In any interactive system, the user operates on *data* (perhaps he only looks at it) by entering *commands* (perhaps he only points at menu items or answers questions). Thus the state of the system must include the following two major components: the *current data environment* (what data is affected by commands entered at this moment), and the *current command environment* (what commands are active). The questions above labelled D and C refer to data and command environments, respectively. The questions labelled T refer to the user's *trail*, i. e. the past and future dialog.

A particularly dangerous version of "insufficient state information" is the deceitful presence on the screen of outdated information. The programmer is aware of the moment he has to write some information on the screen, but he forgets to erase it when it no longer describes the state of the system. The user sees 'current file is TEMP' , when in fact his current data environment is another file. The bad habit of leaving junk lie around the screen is a relic from the days of the teletype, when messages written could not be erased. It should not persist on today's displays.

2.2 Ignoring the user's trail

"We have seen the enemy, and he is us!" The user is by far the most dangerous component of an interactive system - at least 90% of all mishaps that occur during operation are traceable to faulty user actions. Designers of interactive systems must accept the fact that high interactivity encourages trial-and-error behavior. Exhortations to discipline are worse than useless - they are counterproductive. A designer using a CAD system, a writer using a text processor, a programmer debugging his programs - they **must** concentrate on their creative task, and cannot allow a fraction of their conscious attention to be sidetracked, continuously double-checking clerical details. A system should be foolproof enough to absorb most inaccuracies and render them harmless.

It is surprising that virtually all research on reliability and security is directed against either hardware and software failure or deliberate attack by third parties. But the majority of users of interactive systems have no one but themselves to blame when their data is suspect or has been damaged. In cleaning up one's files the most recent version of a document is thrown away and an old one kept instead. You hit **D** instead of **C** on the keyboard, so something gets deleted instead of created. You forget to label a file permanent and so it vanishes. The catalog of plausible errors is different on each system; they all have in common that, as soon as we become aware of the blunder, we gasp: "How could I possibly have done that!". We can and do, at the rate of many oversights a day.

The safeguards built into today's systems to protect the user from his own mistakes are primitive. Are you sure? is the favorite question, followed by a sporadic request to press some unusual combination of keys if the action is really serious. This double-checking is effective only against accidental key pressing, not against an erroneous state of mind. When I delete a file I'm sure that's what I want to do - though I may regret it later.

The most effective protection of the user against his own mistakes is to store as much of the past dialog as is feasible. At the very least a universal command undo must always be active that cancels the most recent command executed. The memory overhead of keeping two consecutive system states is negligible, since these two states differ little - typically by at most one file. The error-prone and time-consuming Are you sure? is unnecessary when the user can always undo the last step. Ideally undo works all the way back to the beginning of the session, but that may be too costly: in order to replay the user's trail *backwards* the system must in general store *states*, not just the *commands* entered, as many operations have no inverses. A practical alternative is for the system to keep a log of commands entered and to allow the user to save the current state as a check point for future use, from which he can replay the session.

3. An interactive system as seen by the user

3.1 The Sites, Modes, and Trails model of interaction

Observation of casual users provides valuable insight into the fundamental design question of how a machine should present itself. Most of the recurring difficulties they encounter are characterized well by the following questions:

- **where am I?** (when the screen looks different from what he expected)
- **what can I do here?** (when he is unsure about what commands are active)
- **how did I get here?** (when he suspects having pressed some wrong keys)
- **Where else can I go and how do I get there?** (when he wants to explore the system's capabilities).

We are beginning to learn that the logical design of an interactive system must allow the user to obtain a convenient answer to the questions above **at all times**. In other words, the man-machine interface must include queries about the state of the system (without changing this state), about the history of the dialog, and about possible futures. This principle is much more important for today's computerized machines, *black boxes* that show the user only as much about their inner working as the programmer decided to show, than it is for mechanical machines of the previous generation, which by visible parts, motion and noise continuously reveal a lot of state information.

In order to answer the user's basic questions in a systematic way, the designer of an interactive system must also design a simple **user's model of the system**: a structure that explains the main concepts that the user needs to understand, and relates them to each other. The major concepts will certainly include many from the following list: the *types of objects* that the user has to deal with, such as files, records, pictures, lines, characters; *referencing mechanisms*, such as naming, pointing; *organizational structures* used to relate objects to each other, such as sets, sequences, hierarchies, networks; *operations* available on various objects; *views* defined on various types of objects, such as formatted or unformatted text; *commands* used to invoke operations; the *mapping of logical commands onto physical I/O devices*; the *past dialog*, how to store, edit, and replay it; the *future dialog*, or help facility.

The list is long and the design of the user's model of the system is an arduous task; but it should be possible to explain the overall structure of the model in half an hour. If the system's behavior then constantly reinforces the user's understanding of this structure, it will quickly become second nature to him. In contrast, if the system keeps surprising the user, it interferes with his memorization process and slows it down. Because questions will always arise during the learning phase, and because the right manual is rarely at the right place at the right time, an interactive system should be self-explanatory: it must explain the user's model of the system to the user, at least in the form of an on-line manual, preferably in the form of an integrated help facility that gives information about the user's current data and command environments at the press of a single key.

The user's model of the system is a *state-machine*: it has an *internal state* and an *input-output behavior.* Components of the state must include:

- **the user's data environment** (data currently accessible)
- **the user's command environment** (commands currently active).

The questions *Where am I?* and *What can I do here?* are then answered by displaying the current data and command environments, respectively [NW 80]. The user must be able to invoke this system display at all times, regardless of which applications program he is currently in, and its presentation must not change the state of the system in any way.

The system must have *universal commands*, which are active at all times. At least the state inquiry commands postulated above must be universal. In a highly integrated system many more commands can be made universal. General dialog commands that are needed in every interactive utility are omitted from the text editor, the diagram editor, the data base query language, and incorporated as universal commands into the system. The consequence on the user's view of the system is that all these utilities "talk the same language", and that in order to become proficient at using a new editor he only has to learn a small number of new commands: the data-specific commands are new, the data-independent commands remain the same.

3.2 Shortcomings of conventional operating systems

Developing good interactive software on existing systems is difficult. Today's programming languages and operating systems often fail to provide the building blocks necessary for processing user inputs and creating graphic output under the stringent real time conditions expected for highly interactive usage.

Today's commercial operating systems are answers to the demands of the computer center, to process a load dominated by batch jobs, on top of which slow-response time-sharing services with textual I/O were added later. Such an operating system provides both too much and too little for the interactive workstation, the computers on which most interactive work is done today. Too much by way of fancy resource management policies designed to accomodate many simultaneous users with different demands; the resulting overhead is wasted on the single-user computer. Too little by way of supporting a dialog with the user. For example, once a program is started, the operating system is typically unavailable to the interactive user, except for an *abort* command. However, in addition to "Control C" there are lots of utilities an operating system contains that should be made available to the end-user at all times, while he interacts with an application program. Directory inspection and manipulation commands, for example, are useful operations when the user wishes to enter a browsing phase, to search his files for useful data. Instead of forcing on the user the irreversible operation of exiting from an editor to inspect his files, the system should make its commands available at all times in a dedicated *system's window*.

3.3 Experimental interactive systems

The concepts described in the preceding sections have been realized in two prototype system XS-1 [Be 82] and XS-2 [St 84], intended as case studies of integrated interactive system for personal computers. Our present interest lies in the systems' kernel, rather than in the application programs that run on it, as an experiment in designing operating systems that support human-computer interaction in the form of the *sites, modes, and trails* model.

The functions and structure of the kernel

In conventional interactive systems the application program is responsible for defining, controlling and displaying its data and its dialog. The operating system does not communicate with the user except for starting or aborting the application program. By contrast, the kernel of XS-1 provides a structured space in which all the data and all commands are embedded, and provides operations to explore this space and manipulate its elements. The function of the kernel is to allow application programs to structure their data in the common space and to leave the dialog treatment to the system. As a consequence, the user interface is mostly determined by the kernel, rather than by the application program, and the user gets the impression that all programs in the entire system *talk the same language*.

In order to achieve this goal, the kernel of XS-1 has four major components:

EXPLORE makes sites, modes and trails visible to the user. It defines and controls all motion on these spaces and handles user requests for screen layout changes by means of universal commands. Explore allows the user to inspect the current system state without changing it.

TREE EDITOR provides universal commands for structural manipulations that are common to most editors regardless of the type of data objects they manage. It is syntax-directed in order to reflect the relationships between different data types.

CENTRAL DIALOG CONTROL is a front-end dialog processor that handles user input at all levels, from key press to command, checks it for syntactic correctness and records it. It directs information to Explore, the tree editor, and to modes.

TREE FILE SYSTEM realizes the data access and manipulation requirements imposed by Explore and the tree editor. It handles data of any size, from a single character to a large file, in a uniform way.

The kernel also contains a screen management package that handles an arbitrary number of windows. It can be considered to be a virtual terminal.

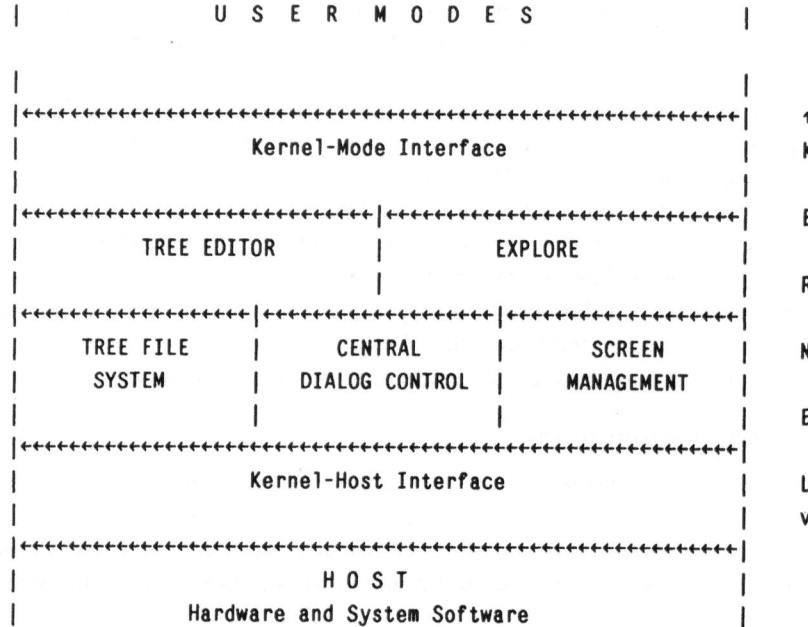

Figure: Components of the XS-1 kernel

Interface between the kernel and modes

Conventional interactive application programs devote much code and time to screen layout and to dialog handling. In XS-1 these activities are centralized in the kernel. A mode presents its data on a virtual screen and lets the screen handling package map the virtual screen onto the physical screen in a window of arbitrary size and position. The kernel handles the dialog and provides universal commands for visualizing the state of the system and for performing structural changes to the site and mode spaces. A sizable portion of the code of typical application programs resides in the kernel; only the mode-specific operations remain to be written by the applications programmer.

The interface between the kernel and modes is syntax-directed with respect to data and to commands: A mode defines the syntax of the site space it works on, and the syntax of its commands. The former drives the tree editor, the latter drives central dialog control. Any collection of data, independently of their size or type, may be attached to a node in the site tree, and any collection of commands may be attached to a node in the mode tree. Data attached to the current site is affected by commands; commands attached to the current mode are active.

Interface between the kernel and host

This interface maps the requirements of the tree file system, of central dialog control, and the screen management package onto corresponding utilities of a conventional operating system, such as schedulers and drivers.

XS-1 is written in Modula-2 and runs on a Lilith workstation. The more recent version XS-2 has also been ported to a VAX equipped with a Bitgraph terminal and to several 68000-based personal computers.

3.4 The controversy about modes

The notion of a *"modeless system"* is one of today's fads. Amusing stories purport to document the disadvantage of systems that behave differently in different modes. For example, that of the user who reaches the conclusion that the computer is down because the terminal echos his key presses without executing any of the commands he enters. The unintentional *insert command* that got the system into the editor's insertion mode has long since been scrolled off the screen, so the user is unaware that he is interacting with a different subsystem than the one he thinks he is.

The major design error behind this story, however, is **not** that the system behaves differently in distinct modes, **but that the insertion mode is hidden.** This type of error stems from the days of printing terminals, on which a periodically appearing line *"now in insertion mode"* is indeed impractical. But on a display terminal a dedicated system status line is conventional. And if the application programmer does not wish to lose a single line on the screen, the system state can be displayed on demand, by way of a universal command which is **not** an insertable character!

How would a "trully modeless system" look? Fortunately we know that, although such systems are disappearing. Early Kanji typewriters for inputting thousands of Chinese characters had thousands of keys. "Trully modeless" means that no physical input device ever has its meaning affected by the system state. As a consequence, every logical command must have its own physical device. This may be practical for special purpose machines such as watches, pocket calulators, or games, but is impossible for general purpose work stations, on which an open-ended spectrum of applications will run.

Interactive systems have a lot of commands! They are not easily counted, as a complex command with many parameters can always be broken into simple commands without parameters, and vice versa. If we count the different decisions that the user may be called upon to make, i.e. we weigh each command with the number of its parameters, we easily reach hundreds of items of information that the user should know (think of a typical BASIC system on a hobby computer). After weeks of daily use, the professional has memorized these hundreds of items in his long-term memory. But the casual user, who must rely on his short-term memory of *"the magical number seven, plus or minus two"* chunks of information [Mi 56], finds himself constantly looking up the manual.

The *"sites, modes, and trails"* model implemented in XS-1 and XS-2 attempts to solve the problem of the "7 + or - 2 chunks in short-term memory" as follows. **The dozen universal commands define a modeless dialog machine** - the user must memorize these application-independent commands once and for all. An interactive application activates its own commands in addition. These are visible in the mode window, so the user need not memorize them. We tend to structure application packages into several modes, typically with less than 10 comands each, whose meanings are related.

4. Is a standard man-machine interface possible?

4.1 The user interface problem in a network of home computers

Once we have mastered the user interface of our first automobile, it only takes minutes to operate the steering wheel, pedals, and buttons of another car. Occasionally we must search for a light switch, but at least the car-driver interface has been standardized to such an extent that we need no manual in order to drive. Contrast this with today's interactive systems, where expertise with one text editor does not guarantee that you can use another without a manual.

The issue of standard man-machine interfaces will become acute with the coming integration of personal computers and telecommunications. I would like to illustrate this point with a study performed at Nippon Telegraph and Telephone. NTT's Information Network System INS [Ki 82] will link home communication terminals to a national fiber optics transmission network of sufficient bandwidth to carry animated pictures. The planned services include electronic mail with sound and pictures, access to libraries and information systems, financial services - whatever comes to mind in the domains of entertainment, education and business at home.

In the rapidly changing market of personal computers, the home terminal cannot be specified ahead of time - most of the needed 30 million terminals would be obsolete before their construction. One will have to accept as home terminals many of the personal computers that will come on the market over the next decades, with properties as yet unknown. The danger of a Babylonian chaos lurks in this scenario. The user who has mastered the terminals in his home and in his office will encounter another machine in the "telephone cabin". Will he be able to operate it without having to study a manual? Whereas the long-lived telephone set made standardized operation possible through essentially identical hardware, the short-lived personal computer will have to rely on a logical standardization as well.

4.2 The scope of existing standards

There is much standardization activity relevant to the problem of the network of home communication terminals, but standard dialog control is has barely been touched. Consider the different system components to be standardized by way of an example: The user wishes to interact with a teaching program obtained from a national library without any loss of quality.

First, some file(s) will have to be transmitted from the library into his home computer. Many standards govern network protocols for data communication. Their multitude is a problem, but a national carrier protected by a monopoly can choose its own, so file transfer is no problem.

When the files have arrived, they must be looked at in several ways. Text poses few problems. The easiest way to standardize text file formats is to limit the opaerating systems to be used to a few that are widely used on personal computers, such as MSDOS, CP/M, or Unix. Graphics is harder, as today's operating systems don't support the concept of a graphics file. Standards are available or underway from the computer field (ACM Core Graphics, GKS) and from telecommunications (Videotex, facsimile). Within the near future, they will be sufficiently stable to serve the purpose of viewing pictures on a home computer.

Text and pictures are only components of an interactive program that the user wants to execute under his control. What standards do we have to govern execution of an interactive program? A superficial answer is: Any programming language standardizes execution, so all we have to do is agree on a few widely used languages such as Basic, Fortran, or Pascal. This will guarantee that a program will execute *somehow*, - with a behavior as criticized in the section on *"Errors in dialog design"*.

The realistic answer is that *standardization of execution control of interactive programs has not even begun yet*. There is substantial experience with individual components of dialog control, but no unity about how to present these to the user.

4.3 The most frequent dialog control commands and their hardware realization

At the Yokosuka Electrical Communication Laboratory of NTT an experiment is underway for standardizing dialog control on a network of heterogeneous home computers. The idea is to freeze half a dozen of the most important dialog control commands, universal commands interpreted by the operating system, not by the application, into hardware. These commands include *motion and selection* (changing data and command environment), *viewing* (control over how much of an object is to be seen), and *operations on the history* (undo). The standard hardware is a *"mighty mouse"*, with five keys and a thumbwheel, tuned to the agility of a person's hand.

The motivation for developing this mouse needs explaining, since it points in the opposite direction of commercial "mouse architecture". The mouse popularized by the Xerox Alto computer has 3 keys, but its successor on the commercial Xerox Star has only 2, and the mouse of the Apple Lisa has a single key. This according to the "theory" that the user gets confused by many keys. However, the software developed for these computers tells a different story: often, different logical commands are multiplexed onto these one or two keys, and coded by such gimmicks as a "fast double click" (different from two slow clicks), or clicking while some key is depressed on the keyboard.

These examples lead to the conclusion that it is *not the number of keys that confuses the user, but the changing semantics assigned to the keys!* If a physical input device is assigned only one function, then we can reasonably use more than one such device. It's the idea of "modeless", not applied to the entire system, but restricted to a few universal dialog control commands.

Physically, the "mighty mouse" has five keys. Thumb and index finger are agile, so they control analog inputs. The other three fingers control 0-1 inputs. The thumb moves with greater freedom than any other finger, so its key can be switched into two different positions; in addition, the thumb glides sideways to turn a thumbwheel.

Logically, the following functions have been assigned to these keys. The thumb key realizes a linear motion command - gas pedal and gear shift in one. Varying pressure causes variable speed forward or backward, depending on the position of the key. A hard click means forward or backward by a unit.

The index finger controls a "show me"-key. Increasing pressure asks for increasing depth of detail, for example through zooming. A hard click selects this object, that is, turns it into the active data environment, the object of the following commands.

An object is identified by moving the cursor to it. For graphical object, built from intersecting lines, an oriented cursor is useful. It determines x, y, and an angle coordinate, which is set by turning the thumbwheel.

The remaining keys are conventional. The middle finger controls a pop-up menu, the ring finger an *"undo"*, the small finger an *"exit"* into the hierarchically next higher data or command environment.

A user must memorize these meanings once and for all - thereafter, he should be able to explore the entire system, even on unfamiliar hardware. In order to master an application program, he will undoubtedly need to learn commands specific to the application- but in order to browse in it and determine whether he really wants to use it, the universal commands of the "mighty mouse" might suffice.

5. Programming the man-machine interface

Programming methodology has emphasized different criteria for judging the quality of software during the three decades of its existence. In the early days of scarce computing resources the most important aspect of a program was its *functionality* - what it can do, and how efficiently it works. In the second phase of *structured programming* the realization that programs have a long life and need permanent adaptation to changing demands and environments led to the conclusion that functionality is not enough - a *program must be understandable*, so that its continued utility is not tied to its "inventor". Today, in the age of interactive computer usage and the growing number of casual users, we begin to realize that *functionality and good structure are not enough - good behavior towards the user is just as important.*

Thus the traditional programmer, trained to analyze systems of great logical complexity in detail, is confronted with an entirely new demand. He is called upon to design expressive pictures and lay them out on the screen, to formulate clear phrases and assemble them into an understandable presentation - skills that demand the creative-artistic flair of a graphic designer and author. It is satisfying to observe that many programmers have recently become concerned with the quality of the man-machine dialog that their interactive programs conduct, as judged from the user's point of view. Well they might, for the computer user population is changing rapidly. With the spread of low-cost single-user computers casual and occasional users abound, and for them the quality of the man-machine interface is crucial - an interactive system is only useful if the learning effort is commensurate with the brevity of the task they want to accomplish. It is for their benefit that computer professionals should start paying as much attention to their communicative skills as writers have always done.

References

[Be 82] G. Beretta, H. Burkhart, P. Fink, J. Nievergelt, J. Stelovsky, H. Sugaya, J. Weydert, A. Ventura, *XS-1: An integrated interactive system and its kernel,* 340-349, Proc. 6-th International Conference on Software Engineering, Tokyo, IEEE Computer Society Press, 1982.

[Ki 82] Y. Kitahara, **Information Network System · Telecommunications in the 21st century,** The Telecommunications Association, Tokyo, 1982.

[Mi 56] G. A. Miller, *The magical number seven, plus or minus two: Some limits on our capacity for processing information,* Psych. Review, Vol 63, No 2, 81-96, March 1956.

[NW 80] J. Nievergelt and J. Weydert, *Sites, Modes, and Trails: Telling the user of an interactive system where he is , what he can do, and how to get places,* in **Methodology of Interaction,** R. A. Guedj (ed), 327-338, North Holland 1980.

[Ni 82] J. Nievergelt, *Errors in dialog design, and how to avoid them,* in **Document Preparation Systems · A Collection of Survey Articles,** J. Nievergelt et al. (eds), North Holland Publ. Co., 1982.

[No 81] D. A. Norman, *The trouble with UNIX,* 139-150, Datamation, Nov 1981.

[St 84] J. Stelovsky, *XS-2: The user interface of an interactive system,* ETH dissertation 7425, 1984.

PROGRAMMING SUPPORT
FOR DATA-INTENSIVE APPLICATIONS *)

Joachim W. Schmidt

Fachbereich Informatik

Johann Wolfgang Goethe-Universität

Frankfurt am Main

Winfried Lamersdorf **)

Fachbereich Informatik

Universität Hamburg

Hamburg

Abstract

Relations are in the process of being accepted as a data structure adequate for a wide variety of applications. On the one hand this is due to the powerful and high level operators on relations, on the other it results from additional services such as recovery management, concurrency control and expression optimization provided by relational systems.

In its first part, the paper presents a database person's view of data definition and data processing, and outlines principles of database programming from a language person's point of view. In its second part, the paper discusses limitations of traditional record-based data models. The extensions proposed are intended for compound data object definition and manipulation.

1.0 INTRODUCTION: DATABASE PROGRAMMING

A data management problem is sometimes called a 'Database Problem' if

- the definition of data objects covers properties of 'real world entities' and their relationships - and the entities are long-lived and large in number;

- the selection of data objects is based on object properties rather than on object identifiers; and

- the operations on data objects are defined and initiated (in parallel) by independent members of some user community.

----

* This paper is an extended version of work from [Schm84], [Lame84].

** Currently with IBM Heidelberg, Scientific Center.

Current algorithmic languages support these requirements only to a limited extent: records define properties of entities but record selection is done by declared names or via references; files and file systems cope with data quantity and longevity and, to some extent, with concurrency but do not support object relationships. These shortcomings stimulated the development of what is now called Database Programming Languages. Examples are ADAPLEX [SmFL81], TAXIS [MBW80], PLAIN [Wass79], Pascal/R [Schm77], or the approach followed by PS-ALGOL [AtCC81].

The main purpose of this paper is to present some of the language constructs found in Database Programming Languages. Furthermore, we want to show how these constructs interact with others which are designed for algorithmic work on data, thereby outlining some of the principles of save and efficient Database Programming. Finally, we will discuss various limitations of the relational model and generalize it to approach complex data object construction and selection.

2.0 A PROGRAMMER'S APPROACH TO THE RELATIONAL MODEL OF DATA

From a programmer's point of view a Database Model can be interpreted basically as an approach to structuring, identifying, and organizing large quantities of variables as required for solving Database Problems.

For traditional reasons, variables in databases are structured as records. What distinguishes, for example, Codd's Relational Model [Codd70] from Hoare's approach to Record Handling [Hoar66] is, in essence, the different ways both approaches deal with record identification; Codd's method has, as we will see, some far reaching consequences.

2.1 NAMING OF VARIBLES AND PARTITIONING OF STATES

A programmer dealing, for example, with persons and houses will define types such as

```
TYPE    streetname  =   string; ...;

        person      =   RECORD
                            ... ;
                            age  :  cardinal;
                            sex  :  ... ; ...
                        END;

        house       =   RECORD
                            ... ;
                            street: streetname;
                            number: cardinal;
                            value : ...; ...
                        END;
```

and declare individual data objects (i.e., variables) such as

```
VAR      This-Person:  person;
         My-House:     house;
```

A **record handler**, expecting large quantities of data objects (i.e., record variables), organizes its state space by collecting all objects of the same type in one, say, **class**; he leaves the problems of record identification to someone else that provides unique references to records:

```
VAR      Persons:            CLASS OF person;
         Houses:             CLASS OF house;
         This-Person:        REF (Persons);
         My-House:           REF (Houses);
```

Finally, **a data modeller** groups data objects of the same type similarly in a set-like structure called **relation**. He starts, however, from the assumption that a property that is capable of identifying an entity, for example a person's name or the address of a house, is so important that it should be modelled explicitly by the type of the corresponding data object:

```
TYPE     personname =  string; ...;
         person     =  RECORD  name:     personname;
                               age:      cardinal;
                               sex:      ...; ...
                       END;

         house      =  RECORD  city:     cityname;
                               street:   streetname;
                               number:   cardinal;
                               value:    ...; ...
                       END;
VAR      Persons:    RELATION OF person;
         Houses:     RELATION OF house;
         This-Person: personname;
         My-House:    RECORD   city:    cityname;
                               street:  streetname;
                               number:  cardinal
                      END;
```

Roughly speaking, one can say that a programmer identifies objects by names, a record handler by references, and a data model through the use of distinguished attribute values, i.e., properties.

Both record handlers and data modellers can easily handle relationships between objects by defining, for example, ownership either through

```
TYPE   houseowner = RECORD ...
                            owner:      REF(person);
                            property:   REF(house);
                            purchasing-date: ...; ...
              END;
```

or through:

```
TYPE    houseowner = RECORD   ...
                            owner:      personname;
                            property:   RECORD  city:    ... ;
                                                street:  ... ;
                                                number:  ...
                                        END;
                            purchasing-date: ... ; ...
        END;
```

Ownership between several persons and houses is represented as above by a class or
a relation of houseowners respectively.

Record handlers and data modellers represent separate schools of database
people: these are, on the one hand, the adherents of the referential data models,
for example, network and hierarchy models, and, on the other, those preferring
associative data models, i.e., the relational one and the derived semantic data
models. This paper concentrates on the relational approach.

2.2 CONSISTENCY OF IDENTIFIERS

Using, for example, strings and cardinal numbers to represent identifiers, as
the relational approach does, is, of course, an open invitation to data
inconsistency, unless specific precautions are taken.

Values of type personname, for example, represent identifiers of persons only
if they are unique within their scope. In other words, the relation:
    VAR Persons: RELATION OF person;

has to fulfil, at all times, the predicate:
    ALL p,p' IN Persons ((p.name=p'.name) -> (p=p')).

Conditions of this kind are often called key constraints and are concerned with
object integrity; we denote them shortly by listing the key attributes within the
relation's type definition:
    VAR Persons: RELATION name OF person;
        Houses: RELATION city,street,number OF house;

While object integrity guarantees that identifiers of relation elements are defined
uniquely, **referential integrity** ensures that identifiers are used properly. Proper
use means that identifiers have to be declared before and as long as they are used;
some string in the owner field of a houseowner record is an identifier of some
person owning that house if and only if there is a record in the relation Persons
identified by that name - otherwise it is just a string of characters. A similar
statement holds for houses identified by their addresses.   In other words, a
relation Ownership with elements of type houseowner has to meet, at all times, the

condition:

```
ALL o IN Ownership SOME p IN Persons ((o.owner=p.name) AND
        SOME h IN Houses (o.property = <h.city,h.street,h.number>)).
```

Codd calls only those database systems **fully relational** that support the two classes of constraints required for object and referential integrity [Codd83].

2.3 VARIABILITY OF STATES

Database Programmers can change the state represented by the elements (i.e., the record variables) of a relation in several ways: either a new value can be associated with some identifier, or a new relation element or tuple (i.e., <identifier, value>-pair) can be introduced, and finally an existing element can be removed.

The Relational Model admits two substantially different perceptions of what a relation is and how a relation is composed of its elements. Both perceptions have their merits.

The set-like perception considers a relation as a set with members of type record constrained by some key condition. It provides a class of assignment operators for element insertion (:+), deletion (:-), and replacement (:&).

A new person, for example, is introduced by:
```
  Persons :+ {<..., 'Klug', 33, ...>};
```

and is replaced by:
```
  Persons :& {<..., 'Klug', 34, ...>};
```

using the person's name, in this case, 'Klug', as the identifier for the record to be updated.

The alternative is an array-like or table perception. In this view an individual relation element is denoted by an array-like selector based on the relation's key, and the previous replacement is equivalent to an assignment:
```
  Persons ['Klug']:= <..., 'Klug', 34, ...>; or shorter
  Persons ['Klug'].age:= 34.
```

When perceived like arrays, relations usually are sparse, that is, there are no values associated with most keys. (A "no value" can be represented, for example, by the empty record, < >). Such 'unassigned' variables do not contribute to the state as represented by the elements of a relation.

An insert in set-like perceptions corresponds to an assignment in the array-like picture, provided the selected element has not been assigned beforehand, and a delete corresponds to the assignment of the empty record to an assigned relation element.

The definition of selectors for relations can be generalized, as in the case of arrays, to selectors denoting more than one relation element, i.e., a subrelation. Since data in a database are provided and processed by different members of a user community, powerful mechanisms for partitioning a relation variable are of particular importance. A discussion of this issue will form part of the subsequent section that extends the basic approach taken in this section through some high level constructs for relations.

3.0 SOME HIGH LEVEL LANGUAGE CONSTRUCTS FOR RELATIONS

The evaluation of expressions with relations as operands requires mechanisms to refer to and operate on all of its elements. The immense cardinalities that may appear with relations does not allow one, however, to step through relations by simply incrementing the key value, as we do with index value of arrays. In addition most of the selected relation elements would be unassigned anyway. With respect to element selection, the set picture itself does not help us at all since there is no selector mechanism defined for sets.

Hence, we start with some primitive access procedures for relations and use them to sketch solutions for some standard problems in database programming.

The uniqueness of key values provides a basis for accessing individual relation elements. Working under the assumption that the set of key values (not the relation!) is ordered we can define some standard procedures that access relations element by element. Procedure low (R,r), for example, accesses that element of the relation, R, which has the lowest key value and returns its value through the record variable, r. Procedure next (R,r) assigns to r the element with the key value next highest to the one provided by r. The Boolean function, eor (R), becomes true if an access fails because of lack of elements.

Each class of database problems addressed in this section will motivate some higher level language construct for relations that abstract from unnecessary implementational details and permit database programs to be more concise and accessible to optimization [Schm77], [ScMa83].

3.1 QUERY EXPRESSIONS

Probably the most frequent and expensive operation on a database is querying, i.e., evaluating logical expressions that have relations as operands.

The case in which a Boolean expression, p(r), is evaluated for some relation element, R[kv], occurs frequently within loops that run over all the elements of a relation:

```
some-s := FALSE;
low (R,r);
WHILE NOT eor (R) DO
  some-s := some-s OR p(r);
  next (R,r)
END;
```

Such statements arise when testing whether an individual data object in a relation makes a Boolean expression true - without knowing the object's identifier. A situation like this sounds strange to an ordinary programmer who always has his variables well under controll, however, it occurs often in an environment with many thousands of data objects shared with other database users. The above coding of our element test is unsatisfactory for several reasons: first, there is no indication that the order in which the program steps through the relation is optimal; second, there should be a loop exit as soon as an element is found that fulfils p; and, finally, the statement sequence is too long.

From predicate logic we know that the above program computes the same result for s as given by the first-order predicate:

SOME r IN R (p(r)).

The existential quantifier introduces a variable, r, that denotes arbitrary elements in relation R; the predicate becomes true if and only if at least one element of R fulfils p.

In the given context, a predicate denotes more directly what we mean, and, ideally, its evaluation requires access to only one relation element that fulfils p and it is left up to some clever execution model to approach that optimum.

Introduction of universal quantification, ALL r IN R (p(r)), can be justified in an analoguous way.

A variation of the above case arises when we are interested in the values of all those relation elements that fulfil some selection predicate.

Relational queries generalize expressions consisting only of selected relation elements, {R[kv1], ..., R[kvn]}. Instead of selecting elements that match a specific key condition, relational queries ask for all elements of a relation that fulfil an arbitrary selection predicate, p:

```
result := { };
low (R,r);
WHILE NOT eor (R) DO
  IF p(r) THEN result :+ {r} END;
  next (R,r)
END;
```

Following similar arguments as with Boolean queries, we replace the above statement sequence by the relation-valued expression:

{EACH r IN R: p(r)}.

The quantifier EACH introduces the variable r that denotes arbitrary elements in relation R; the selection phrase EACH r IN R: p(r) selects all the elements that fulfil p; the relation constructor, {...}, finally turns the selected elements into a relation.

In their most general form relational queries introduce several variables, r, s,..., which run over various relations, R, S,...; they allow first-order selection predicates over r, s,..., and admit record constructors to structure the resulting relation elements:

{<..,r.f,..,s.g,..> OF EACH r IN R, EACH s IN S,...: p(r,s,..)}.

Note that relation expressions can be nested and that they can be combined freely with first-order predicates.

With respect to their selective power the above class of relational expressions is equivalent to what Codd called a **relationally complete** query language [Codd71].

3.2 REPETITION STATEMENTS

The two classes of problems, element test and element selection, occur frequently in database programming; they are, however, special cases. In general, an arbitrary statement, S, is executed for all relation elements selected by some predicate:

```
low (R,r);
WHILE NOT eor (R) DO
  IF p(r) THEN S END;
  next (R,r)
END;
```

If the order in which relation elements are processed is irrelevant, element access by increasing key value is an unnecessary and often costly decision. Often any order will suffice as long as statement S is executed once and only once for each element of R that fulfils predicate p, and the decision on the order can be left up to the implementation. This freedom is provided by applying the principles of control abstraction, and allowing a selection phrase, i.e., EACH r IN R: p(r), for loop control:

```
FOR EACH r IN R: p(r) DO S END;
```

Nested loops with a common selection criterium, that is,

```
FOR EACH r1 IN R1: TRUE DO
    FOR EACH rn IN Rn: TRUE DO
       IF p(r1,...rn) THEN S END
    END
END;
```

can be replaced by one loop controlled by one compound selection:

```
FOR EACH r1 IN R1, ... EACH rn IN Rn: p(r1,...rn) DO S END;
```


3.3 SELECTED VARIABLES

In practice several users contribute to the data integrated into a single relation of some database. Consequently, individual users often do not require access to a complete relation variable but only to selected parts of it. In this section we will extend the notion of selected relation elements introduced above and outline the concept of generalized selected relation variables.

Let us assume that R is a relation with a key composed of two attributes, k1 and k2. If we switch from the array-like perception of relations to the set-like view, the assignment:

```
R[kv1,kv2] := <...,kv1, ..., kv2,...>;
```

converts into the assignment:

```
R := {EACH r IN R: NOT (<r.k1, r.k2> = <kv1, kv2>), <...,kv1,...,kv2,...>}.
```

Assignment of a record variable, rec, to a selected relation element:

```
R[kv1, kv2] := rec;
```

is equivalent to the above assignment to the entire relation, R, controlled by a test:

```
IF <kv1,kv2> = <rec.k1,rec.k2> THEN
   R := {EACH r IN R: NOT (<r.k1,r.k2> = <kv1,kv2>), rec}
ELSE <EXCEPTION:...>
END;
```

As a first generalization, the notion of an element selector can be extended, as for arrays, to a selector for subrelations. The following statement assigns a relational expression, rex, to a subrelation variable selected from selection variable R by a truncated key list:

    R[kv1] := rex;

Its semantics can be defined by an equivalent assignment to the entire relation:

```
IF ALL x IN rex (x.k1=kv1) THEN
   R := {EACH r IN R: NOT (r.k1=kv1), EACH x IN rex: TRUE}
ELSE <EXEPTION:...>
END;
```

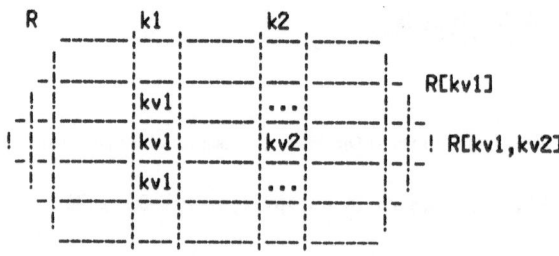

Key-Based Relation Selection

The above selectors are based on specific selection predicates that depend on keys and key values only. We now want to generalize selectors and permit, as with query expressions, arbitrary predicates for relation selection. The linguistic support for defining generalized selectors is provided by a language construct called a **selector generator**. We can use this generator to redefine, for example, the above selector for relation elements:

```
SELECTOR sk FOR rel: RelType (kf1:k1type; kf2:k2type): RecType;
BEGIN EACH r IN rel: (r.k1=kf1) AND (r.k2=kf2) END sk;
```

A selector definition introduces a selector name, may have parameters, and binds a selector to a formal relation of a given type. It includes the definition of a selection phrase, which introduces a selection predicate. Finally it gives the type of the selected data object.

The use of a selector, for example, R[sk(kv1,kv2), requires that the selected relation, R, is of type RelType, as required by the selector definition, sk. The denotation R[sk(kv1,kv2)] is equivalent to the shorthand R [kv1,kv2] used above.

In its most general form a selector is based on an arbitrary selection predicate, p:

```
SELECTOR sp FOR rel: RelType (...): RType;
BEGIN EACH r IN rel: p(r) END sp;
```

An **assignment**

```
R [sp(...)] := rex;
```

is equivalent to:

```
IF ALL x IN rex (p(x,...)) THEN
    R := {EACH r IN R: p(r,...), EACH x IN rex: TRUE}
ELSE <EXCEPTION>
END;
```

and the value of a selected variable, R[sp(...)], is equal to that of the relational **expression**:

```
{EACH r IN R: p(r,...)}.
```

Note that the selection phrase used in a selector definition, SELECTOR sp ...; BEGIN EACH r IN R: p(r,...) END sp, is the same as that already used in relation expressions, {EACH r IN R: p(r,...)}, and for loop control, FOR EACH r IN R: p(r,...) DO... .

In database literature selectors and selected relation variables are often called views [Ston75]. In subsequent sections we will demonstrate how selected relation variables can be used to support database integrity and concurrency.

4.0 SUPPORT FOR DATABASE INTEGRITY, RECOVERY AND CONCURRENCY

In some respects database users take their database for the "reality" it is supposed to describe and, therefore, have to be very concerned with data integrity. There are numerous reasons why the integrity of a database can get lost, however, there are some ways to keep database integrity under control.

First of all, programs may be incorrect and provide unintentional database input or output. Database programmers, being responsible for the data of a whole user community, are well advised to make use of whatever the state of the "Art and Craft, Science and Logic of Programmming" ([Hoar69], [Grie81], [Reyn81], [Hehn84]) offers in order to improve program correctness.

Some cases of unintentional input are detected because they violate one of the integrity constraints that are defined explicitly on the database, for example, a key constraint. In such a case subsequent statements and some previously executed ones usually become obsolete. Situations like these and similar ones caused by hard- or software malfunction call for recovery mechanisms that re-establish previous database states. Furthermore, concurrent execution of database programs requires careful control of program dependencies to avoid additional problems with data integrity.

4.1 STATEMENTS FOR DATABASE INTEGRITY

The typical sequence of tasks a database program has to perform can be characterized as follows:
- local pre-processing of what is going to become the database input, and subsequent transformation of the database into some new consistent state; and
- local evaluation of some consistent database state, and subsequent post-processing of what happens to be the database output.

Thus, the interactions between a program and database imply a structure of database programs where the operations on the database are concentrated in statement sequences, S, that begin and end in a consistent database state. We can express this by annotating a database program, i.e., by placing assertions (integrity constraints, ICs) around statements: {IC} S {IC}.

A database programmer usually knows — or should know — precisely what specific condition, Q, the output state of S has to fulfil. Furthermore, following the approach of Hoare [Hoar69] and others, the postcondition, Q, should be taken as a starting point for developing the statement sequence, S, so that it guarantees the postcondition, provided execution begins in some precondition, P: {P} S {Q}.

Of course, any postcondition has to imply the integrity constraints defined on the database, $Q \longrightarrow IC$, and any precondition (in particular its part that refers to the database) has to be shown to be a consequence of the postcondition established by the preceding statement sequence.

The following sketch of an annotated program includes explicitly the initial values, Rv, ..., of the relation variables, R, ..., used by statement sequence S:

$$\{ . . . \} \ S^{j-1} \ ; \ \{Q^{j-1}\} \ ...$$

$$\{ \ P^j \ \& \ ((R^j = Rv^j) \ \& \ ... \ )\} \ S^j \quad ; \ \{Q^j \ \}$$

with $Q^{j-1} \longrightarrow IC$; $Q^j \longrightarrow IC$; and $Q^{j-1} \longrightarrow P^j \ \& \ (...)$.

The omissions (...) between S^{j-1} and S^j refer to statements that work exclusively with local program variables and not with the database.

We will use pre- and postconditions to discuss some of the consequences that database recovery and concurrency impose on database programming.

4.2 ACTIONS FOR DATABASE RECOVERY

Provided S fulfils {P} S {Q}, we know that the following is true:

"if execution of S is begun in a state satisfying P, then it is guaranteed to terminate in a finite amount of time in a state satisfying Q" [Grie81].

This interpretation says nothing about the state of the variables referenced by the statements of sequence S if, for one reason or another, the actions specified by S cannot be executed.

Database Programming Languages should enable a programmer to control actions on a database even under circumstances of failure. This may be achieved by converting any statement sequence, S, into an action sequence, A, ([KMPR83], [RRUZ83]) with consequences that can be expressed best in terms of pre- and postconditions:

if   {P & ((R=Rv) & ...)} S {Q}   holds for statement sequence S,
then {P & ((R=Rv) & ...)} A {Q} V {((R=Rv) & ...)and NotA}   holds for the
corresponding action sequence, A. The predicate NotA indicates that action
A has not been executed.

In other words, the postcondition of action sequences, A, is that of the underlying statement sequence, S, or it is, due to recovery actions in cases of failure, given by the re-established initial values of the database variables referenced by S.

Thus, database recovery leads to disjunctive postconditions of action sequences and, therefore, has an effect on database programming. The precondition of action sequence, Aj, now has to follow from a weaker postcondition:

$$Q^{j-1} \ \lor \ ((R^{j-1}= Rv^{j-1}) \ \& \ \ldots) \ \longrightarrow P^j \ \& \ (\ \ldots \ ) \ .$$

Converting statement sequences into action sequences not only maintains database integrity,

$$Q^{j-1} \ \lor \ ((R^{j-1} = Rv^{j-1}) \ \& \ \ldots) \ \longrightarrow IC,$$

but also extends database validity to cases that require database recovery.

4.3 TRANSACTION FOR DATABASE CONCURENCY

Up to now we assumed that all action sequences are part of the same database program, U. Going one step further and converting action sequences, A, into transactions, T, allows two transactions, T_U^j and its successor $T_{U'}^i$, to originate from independent database programs, U, U'. The consequences for database programming are obvious: the precondition of $T_{U'}^j$ has to follow from the postcondition of some T_U^i

$$Q_U^i \ \lor \ ((R^i = Rv^i) \ \& \ \ldots \ ) \ \longrightarrow P_{U'}^j \ \& \ (\ \ldots \ ) \ .$$

Since concurrency models currently in use in databases exclude explicit communication between programs, U and U', there is very little the precondition of transaction $T_{U'}^j$ can learn from the postcondition of its predecessor, T_U^i . All one knows for sure is that T_U^i re-establishes the integrity constraints:

$$Q_U^i \ \lor \ ((R^i = Rv^i) \ \& \ \ldots \ ) \ \longrightarrow IC$$

and, therefore, $T_{U'}^j$ can rely upon them:

$$IC \ \longrightarrow \ P_{U'}^j \ \& \ (\ \ldots \ ) \ .$$

In the context of a Database Programming Language the syntax of transactions [MRS84], [Reim84] may look like

```
TRANSACTION T(...);
IMPORT Rel[sp(...)], ... ;
BEGIN
      S
END;
```

The notion of a selected relation variable R[sp(...)] as introduced in section 3.3 has been used to indicate that a transaction may demand access only to selected elements of a relation

The weakening of conditions after actions and before transactions forces database programs to re-strengthen them through the explicit use of conditional statements. This is one of the reasons why Database Programming Languages need conditionals supported by powerful Boolean expressions, for example, first-order predicates.

5.0 REQUIREMENTS FOR COMPOUND OBJECT MODELLING

While the above constructs aim for improving the usability of current database models, other research efforts try to overcome the limitations of traditional record-based data models [Kent79]. These limitations become especially obvious when modelling data objects with highly varying structures which are semantically inter-related [Codd79], as required, for instance, in CAD/CAM [Halo82], information retrieval [ScPi82], or office applications [GiTs83].

Modelling advanced data-intensive application requires primitives for representing compound data objects adequately with respect to all their structural as well as operational aspects. An important criterion for the adequacy of a data model is the coherence of its modelling primitives with the underlying semantic structures of an intended class of applications. In office systems, for example, data objects to be specified are characterized by a great variety of structured components related by complex inner relationships.

In comparison to the limited set of tools for conventional database modelling, data modelling support provided by modern programming languages can be characterized by a great variety of modelling concepts (i.e., data type generators) which can be combined freely [Zill84]. The 'recursive' data model subsequently presented integrates data definition and abstraction mechanism from modern programming languages with high-level data structuring, selection, and management concepts from current database technology.

5.1 SEMANTIC MODELLING PRIMITIVES

In our approach to compound object modelling we provide support for

A) the **construction** of compound data objects from elementary ones or from those already constructed;

B) the **selection** of specific object components from compound data objects; and

C) the **recognition** of the construction rules -or types- underlying some compound data object instance.

For example, in an office environment we allow for

A) the construction of a 'letter' instance from components as, for example, names, addresses, dates, and textual fragments;

B) the selection of single letter components as, for example, the sender's address or the letter's textual content; and

C) the recognition of a given document as a letter (as opposed to a form, a memo, etc.).

5.2 THE RELATION MODEL REVISITED

The relational data model supports the above modelling primitives in a more limited way. Its basic constructor mechanism allows for the representation of tuple or record structures, i.e., for data objects composed of a fixed number of atomic data elements. By a second constructor mechanism a varying number of tuples can be related by collecting them in a set or relation structure. Those elementary structuring mechanisms meet the requirements of simple or simplified applications only. As soon as application modelling requires notions such as order, repetition, or nesting the resulting data objects cannot be declared directly by the constructors of the relational model. Instead, complex structures have to be implemented utilizing the representational means of the relational model, i.e., the possibility of

- dynamic allocation and de-allocation of records by insert and delete operations on relations;

- associative identification of individual records by key values, and

- dynamic alteration and association of key-identified records by assignment.

The need for using the relational model on the "lower level of implementation" and not on the "higher level of declaration" increases, however, the complexity of a database program considerably.

For the selection of individual data objects from a data base the relational model provides powerful query languages equivalent to first-order predicative set selection. If the data objects to be selected are spread over more than one relation, selection requires subsetting of the Cartesian product of the relations involved, and severe consistency and performance issues may be raised. Finally, if the number of relations involved in a query problem depends on the value of a relation (e.g., on its emptiness or on the existence of specific relation elements), traditional query languages are not sufficient to express such a query. Those problems, for example the derivation of a family tree from a parents relation, are recursive in nature and require extensions of traditional database query languages. Database Programming Languages, such as Pascal/R [ScMa80], that augment a relationally complete data language by an algorithmically complete set of language constructs, provide, of course, a framework powerfull enough to program appropriate solutions – however, again on the expense of an increased program complexity.

Recognition of the construction rules (or types) of data objects is necessary for any save data processing system. While the type of an individual relation element or tuple can be determined easily, the construction rules of a set of tuples, i.e., a relation (including its key and other constraints) are difficult to determine or even indecidable [KSW81], [Klug 80].

These limitation of the relational data model lead directly to what we call a "recursive data model", providing extended concepts for the construction selection, and recognition of complex data objects.

5.3 RECURSIVE DATA MODELS

Recursive data models [Lame84] are based on simple as well as recursively defined data types [Hoar75] and extend the modelling tools of classical data models by allowing representations of variable length structured data objects, possibly nested in a varying depth.

A traditional type definition is based on a single data structure or type generator from a set of data structures predefined by the data model, for example, RECORD or RELATION. Each such data structure provides a fixed value set for each type. A recursive data type, RecursiveType, however, is defined by a set of user-defined "structure generators", Gen_i, used to generate instances of that recursive type. Structure generators are based on limited sets of component types, CompType_ij, which may, recursively, contain other recursive data types (including the data type to be defined), or consist of simple data types as known from conventional, high-level programming languages. Single components can be identified by elementary 'component selectors', sel_ij.

```
RecursiveType =   (Gen_1 (sel_11:CompType_11;...; sel_1n: CompType_1n)
                  ¦ . . . ¦Gen_i (...)¦ . . .
                  ¦Gen_k (sel_k1: CompType_k1;...;sel_km: CompType_km) );
```

defines a data type whose value set consists of all hierarchically structured data values which can be generated by (in general nested) application of the generators 'Gen_i' to components of type 'CompType_ij'.

A recursively defined data type provides concepts for all three basic semantic modelling primitives listed above by introducing:

A) a set of structure generators, Gen_i, to construct recursive object from components (which may, in turn, be constructed by some structure generator, or denoted by conventional programming language expressions);

B) for each structure generator a set of elementary component selectors, sel_ij, to select a component from a recursive object generated by the generator 'Gen_i'. Syntactically, component selectors are enclosed in square brackets and follow the object to be selected. For a recursively defined object, for example rcsvar, generated by 'Gen_i (...,compval_ij,...)', we get

$$\text{rcsvar [sel\_ij] = compval\_ij.}$$

Component selectors identify component variables (i.e., they may be used in an expression context as well as on the left-hand side of an assignment statement and also nested). Nested component selectors are written one after another and evaluated from left to right;

C) a set of Boolean characteristic functions, is-Gen_i, to recognize a recursive value as being constructed by some structure generator 'Gen_i'. For a recursive variable, for example rcsvar, whose value was generated by the structure generator, 'Gen_i(...)',

$$\text{is-Gen\_i (rcsvar) = TRUE.}$$

In terms of a recursive data model 'DocumentType' can be defined as follows:

```
DocumentType = (Letter (sender, receiver: NameType;
                        from, to: AddressType;
                        date_mailed, date_received: DateType;
                        content: TextType)
                : Form (...) : Memo (...) : ... );
```

Recursively defined data types comprise complex data objects which are, in different ways, generated by (in general nested) applications of the structure generators defining the respective data types. Our recursive approach generalizes traditional (record-based) data models, restricted to data objects with an identical structure [Kent79], to more advanced data models that allow for the definition and manipulation of data objects sharing an identical structuring concept. In other words, recusive data models are based on a limited number of modelling primitives and gain their power by allowing for their orthogonal combination.

6.0   EXTENDED CONSTRUCTS FOR COMPOUND OBJECT DEFINITION

For a more convenient notation of frequent classes of compound objects we augment our recursive data model by a more abstract layer of structural and operational modelling concepts.

6.1   ABSTRACT OBJECT REPRESENTATIONS

In general, recursive data structures are powerful enough to express all of the following data object representations. The following data structuring mechanisms serve only as shorthand notations for those structures occurring frequently in compound object modelling.

The management of varying numbers of application objects is supported by a 'set' type for **object collection**. Sets have a varying cardinality, and element duplication has no meaning. Instead of introducing a set type explicitly by the recursive definition

```
SetType = (Empty ( ) ¦ Insert_element (new: ElementType,
                                        old: SetType)
           ¦ Delete_element (...) );
```

we provide as a syntactically shorter, semantically equivalent alternative a predefined type generator for sets that includes the usual set operators in infix notation (:+, :-):

```
SetType    = SET OF ElementType.
```

Set instances can be generated by applying a standard set constructor 'SetType {Elem_1,...Elem_n}. (The syntax for complex value constructors follows that of typed generators as, for example, in Modula-2 [Wirth82].) Set restriction is based on first-order predicate expressions following the line of first-order query languages:

```
SetType {<element_id> IN <set> : <predicate>}.
```
For example, a file cabinet may be perceived as a set of drawers:
```
FileCabinetType = SET OF DrawerType.
```

Consequently, cabinet1 :+ cabinet2; integrates the drawers of cabinet2 into the set of drawers represented by variable, cabinet1, and FileCabinetType {draw IN cabinet1 : <predicate>} selects a certain subset of drawers from 'cabinet1'.

Object identification is another concept essential for data-intensive applications. It can be represented by 'mapping' object identifiers into the set of objects to be identified. Maps are sets of pairs ∈ (Domain x Range) where 'Domain' and 'Range' are sets, too, and no two element pairs have the same domain value; syntactically:

```
MapType   = (DomainType-->RangeType).
```

Map instances are created using a map generator 'MapType {d_1-->r_1, ...,d_n-->r_n}'. The operatons on maps are 'DOM' (domain value set), 'RNG' (range value set), ':+'(map extention, i.e., inserting new pairs), ':&'(map update, i.e., overwriting existing pairs), ':-' (map reduction,i.e., deleting existing pairs) and 'map [argument]' (map application, i.e., identifying an object by some argument).

For example, the above 'DrawerType' can be modelled as a map from file identifiers to the corresponding files in the drawer (file identifiers are regarded as unstructured 'tokens', only subject to equality tests):

```
DrawerType   = (FileIdType-->FileType);
FileIdType   = TOKEN.
```

Then, for example, 'drawer :+ DrawerType {file_k--> file} extends the content of a given drawer by a new file which is identified by 'file_k'.

Note, that relations can be interpreted as maps of a specific type:
```
TYPE RelType  = ( KeyType --> ElemType);
      ElemType = RECORD ... key : KeyType; ...END;
VAR   Rel : RelType;
```

that always meet the condition

```
Rel[keyval]. key = keyval.
```

Another important concept supported by recursive types is that of **object ordering**. It leads to a 'list' type data structuring mechanism:

ListType = LIST OF ElementType.

Single list instances are generated by a typed standard generator
for lists, 'ListType {...}'.

Together with the 'list' type come the usual list operation primitives as, for instance, the element selectors 'LAST' (yields the most recently added element), 'REST' (yields the list without the last element), 'list[i]' (selects the i-th element), and the operators 'ELEMS' (returns the set of all list elements), 'LENGTH' (returns the number of elements), 'VOID' (tests for zero elements), ':+' (appends r.h.s. elements to l.h.s. list variable) etc.

For example, a data type by which we can file documents can now be modelled as

FileType = LIST OF DocumentType.

Given a 'file' variable, 'file[j]' selects the j-th document from that file (if existent), and 'file [LAST]' selects the most recently added one; FileType {doc_1,...,doc_n} generates a file instance from given document instances, doc_1,...,doc_n.

6.2 OFFICE OBJECTS EXAMPLE

In the following, we complete the example representation of a file cabinet containing files and documents that define part of an office environment:

```
TYPE   FileCabinetType = SET OF DrawerType;
       DrawerType      = (FileIdType --> FileType);
       FileType        = LIST OF DocumentType;
```

```
DocumentType     = (Letter (sender, receiver: NameType;
                            from, to: AdressType;
                            date_mailed, date_received: DateType;
                            content: TextType)
                  ¦ Form  (type_no, serial_no: INTEGER;
                            entries: (EntryIdType --> EntryType)
                  ¦ Memo  ( ...)          ¦     . . .     );
```

Components of letters, forms, memos, etc. may be defined recursively, too:

```
NameType         = Name (first, middle, last: WordType);
AddressType      = Address (organization: LIST OF WordType;
                            street, city, state, country:
                            WordType;
                            no, zip: INTEGER);
DateType         = Date (year: (1950...1990); month: (1...12);
                            day:  (1...31) );
EntryType        = Entry (descr: TextType; content: ContentType;
                            domain: SET OF ContentType);
ValueType        = ...;
FileIdType       = TOKEN;
FieldIdType      = TOKEN;
EntryIdType      = TOKEN;
```

A more complex recursive data type is 'TextType' comprising component types for paragraphs, sentences, words, etc.:

```
TextType         = LIST OF ParagraphType;
ParagraphType    = ( Titled_para  (title: SentenceType;
                                    cont: ParagraphType)
                  ¦ Untitled_para (para: LIST OF SentenceType));
SentenceType     = Sentence (elems: LIST OF ElementType;
                            end_mark: ('.', '!', '?', ...));
ElementType      = ( Word (WordType)
                  ¦ Mark (',', ';', ...));
WordType         = LIST OF CHAR;
```

Finally, we are able to declare a single compound variable representing a file cabinet with all its structured components by:

```
VAR file_cabinet : FileCabinetType;
```

7.0 EXTENDED CONSTRUCTS FOR COMPOUND OBJECT OPERATION

In the proposed data model the operational primitives of the extended object types are the basis for modelling the semantics of office procedure. They provided mechanisms for compound data object construction, decomposition, and recognition. More powerfull operations on recursive objects can be defined (recursively) in terms of the elementary operators.

7.1 DATA OBJECT SELECTION

For example, elementary selection of object components can be modelled in terms of the operational primitives as demonstrated by the following operations on a 'filecabinet' variable as declared above:

"select the drawers from a file cabinet which contain a 'file_k'":
      DrawerType { draw IN file_cabinet:
                        file_k IN DOM draw }
"select the 4th document from 'file_k' in a given drawer":
      drawer [file_k] [4],
"is that document a form?":
      is-Form (drawer [file_k] [4]),
"select the content of the entry marked 'confidential' of that form":
      drawer [file_k] [4] [entries] [confidential] [content]

A more complex example selects "the latest letter(s) from a sender 's' contained in a file 'f'":

      SET OF DocumentType { doc IN ELEMS f :
                        (is-Letter (doc) AND (doc[sender] = s)
                        AND ALL doc' IN ELEMS f (is-Letter (doc')
                        AND (doc' [sender'] = s ==>
                        (doc [date] >= doc' [date]))) }

Since component selection is frequent in structured data object applications and, as shown by the last example, may be complex, it is supported additionally by 'dedicated component selectors'. Similar selector or view mechanisms have been proposed for the relational database model by [Ston75], [Rous82], and [MRS84], and are defined for the recursive data model in [Lame84].

The previous component selector example can be expressed recursively using a generic, parameterized selector mechanism. The definition of the selector 'last_letter_from' is based on the fact that the latest letter was filed last (with unary list operators binding stronger than selectors):

```
SELECTOR last_letter_from FOR file: FileType (s: NameType) :
                                             DocumentType;
    IF VOID file
        THEN <EXCEPTION: message ("there is no letter from" s "in" file)>
        ELSE IF is-Letter (f [LAST]) AND f [LAST] [sender] = s
                THEN SELECT f [LAST]
                ELSE SELECT f [REST] [last_letter_from (s)]
    END last_letter_from.
```

An application of this selector to file 'f', 'f [last_letter_from (s)]', returns either the required document or raises the defined exception.

7.2 DATA OBJECT MANIPULATION

Data objects are represented by variables as, for example, demonstrated in section 6.2. . The simplest case of data object manipulation is given by the selection of an object or an object component, and by a subsequent replacement by a new object. For example, setting the receiving date of the last letter from a sender in a given file to '84-7-14' is expressed by:

```
file [last_letter_from (sender)] [date_received] := Date (84,7,14).
```

Insertion, deletion, or update of object (or their components) is expressed via set-oriented update operators and operands constructed by the extended operational primitives. For example:

```
"insert a file identified by 'file_i' into a given drawer":
    drawer :+ FileType {file_i --> file}
"delete some values from the domain of a given data field" :
    field [domain] :- SET OF ValueType {value_1, ..., value_n}
```

"replace the last document in a file by a new one" :

    file [Rest] :& new_document.

    More powerful query and data manipulation language constructs may include multi-component selection mechanisms as well as extended manipulation functions; furthermore, control structures may be defined, in terms of sets, lists, and maps of object components.

8.0 CONCLUDING REMARKS

    The programming support for data-intensive applications discussed in this paper is under development. Integrated database programming languages, for example Pascal/R [ScMa80], TAXIS [MBW80], or GALILEO [ACO84], exist in prototype versions. Languages that are less procedural than the above examples and show more of a specification-like or AI-like flavor are under design [BGMV84].

    The choice of the recursive data structuring machanisms proposed in this paper is influenced by experience gained by applying 'META IV', the meta-language of the semantic specification method 'Vienna Development Method' (VDM) [BjJo78] in a database context (see, e.g., [BjLø82], [LaSc80]). A prototype implementation of the recursive data model has been implemented [Saun84], [LaSc84] using a compiler writing system and the database programming language Pascal/R [Schm77]. The prototype maps our recursive approach to complex object construction, selection, recognition, and manipulation down to the data objects of a conventional relational database system [ScMa80]. Other approaches to complex object modelling are reported by [HaLo82], [ScPi82].

9.0 REFERENCES

[ACO84]   Albano,A., Cardello,L., Orsini,R.: GALILEO: A Strongly Typed
           Interactive Language. To appear in ACM TODS.

[ACC81]   Atkinson,M.P., Chrisholm,K, Cockshott,P.: The New Edinburgh Persistent
           Algorithmic Language. University of Edinburgh, Department of Computer
           Science, CSR-90-81, August 1981.

[BGMV84]   Borgida,A., Greenspan,S., Mylopoulos,J., Vassiliou,J.: The Conceptual
           Modelling Language CML, internal document.

[BjJo78]   Bjørner,D., Jones,C.B. (Eds.): The Vienna Development Method: The
           Meta Language. Lecture Notes in Computer Science, No.61,
           Springer-Verlag, Berlin Heidelberg New York, 1978.

[BjLø82]   Bjørner,D., Løvengreen,H.: Formalization of Database Systems – and a
           Formal Definition of IMS. Proc. 8th Int. Conf. on VLDB, Mexico
           City, September 1982.

[BMS84]   Brodie,M.L., Mylopoulos,J., Schmidt,J.W. (Eds.): On Conceptual
           Modelling: Perspectives from Artificial Intelligence, Databases, and
           Programming Languages, Springer-Verlag, Berlin Heidelberg New York,
           1984.

[BMS84]   Brodie,M.L., Mylopoulos,J., Schmidt,J.W. (Eds.): Proc. Symp. on
           Conceptual Modelling: Perspectives from Artificial Intelligence,
           Databases and Programming Languages. Intervale, New Hampshire, June
           1982, Springer-Verlag, Berlin Heidelberg New York, 1984.

[Codd70]   Codd,E.F.: A Relational Model of Data for Large Shared Data Banks.
           CACM Vol.13, No.6, June 1970.

[Codd71]   Codd,E.F.: Relational Completeness of Data Base Sublanguages. Courant
           Computer Science Symposia, No.6, Prentice-Hall, May 1971.

[Codd79] Codd,E.F.: Extending the Relational Database Model to Capture More Meaning. ACM TODS, Vol.4, No.4, December 1979.

[Codd83] Codd,E.F.: "Foreword" of [ScBr83].

[GiTs83] Gibbs,S., Tsichritzis,D.: A Data Modelling Approach for Office Information Systems. ACM Transactions on Office Information Systems, Vol.1, No.4, October 1983.

[Grie81] Gries,D.: The Science of Programming. Springer-Verlag, Berlin Heidelberg New York, 1981.

[HaLo82] Haskins,R.L., Lorie,R.A.: On Extending the Functions of a Rela tional Database System. Proc. ACM SIGMOD Int. Conf. on Management of Data, Orlando, Florida, June 1982.

[Hehn84] Hehner,E.C.R.: The Logic of Programming. Prentice-Hall, 1984.

[Hoar66] Hoare,C.A.R.: Record Handling. In F. Genuys (Ed.): Programming Languages. Academic Press, 1968.

[Hoar69] Hoare,C.A.R.: An Axiomatic Approach to Computer Programming. CACM Vol.12, No.10, October 1969.

[Hoar75] Hoare,C.A.R.: Recursive Data Strustures. International Journal of Computer and Information Science, Vol.4, No.2, 1975.

[Kent79] Kent,W.: Limitations of Record Based Information Models. ACM TODS, Vol.4, No.1, March 1979.

[Klug80] Klug,A.: Calculating Constraints on Relational Expressions. ACM TODS, Vol.5., No.3, September 1980.

[KMPR83] Koch,J., Mall,M., Putfarken,P., Reimer,M., Schmidt,J.W., Zehnder,C.A.: Modula/R Report (Lilith Version), ETH Zürich, Institut für Informatik, Februar 1983.

[KSW81] Koch,J., Schmidt,J.W., Wunderlich,V.: Type Derivation for First Order
         Relational Expressions. Techn. Report No.79/81, Fachbereich
         Informatik, Universität Hamburg, June 1981.

[Lame84] Lamersdorf,W.: Recursive Data Models for Non-Conventional Database
         Applications. Computer Data Engineering Conference (COMPDEC), IEEE
         Computer Society, Los Angeles, April 1984.

[LMS84] Lamersdorf,W., Müller,G., Schmidt,J.W.: Language Support for Office
         Modelling. Proc. 10th Int. Conf. on VLDB, Singapore, August 1984.

[LaSc80] Lamersdorf,W., Schmidt,J.W.: Specification of Pascal/R. Technical
         Reports No.73 and 74, Fachbereich Informatik, Universität Hamburg, July
         1980.

[LaSc84] Lamersdorf,W., Schmidt,J.W.: Specification and Prototyping of Data
         Model Semantics. Proc Working Conf. on Prototyping, Namur, Belgium,
         Springer-Verlag, Berlin Heidelberg New York, 1984.

[MRS84] Mall, Reimer,M., Schmidt,J.W.: Data Selection, Sharing and Access
         Control in a Relational Scenario. In: [BMS84].

[MBW80] Mylopoulos,J., Bernstein,P., Wong,H.K.T.: A Language Facility for
         Designing Interactive Database-Intensive Applications. ACM TODS,
         Vol.5, No.2, June 1980.

[Reim84] Reimer,M.: Transaktionen in Datenbankprogrammiersprachen. ETH Zürich,
         Dissertation, Nr.7553.

[Rous82] Roussopoulos,N.: View Indexing in Relational Databases. ACM TODS,
         Vol.7, No.2, June 1982.

[Reyn81] Reynolds,J.C.: The Craft of Programming. Prentice-Hall, 1981.

[RRUZ83] Rebsamen,J., Reimer,M., Ursprung,P., Zehnder,C.A.,Diener,A.: LIDAS -
         The Database System for the Personal Computer Lilith. Proc. INRIA
         Workshop on Relational DBMS Design, Implementation, and Use on
         Micro-Computers, Toulouse, February 1983.

[Saun84] Saunus,L.: Adaptive User Interfaces for Relational Systems Utilizing
          Compiler Writing Techniques (in German). Diploma Thesis, Fachbereich
          Informatik, Universität Hamburg, 1984.

[ScBr83] Schmidt,J.W., Brodie,M.L.: Relational Database Systems: Analysis and
          Comparison. Springer-Verlag, Berlin Heidelberg New York, 1983.

[Schm77] Schmidt,J.W.: Some High Level Language Constructs for Data of Type
          Relation. ACM TODS, Vol.2, No.3, September 1977.

[Schm84] Schmidt,J.W.: Database Programming: Language Constructs and Execution
          Models. In: Amman,U. (Ed.): Proc 8th GI Fachtagung on "Programming
          Languages and Program Development", ETH Zürich, Switzerland, Informatik
          Fachberichte, Vol.77, Springer-Verlag, Berlin Heidelberg New York,
          1984.

[ScMa80] Schmidt,J.W., Mall,M.: Pascal/R Report. Universität Hamburg,
          Fachbereich Informatik, Report No.66, Januar 1980.

[ScMa83] Schmidt,J.W., Mall,M.: Abstraction Mechanisms for Database
          Programming. Proc. ACM SIGPLAN Symp. on Programming Language Issues
          in Software Systems, ACM SIGPLAN Notices, Vol.18, No.6, June 1983.

[ScPi82] Schek,H.J., Pistor,P.: Data Structure for an Integrated Data Base
          Management and Information Retrieval System. Proc. 8th Int. Conf.
          on VLDB, Mexico City, September 1982.

[SmFL81] Smith,J.M., Fox,S., Landers,T: Reference Manual for ADAPLEX. Computer
          Corporation of America, Cambridge, January 1981.

[Ston75] Stonebraker,M.: Implementation of Integrity Constraints and Views by
          Query Modification. Proc. ACM SIGMOD Conf., San Jose, May 1975.

[Wass79] Wasserman,A.I.: The Data Management Facilities of PLAIN. Proc. ACM
          SIGMOD Conf., Boston, May 1979.

[Wirt82] Wirth,N.: Programming in Modula/2. Springer-Verlag, Berlin Heidelberg
          New York, 1982.

[Zill84] Zilles,S.N.: Types, Algebras, and Modelling. In: [BMS84].

KNOWLEDGE-BASED AND EXPERT SYSTEMS: REPRESENTATION AND USE OF KNOWLEDGE

I. Hofmann, H. Niemann, G. Sagerer
Institut fuer mathematische Maschinen und Datenverarbeitung
Lehrstuhl fuer Mustererkennung
Martensstr. 3
University of Erlangen-Nuernberg (FRG)

ABSTRACT

The rapid development of data processing, data management, and artificial intelligence techniques made it possible to design systems based on a large amount of expert knowledge. The following sections will give an overview of this area. First the combination of data processing, building large information systems, and integrating knowledge-based systems will be illustrated. A definition of the terms 'knowledge-based systems' and 'expert systems' is suggested by choosing examples among the growing number of existing systems, and by specifying components or modules of such a system. One chapter will be dedicated to a system for the automatic analysis of heart scintigrams developed at the University of Erlangen. The last chapter reports on experiences, possibilities for use and further expansions.

NATO ASI Series, Vol. F19
Pictorial Information Systems in Medicine
Edited by K. H. Höhne
© Springer-Verlag Berlin Heidelberg 1986

1. INTRODUCTION

There has been a rapid development in computer-based management and processing of information. The way led from simple programs to complicated program systems. So called information systems are concerned with storing, managing and evaluating large amounts of data which are found in administration, economy or control, for instance. Since data to work on are getting more complex, this requires a good deal of data management using complex strategies for using the stored information. Up to now those special program systems are concerned with alphanumeric data most of the time. The scale runs from database systems up to systems, which contain methods and offer the possibility of program synthesis [GRE 79], and further on systems for evaluating data [HAR 68, LOC 78]. An important expansion of information systems are pictorial information systems (PIS), which work on pictorial data, additionally. The question raises how to combinate digital and perhaps analog images with alphanumeric and graphical information. The resulting problems are:

- accuracy, and on the other hand, fuzzy data
- availability and reliability
- time variant data
- access time.

So, the representation of data in the system and to the user is one major objective. The storing of information and its management has occupied many database administrators for the last 20 years. All these problems of data independence, reducing redundancy, avoiding inconsistency, integrity, data protection and privacy (which is very important if there are confidential data) have been the reason for intensive research in this field. Among all publications we will cite only one example, [DAT 81].
Further developments in network technology support data distribution between different departments, even between countries. The result in doing research affecting (pictorial) information systems, should be modular systems which allow easy integration of new developments and expansions as well as easy modifications.
Another important aspect in building an information system is to look at the environment in which it will be used. Each special application

involves specific problems in planning, design, and realisation of such a system. The following chapters are restricted to pictorial information systems in medicine (see [DUN 79], for instance) which, especially in the future, will contain one component called knowledge-based systems (KBS) or expert systems (ES) (Fig. 1.1).

Besides administrative data a large amount of pictorial information occurs in departments of radiology or nuclear medicine, for instance. Those data as a whole are sensitive, fuzzy, time critical and confidential; failures of every kind may be fatal. The resulting PIS should fulfill the following functions, besides those characterising common information systems:

1. image acquisition: cameras, for instance

2. image distribution: connecting different departments by installing image transmission networks

3. image redisplay

4. image archiving: centralized databases

5. image interpretation at two levels of abstraction

   - image processing

   - image analysis, image understanding

6. control and management of pictorial, alphanumeric and graphical data.

There is the possibility to distribute these tasks on distinct subsystems (Fig. 1.1):

1. a pictorial archiving and communication system (PACS) which should fulfill all functions except '(5) image interpretation at two levels of abstraction'. However, some methods for image processing should be included in order to support diagnosis and therapy by image displaying of high quality.

2. one or more KBS or ES for an automatic interpretation of data,
    either as a system for consultation in therapy or as a system for
    deriving medical evidence.

    The output (text, images, graphics, statistics,...) is given back
    to the PACS or immediately to the user.

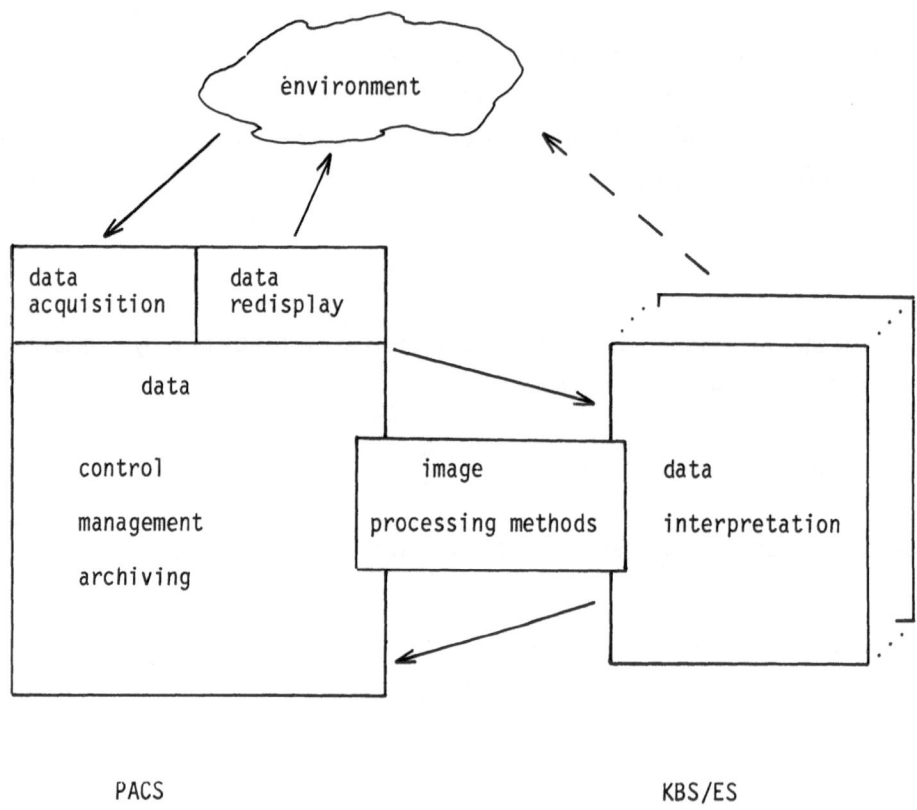

PACS   KBS/ES

Fig. 1.1   Principle structure of a PIS

2. AN OVERVIEW OF KNOWLEDGE-BASED AND EXPERT SYSTEMS

The development of expert systems started in the 1960s by trying to simulate human perception, thinking and learning ([FEI 61], for instance). Considerable success has been achieved in the meantime, but most of the existing systems are experimental prototypes used in the laboratories of universities and other research institutes. The area of ES investigates methods and techniques for constructing man-machine systems with specialized problem-solving expertise in different applications. One may distinguish [HAY 83]

1. interpretation systems, which explain observed data by assigning to them symbolic meanings describing the situation or system state accounting for the data,

2. monitoring systems, which compare observations of system behaviour to features that seem crucial to successful plan outcomes,

3. debugging systems, which support correction of a diagnosed problem

4. repair systems, which develop and execute plans to get rid of diagnosed problems,

5. planning systems, which design actions,

6. design systems, which construct descriptions of objects in various relationships with one another and verify that these configurations conform to stated constraints,

7. prediction systems, which typically employ a parametric dynamic model with parameter values fitted to the given situation,

8. diagnosis systems, which relate observed behavioural irregularities with underlying causes,

9. instruction systems, which train the students' behaviour and show weaknesses in their knowledge, and at last

10. control systems, which govern the behaviour of another system.

Existing systems

The following overview doesn't claim to be complete, it is directed to give some examples of what has been done up to now whether in mineral prospecting [GAS 80], chess [WIL 80] or, in more details, describing some existing systems.

MACSYMA [MAR 71] is a mathematical system. It performs differential and integral calculus symbolically and excels at simplifying symbolic expressions by reformulation.

DENDRAL [LIN 80] analyzes mass spectrographic, nuclear magnetic resonance data to infer the plausible structures of an unknown compound. By systematically generating all plausible structures, it finds even those candidates that human experts overlook occasionally. The additional system META-DENDRAL adds knowledge to DENDRAL which had been accumulated during an analysis.

R1 [MCD 80] serves for configuring DEC VAX computer systems. It is used industrically. The skill of this system is astonishing.

HEARSAY II [ERM 80] is capable of understanding connected discourse from a 1000 word vocabulary.

PROSPECTOR [DUD 81] examines mineral deposit relationships. It contains about a dozen knowledge bases for difficult kinds of deposits. PROSPECTOR uses a production system to encode the knowledge gleaned from experts. It consists of a data base of assertions, a set of production rules of the form <antecedent> ==> <consequent>, a control program deciding on an inference strategy, and an inference mechanism to propagate the effects of input data. The probabilistic production system has two special restrictions [KON 79]:

1. no variables are bound in the production rules

2. there are no inference loops like A=>B, B=>C, C=>A.

Given these properties, the set of productions can be converted into an inference net without any loops. Productions are indicated by edges. The leaf nodes represent input assertions to which probabilities are assigned by an input source (or the user). The control strategy is searching a path from the top of the inference net downto an input leaf node and is asking for a probability (goal-directed search by establishing a query order for the leaf nodes). The other way round, if the probabilities of the leaf nodes are known when starting the system, inferencing becomes data-directed from the input leaf nodes.

An example for an inference net is shown in Fig. 2.1

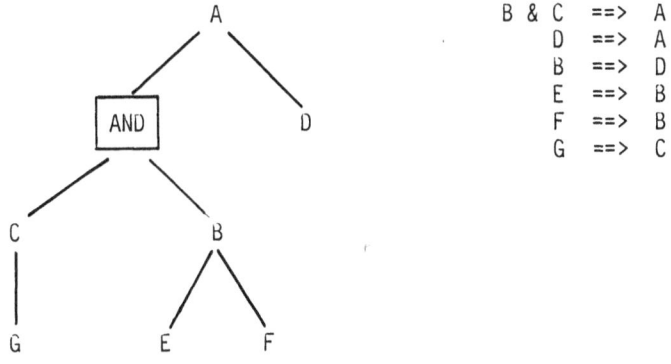

B & C	==>	A
D	==>	A
B	==>	D
E	==>	B
F	==>	B
G	==>	C

Fig. 2.1 Inference net

PATREC [MIT 84] is a subsystem of an automated system MDX for diagnosing liver diseases by examining the cholestasis syndrome. PATREC allows database reasoning by using inferential knowledge, embedded in the underlying knowledge base, in order to generate answers when corresponding data are not explicitly stored in the database. Temporal aspects are also observed.

MDX [MIT 79] has three components: the diagnostic system, a patient data base (PATREC), and a radiology consultant. The medical knowledge needed is represented in a scheme called conceptual structures organising knowledge in the form of production rules or more complex procedures. This structure is mainly hierarchical with successors of a conceptual node representing further refinements of a concept. Each conceptual node can be viewed as an expert in its conceptual area able to turn over control to selected subconcepts which possess more detailed knowledge in a subarea. A set of procedures is associated with each node in order to decide whether the concept is applicable to actual data. The underlying structure is represented as a tree, Fig. 2.2, for instance. The concepts are refined

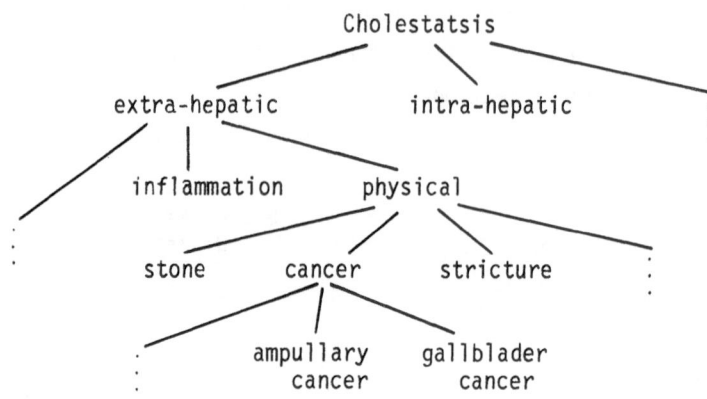

Fig. 2.2 Conceptual structure of a portion of 'Cholestasis' [CHA 79]

due to a function of further learning. In this way an expansion of the knowledge base can be performed easily. The second advantage is an effective use of knowledge by facilitating the access. Because each concept may have different knowledge there is no uniform mechanism operating on different concepts. Nodes of higher degree in the tree have to search below in order to fulfill their task. Control is distributed systematically to other experts so that each one is able to solve one part of the refined problem. The system starts working on some initial data and asks for additional ones, if needed during refinement. These are used by an expert controlling the process to re-compute only what was affected within its domain. By passing control back and forth, each expert determines the relevance of the data and acts accordingly but only if it gets the control.

CRYSALIS [ENG 79] is a knowledge-based system whose goal is to infer the three-dimensional structures of proteins from x-ray crystallographic data. Many diverse sources of knowledge contribute to the inference of a protein structure from an electron density map (EDM) and associated data. The EDM describes the electron cloud surrounding the molecule. The knowledge base is structured hierarchically, and divided into domain knowledge, task knowledge and strategy knowledge. The knowledge sources comprise the formal and informal procedures expressed by rules being appropriate to use if particular events occur. The correspondence between events and rules is established by another set of rules, the task rules for method selection. These are used to decide which knowledge source or sequence of knowledge sources is to be activated in order to perform one of the typical tasks in building the structure of a protein. Once a task is completed or if the task fails, the system passes control to a higher level (strategy level) to determine the next step. Strategy knowledge is also expressed by rules which make use of the current state of the knowledge sources and the event list. The conditions of strategy rules refer to global features of the current hypothesis, such as the presence of solved and unsolved regions along the amino acid sequence.

The following systems are found in the area of medical diagnosis and therapy.
MYCIN [SHO 76, BUC 84] is one of the first, and one of the most successful ES. It addresses the problem of diagnosing and treating infectious blood diseases.
An ES PUFF [KUN 78] for diagnosing pulmonary diseases is in routine use in laboratory. Approximately 95% of its reports are accepted without modification.
CASNET [WEI 81] is an ES for consultation in the diagnosis and treatment of glaucoma. The knowledge is represented in a network of dysfunctional states with well defined etiologic and ending states. Evidence about the likelihood of these states can be obtained from another network relating states to tests. Diseases are defined by paths in the net. The system calculates the status of each state, based on the test values. This computation is repeated as soon as new data are entered. It also computes the weight of each node which is independent of its status, but depends on those nodes occuring in the same path. Diseases are hypothesized and verified on the basis of which paths are the most likely or potentially ones able to account for all confirmed states.
CADUCEUS [POP 81] contains 85% of all relevant knowledge concerning internal medicine. A semantic network accents relationships between symptoms and diseases which are distinguished by sophisticated strategies.
The experience in the development of ES has led to some tools for building those systems, like EXPERT [WEI 81], EMYCIN [MEL 81] or TEIRESIAS [DAV 80], for instance.

This survey documents the breadth and significance the area has attained in less than two decades, supporting the impression that, while maturing rapidly, it can expect considerably more progress in the years to come.

Definition and features of an ES

One major objective will be the definition, what an ES or a KBS should be, and what it should look like. In [MCG 82] an expert program is defined as:

> "Extensive experience enables humans to exhibit expert performance in many tasks, even though no firm scientific or calculational base exists and the knowledge does not exist in any explicit form. So-called expert artificial intelligence programs attempt to exhibit equivalent performance by acquiring and incorporating the same knowledge that the human expert has."

[ERM 83] tries to draw the line between KBS and ES by:

> A knowledge-based system is an intelligent computer program that uses knowledge and inference procedures to solve difficult problems;

> an expert system is a knowledge-based system that performs at or near the level of human experts.

Both definitions mention some features of KBS or ES:

1. intelligence:
   The term 'intelligence of humans' should be understood as the "general mental ability due to the integrative and adaptive functions of the brain that permit complex, unstereotyped, purposive responses to novel or changing situations, involving discrimination, generalisation, learning, concept formation, inferences, mental manipulation of memories, images, words and abstract symbols, education of relations and correlates, reasoning and problem solving" [MCG 82]. However, the restrictions concerning human intelligence [ENC 62] are valid also to the 'intelligence of machines', to the highest degree:

(the features of intelligence)..."may develop to a considerable extent in all aspects of one of the forms of ability without equivalent development in the others, though there is also some tendency for all of them to be related as different manifestations of one general intelligence."

Therefore, an intelligent behaviour of a program is characterized by owning at least some of the properties mentioned above, and simulating human behaviour in some fields of application.
The following points describe some of the more important features in details.

2. efficiency:
   Expert behaviour seems to demand that blind search through large numbers of hypotheses should be avoided in favour of quick elimination of many possibilities in each inferential move.

3. difficulty and complexity:
   The problem to be solved automatically should be of a certain complexity and difficulty to require an expert. This point parallels the notion that the more complicated the domain, the more expertise one can (or should) attain. A system will be restricted to narrow, specialized domains up to now.

4. type of task:
   Different tasks may substantially change the picture of a system's architecture. The architectural implications of the task requirements are determined decisively by the available knowledge, knowledge representation schemes, and the employed strategies.

5. expertise:
   All expert level of performance depends upon the ability of a system to access large amounts of information and to refine continuously that information in the face of new data (fragmental, incomplete and inexact knowledge). Information might be factual knowledge, heuristic knowledge like information or understanding acquired through experience, and judgmental knowledge. Knowledge which can be represented in an automatic system is, most of the

time, restricted to constraints and associations between objects and events occuring in the real world. This has to be combined with an opportunistic behaviour to exploit knowledge at the right time.

6. reformulation:

One of the critical tasks that real experts do and that ES are just beginning to approach is to take a problem stated in some arbitrary initial form and convert it into the form appropriate for processing by expert rules. The success of both reformulation and processing depends upon a correct model of the world of a special application.

7. reasoning:

A central premise of ES work concerns the kind of information with which the system reasons. An appropriate knowledge representation scheme facilitates this task. The process of reasoning is connected to the process of retrieval which involves sorting out large quantities of knowledge to decide which ones are relevant to what is being done. This includes high flexibility in the behaviour of a program system which is one of the most important factors distinguishing work on ES from simply high-quality special-purpose programming.

8. reasoning about self:

The performance of an ES can be improved by supplying various sorts of knowledge about knowledge used in the system and different strategies used by the system. If there are provided explanations or justifications for conclusions reached, a system becomes transparent to the user which might be urgent as the cost of a wrong decision and the frequency of unexpected results increase.

9. speed:

The speed at which a decision is reached, is one important factor to determine the quality of a system. For instance, applications in medical diagnosis work on time-critical data. The worst case would be an accurate medical evidence slowly arrived at while the patient died in the meantime.

Architecture

These features are to be completed by an architecture of the ES well adapted to a specific application. There are some components which are contained in almost every ES but there are others which occur only in some of them. Every existing ES contains an user interface which realises problem-oriented communications between the user and the ES. In general, there will be a language processor which parses and interprets user questions, commands, and desired information. Answers, explanations, and justifications given by the system are represented in an understandable form.

A memory for storing intermediate decisions and hypotheses that the ES manipulates, as well as final results reports the way of getting a solution of the problem. These results are the basis to any search strategy and control.

One central part of an ES is the knowledge base. It contains facts, heuristics, and rules concerning one specific domain. The quality of a system depends on this component to a high degree.

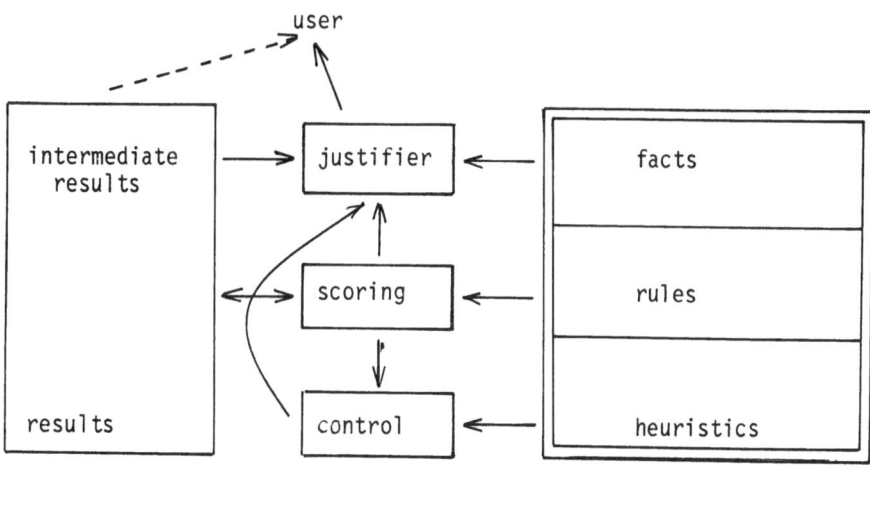

Fig. 2.3   Representation of an ES

If the knowledge base consists of a large amount of rules, an interpreter is used to choose a corresponding knowledge base rule and to apply it to intermediate results. Generally the interpreter validates the relevance conditions of the rule, binds variables in these conditions to particular solution elements, and then makes those changes to the memory that the rule prescribes.

A similar function is performed by one module 'control'. It uses some plans and search strategies to derive the solution of the problem from input data. The justifier explains the actions of the system to the user. To do this, the justifier uses a few general types of question-answering plans. These typically require to trace backward along the paths in the intermediate results' memory from the questioned conclusion to the intermediate hypotheses or data that support it. The last component, existing in almost every system, is a module for scoring decisions. Most of the time a numerical adjustment scheme is used to determine the degree of believe in each potential decision. This scheme attempts to ensure that plausible conclusions are reached and inconsistent ones are avoided.

Fig. 2.3 shows an idealized representation of an ES, which is a combination of the approaches in [NIE 81] and [HAY 83].

Conclusion:

There are a large number of ES and KBS known from literature documenting intensive research. However, there is a lack of pictorial systems to notice. The problem is to acquire the necessary knowledge about understanding images and scenes, and to store it in an appropriate way.

3. REPRESENTATION AND USE OF KNOWLEDGE

Knowledge acquisition

Expert knowledge provides the key to expert performance while knowledge representation and inference schemes provide the mechanisms for its use. A representation of knowledge is a combination of data structures and interpretive procedures that, if used in the right way in a program, will lead to "intelligent" behaviour. The process of extracting knowledge from an expert or source of expertise and transferring it to a program (ES) is an important and difficult problem. It can reduce the costs of knowledge reproduction and exploitation and it can make private knowledge available for public test and evaluation, but first it involves problem definition and refinement, as well as representing facts and relations acquired from an expert. The expertise to be elucidated is a collection of specialized facts, procedures, and judgmental rules about the narrow domain area rather than general knowledge about the domain, or commonsense knowledge about the world. The transfer and transformation required to represent expertise for a program may be automated or automated partially in some special cases. However, up to now there is the need for a dialog between the expert and the system designer. Most of the time a person is required to communicate with the expert and the program. One of the most difficult aspects is to help the expert to structure the domain knowledge, to identify and formalize the domain concepts. The knowledge engineer and domain expert work together closely to define the problem. So the knowledge relevant to solve the problem is isolated and verbalized. One of the most troublesome difficulties is representation mismatch, the difference between the way a human expert normally states knowledge and the way it must be represented in the program.

A knowledge base is restricted to be a model of a slice of reality. Knowledge consists of symbolic descriptions that characterize the definitional and empirical relationships in a domain, and of procedures for manipulating these descriptions. A knowledge base needs organisational principles to be understandable as a unit.

For any representation formalism it is not possible to show that one scheme will be better than another. A choice is dependent on the special area of application. But each scheme has been successfully used in a variety of programs that do exhibit intelligent behaviour. However, there are some features of knowledge representation to consider: modularity, flexibility, understandability, and completeness to a certain extent.

Knowledge representation

The problem in knowledge representation is to use appropriate representation schemes which allow the combination of different types of knowledge. One major difference is between declarative and procedural knowledge. The former describes facts, objects or events, and their interrelationships (static knowledge). Dynamic knowledge like actions on objects, time variance, or heuristics, for instance, is represented as procedural knowledge.

The following section will represent some well-known knowledge representation schemes:

PREDICATE LOGIC [JOK 79, BAL 82] is a system for expressing propositions and for deriving consequences of facts. There are used expressions of the form

$$P(t_1, \ldots, t_n) \qquad n \geq 1$$

which are called atoms having n arguments called terms. Terms are either constants, variables or functional terms which are expressions of the form

$$f(t_1, \ldots, t_m) \qquad m \geq 1$$

with t_i terms. Based on atoms it is possible to represent knowledge in sentences which are a collection of clauses

$$A_1, \ldots, A_k \quad \texttt{<---} \quad B_1, \ldots, B_p$$

where $A_1, \ldots, A_k$ are called conclusions of the clause and $B_1, \ldots, B_p$ are called the conditions. Both conclusions and conditions are atoms. The conclusions of a simple clause are a disjunction of alternatives, whereas

the conditions are a conjunction. A clause without conditions is an unconditional assertion, a clause without conclusions is used as a negation.

For instance [DEL 79], the sentence "John gives the book to Mary" might be represented by using a three argument predicate

$$GIVE(John, book, Mary) \quad \texttt{<---} \quad .$$

The representation of "John gives the book to everyone he likes" is given by

```
    GIVE(John,book,x) <--- LIKES(John,x)
```

with x variable and <--- representing the logical connection "if". The
clause

```
    <--- GIVE(x,y,John),LIKES(John,y)
```

denies that anyone gives John anything which John likes.
 In addition to using symbols for constants and predicates, the predicate
calculus allows for functions and logical connections, also expressing
sentences involving the universal (to all x ...) and existential (there is
one x ...) quantifier. Even the derivation of new facts from old ones can
be mechanized.
 The semantics of a set of logic statements is specified completely by the
rules of inference. A formula can be given an interpretation by assigning
a correspondence between the elements of the language and the entities and
relations of the domain of discourse.
 Predicate calculus is a widely studied formal language of symbol
structures that can be used for representation. The advantages of
predicate calculus systems are given by their modularity and generality.
Each fact is valid, independent of whatever else is in the system. The
notation is general, not tailored to any particular sort of knowledge. The
most important feature of logic and related formal systems is that
deductions are guaranteed to be correct and complete: if there is enough
knowledge in the system to proof something, most theorem provers will get
to it eventually. But the demands of generality make these systems
inefficient in a combinatorial way. Incomplete knowledge and beliefs are
not to be represented, so the introduction of new axioms can invalidate
old theorems. Classic symbolic logic lacks tools for describing how to
revise a formal theory dealing with inconsistencies caused by new
information, and for dealing with imprecise knowledge.

PROCEDURAL REPRESENTATION schemes store the knowledge about the world in
procedures that specify entities, and define or perform the transitions
from one well-specified situation to another state. The disadvantage is,
that knowledge is not stated explicitly, and thus is not extractable
typically in a form that humans understand easily. The consequent
difficulty in verifying and changing procedural representations is the

major flaw of these systems. One major advantage is the facility for representing heuristic knowledge, especially domain-specific information that might permit more directed deduction processes.

Besides knowledge utilization, PRODUCTION SYSTEMS [WIN 77] offer the possibility for knowledge representation. The central ideas are condition-action pairs:

IF(condition occurs) THEN { action(s), conclusion(s) } .

The IF-term specifies a condition which has to be met by actual data if the rule is applicable. In this case the THEN-term is activated. The context is the focus of attention of the production rule. The actions of the production rules can change the context, so that other rules will have their condition parts satisfied. Production rules have been found useful as a mechanism for storing and using information. The latter is valid because a rule describes how to apply knowledge.
 Great advantages of production systems are the ability to infer new facts from what it has been told already, and the possibility to expand easily the knowledge base by simply adding new rules. This involves creating dynamically new symbol structures from old ones. There are obvious qualities like modularity, uniformity and a certain naturalness. However, by growing data bases a procedural approach like a production system will result in inefficiency and a lack in lucidity.

FRAMES [MIN 75, TSO 80, WIN 83] are complex data structures to whom are attached different kinds of information: declarative and procedural knowledge in predefined internal relations. Some of this information is about how to use the frame, what one can expect to happen next, or what to do if these expectations are not confirmed. There are some slots that must be filled with specific instances, data or default values. A frame, once evoked on the basis of partial evidence or expectation, would first direct a test to confirm its own appropriateness. Thereby, using knowledge about recently noticed features, relations and plausible subframes, the frame tries to match itself to the data it discovers. Next, it could request information needed to assign values to those slots that cannot retain their default assignments. In case of failure, it could transfer control to a more appropriate frame by observing transformations which relate frames, and similarities. Frames are a favourable mechanism to deal with hierarchies and inheritance of properties. There are some similarities to network schemes.

The research on the mentioned representation schemes introduced additional languages and special developments for simplification of the design of a knowledge base (FOL [WEI 80], PLANNER [HEW 72] or FRL [GOL 77], for instance).

One important representation scheme is a RELATIONAL DATABASE. Entities are interconnected by relations. Each relation is determined by a number of attributes as arguments, some of which are distinguished as so-called keys, to identify sets of attributes and relations. The basic property of these structures is that the data are stored in an associative format by means of tables. The advantage is to describe logical relationships between data by defining different views and subviews. But there are difficulties to observe if connecting attributes of different domains in dependence on conditions. There often will be the need for the introduction of new relations. The same consequences will result by representing many-to-many-correspondences. Another problem is the representation of heuristics which often cannot be described as relations. A detailed introduction on this area is given in [DAT 81].

The last representation mechanism mentioned in this context are SEMANTIC NETWORKS [HAN 78, FIN 79, BAL 82, MYL 83]. Such a net consists of nodes, representing objects or events, and links between the nodes, representing their interrelations. Both the nodes and the edges can be labelled.

Relevant facts about an object can be inferred from the nodes to which they are linked directly, without a search through a large database. A property inheritance hierarchy is established in the net with the help of the IS-A-relation. The interpretation (semantics) of net structures, however, depends on the program manipulating them; no conventions exist about their meaning. This mechanism shows the static aspects of knowledge (declarative knowledge), as well as dynamic knowledge (procedural knowledge) by attaching procedures to nodes or links. The development of semantic networks had been in parallel to that of frames.

Control and use of knowledge
But all knowledge of the world is worthless if one has no possibility to
use it efficiently. So the second major part of on ES is a module for
control and use of knowledge.

Many problem-solving systems are based on the formulation of problem-
solving by search NIL 80 . A description of the desired solution is called
a goal and the set of possible steps leading vom initial conditions to a
goal is viewed as a search space. The choice of a search method is affected
by many characteristics of a domain, such as the size of the solution
space, errors in the data, and the availability of abstractions.
 State-space search: This representation of problem-solving uses states and
operators. A state is a data structure comparable to a snapshot of the
problem at one stage of solution; operators change one state into another.
The idea is to find a sequence of operators that can be applied to a
starting state until a goal state is reached.
 A lot of search strategies had been developed which may be distinguished
mostly in

(a) Blind search: Blind search methods can be goal-directed (involve
     searching backward from the desired end state to the initial
     conditions), data-directed (starts from the given initial conditions
     towards the goal) or bidirectional. In order to handle alternatives on
     the way to a solution one has also to decide whether to perform depth-
     first or breadth-first search.

(b) Graph search algorithms for finding the optimal solution of the given
     problem in dependence on the chosen criterium for being optimal: The
     method mentioned in (a) is limited by the amount of time and storage
     available. This illustrates the phenomenon of combinatorial explosion
     arising when problem-solvers lack sufficient knowledge to guide an
     inferential process. For many applications it is possible to find
     domain-specific information guiding the search process and reducing the
     amount of computation. One form of this kind of search is to direct the
     search in a ´best-first´ order by using a domain-dependent evaluation
     function to estimate the ´distance´ of the chosen search path to the
     goal. Those functions are said to be well behaved if they indicate
     reliably and monotonically an optimal path to a goal. During a search

process there will be nodes to which some but not all applicable operators have been applied; the node is expanded partially, reserving the possibility of further expansion at a later state in the search. The problem arises to assess the search state by the specific evaluation function.

[NIL 80] introduces the so-called A*-algorithm, which is proven to be optimal in the sense of finding the best solution in minimal time by estimating costs on the way to the goal.

Another graph-search algorithm is known as dynamic programming [RAB 78]. This method is used to solve matching problems. The transitions needed to match actual input data against an idealized (optimal) prototype are represented in a network with edges describing the transitions from one state (node) to another. Each node is weighted by an estimate of costs on the solution path. The path of minimal costs, that is the minimal distance between input data and the prototype, describes the optimal solution.

A general idea for reducing search in many kinds of problem-solving is searching an abstracted representation of the search space. This amounts to hierarchical problem-solving through various levels of abstraction.
This method is called problem reduction. A problem is divided into independent subproblems until a problem is reached which can be solved immediately. An appropriate representation scheme are AND/OR trees which on each level have only nodes of either the type AND or the type OR for connecting the relevant subproblems. An AND node only has OR nodes as successors, an OR node only has AND nodes as its successors. It may happen that there is necessity for a general AND/OR graph instead of a "simple" tree.

Besides search algorithms problem-solving could be achieved by symbolic reasoning, that is making and retraction of assumptions. Assumptions and further inferences based on them must be regarded as tentative and might be revised given new information. Dependencies among a current set of beliefs might effect on more than one assumption if new knowledge is accommodated. This method is called backtracking or tracing.
The domain of resolution theorem proving has always equated problem solving with deductive logic. There has to be some replacement of "axiom sets" by additional inference rules which allow reasoning moving foreward

in macrosteps. This is not limited to applications in mathematical logic. The general idea is to accumulate knowledge, until the values of all desired quantities are known. If there is the possibility that during problem solving a statement accepted for being correct may lose its validity, reasoning becomes difficult. In a mathematical sense there will be the need for modal or other non-monotonic logic. That is offering the assertion of certain propositions without direct proof, but on the basis of their consistency with the remaining propositions, so that there is no reason for them not to be true. An example would be "if there is no known reason why an animal with wings should be no bird, then it is a bird". An informal introduction to this area gives [SIM 83].

In the case of procedural systems one will meet the great advantage of being directed in the problem-solving activity in the sense that irrelevant knowledge is not used and unnatural paths are not followed (instructions for the use of knowledge). No explicit means of procedural control are provided. The system is thus solely event-driven and directly guided through its database [GEO 82].

In many cases, the limitations of computer time and storage show the need for heuristics to reduce the search space.
One method is to define special cost functions using commonsense or probelm-specific a-priori-knowledge.
Another possibility is the development of schemes for constructing and executing plans. The basis of these schemes is to construct a global plan constraining the set of possible paths in the search space, and then searching for a solution in this constrained solution space.
If the use of knowledge and control is performed by state-space-search or problem reduction, there is the possibility to reduce the search space by pruning. In many cases it is infeasible to insist on finding the optimal path because search effort would exceed any reasonable bound. The consequences are to bound the effort by passed computer time or depth of search, for instance. Of course, any of the pruning techniques may result in a nonoptimal path or the missing of the path, but most of the time they will succeed, if a reasonable cost function is available.

All strategies, algorithms and access mechanisms for the use of knowledge and control are strongly dependent on the knowledge representation scheme and vice versa. A graph-like structure of the knowledge base yields for graph-search algorithms, intending to use an interpreter of rules a production system will result, and so on. This decision is to be made by the knowledge engineer of the system with respect to the application area.

4. A SPECIAL APPLICATION: A PICTORIAL KNOWLEDGE-BASED SYSTEM

The last chapter will be dedicated to a knowledge-based system for the automatic analysis of nuclear image sequences describing one cycle of the beating human heart. Those sequences are Tc-99m gated bloodpool studies triggered by the ECG, serving as input to a system which automatically derives a medical description of the motional behaviour of the human heart, especially of the left ventricle. This system has been developed at the University of Erlangen (FRG). A prototype version has been completed, and, therefore, first results are available.

System architecture
The principal architecture of the system is shown in Fig. 4.1. It is based on the approach proposed in [NIE 81].
The system consists of five major modules, each of which is concerned with a particular task which is to be described in the following paragraphs.

Input image sequences
Input to the system developed at our institute are scintigraphic image sequences showing the radiation and distribution of the radionuclide Technetium-99m. The two-dimensional intensity distribution in an image corresponds to the three-dimensional radiation distribution in the bloodpool. In this way, the interior space of the human heart is visualised.
A sequence of images triggered by the ECG is taken for describing the

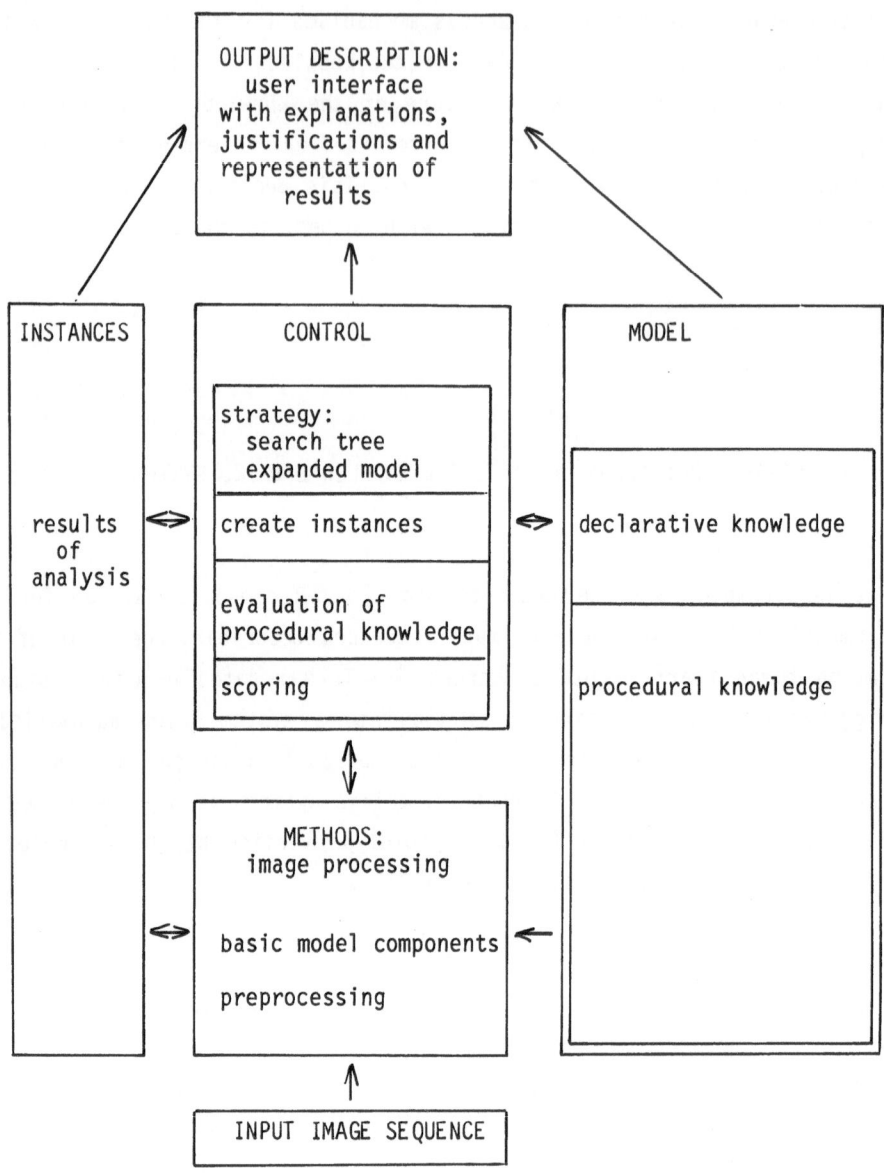

Fig. 4.1 Architecture of the system

motility of the heart and for medical diagnosis. Each sequence consists of
n images, 12≤n<32, with a spatial resolution of 64x64 pixel and an
intensity resolution of 8 bit. The images are taken in LaO 45 (Fig. 4.2).

The main application of scintigraphic image sequence analysis are in diagnoses and supervision of patients after heart surgeries.

Module METHODS

The module METHODS comprises several submodules for image processing and extraction of basic model components from an image sequence. At the beginning of an analysis this module is activated by the module CONTROL. The following sequence of operations is carried out: intensity normalisation, median filtering, thresholding for separating the heart from the background, locating the septum for dividing the heart into left and right ventricle, and segmentation of the left ventricle. Some of these operations are shortly explained.

Since the quality of those images is poor, a smoothing operation

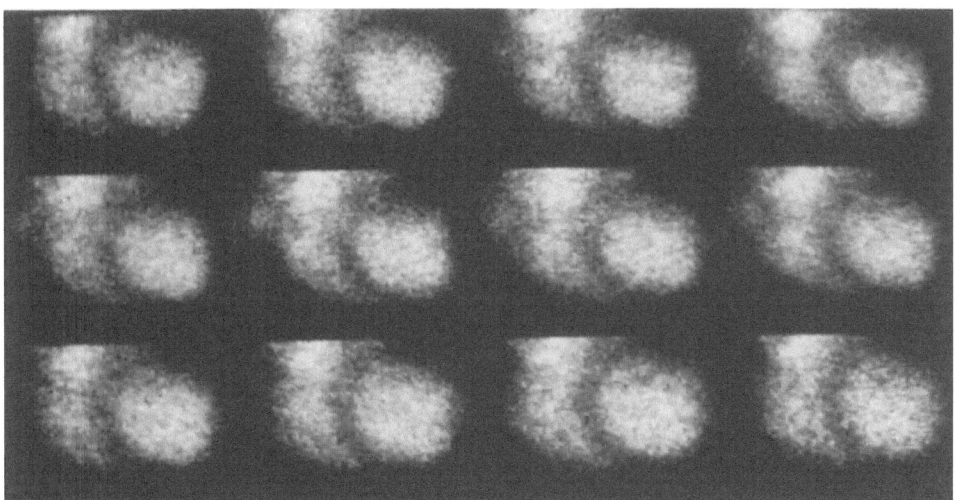

Fig. 4.2 Input image sequence

is performed first. It is done by a median filter of size 7x7. The window size has been determined heuristically based on subjective inspection. The sequence of Fig. 4.2 is shown again in Fig. 4.3 after median filtering. This operation improves the subjective quality of the input data and facilitates further processing steps.

The major goal is to outline the contours of the heart, especially of the left ventricle, and to perform segmentation of the left ventricle.

318

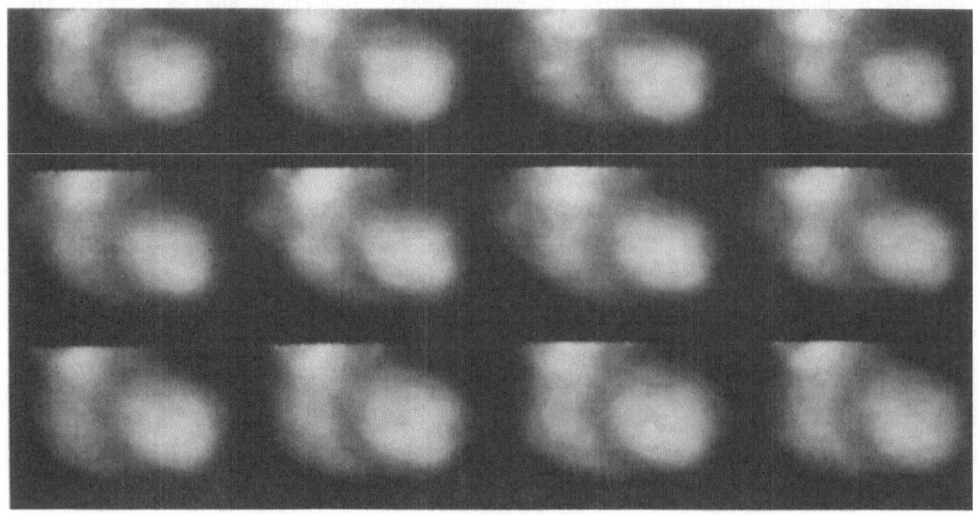

Fig. 4.3 Image sequence after smoothing

There are several approaches to the automatic detection of the left
ventricular contour known from literature. They are based on thresholding
([BUN 82]), or a certain kind of gradient ([GER 81, GOR 81, HAW 81]). The
results using a modification of [GER 81] proved better results than the
others, especially thresholding.

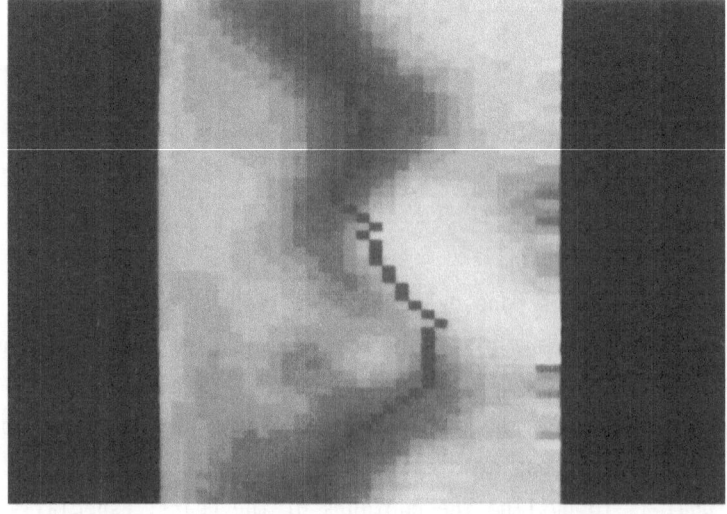

Fig. 4.4 Optimal contour in polar space

Our approach is based on a polar coordinate transform, using the a-priori-knowledge that the left ventricle is of circular or elliptical shape, a local edge detector and a contour following procedure based on dynamic programming. The center of the polar coordinate transform is a point inside the left ventricle which is detected by means of a heuristic procedure. Fig. 4.4 and Fig. 4.5 show the results of contour detection applied

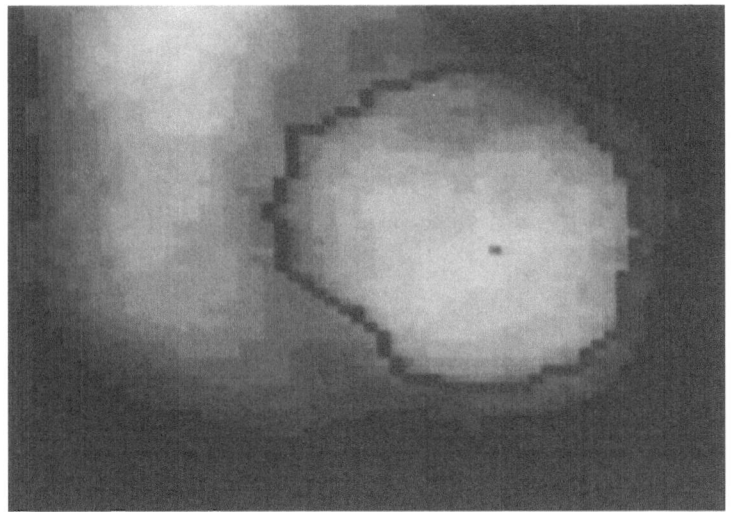

Fig. 4.5 contour of the left ventricle

to the first image of the sequence shown in Fig. 4.3. The polar transform, and then application of the local edge detector is shown in Fig. 4.4, Fig. 4.5 gives the contour of the left ventricle after a backtransformation into x,y-space.

After contour extraction, segmentation of the left ventricle on each image of the sequence is performed. A local motility description is based on such an operation. Several approaches are known from literature, [SIL 80] or [BRO 78], for instance.

The combination of two kinds of segmentation, supplementing each other, has proved best results. A segmentation into n sectors is done, each of which is given by an angle 2π/n.

Another segmentation is guided by medical criteria:

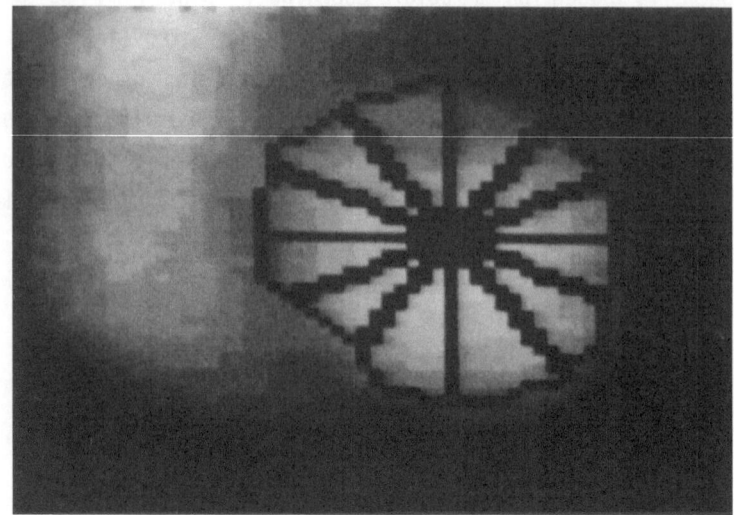

Fig. 4.6 Segmentation using sectors

S : septal, the segment left to the center,

PL: posterolateral, the segment right to the center,

B : basal, the segment above the center,

IA: inferioapical, the segment below the center.

The anatomical definition of those segments is given by a certain range of occurrence, and an expected direction of the left ventricular contour.
The combination of both types of segmentation allows for testing effects of the weak motility of one segment on its neighbours.
Fig. 4.6, 4.7 show the results of segmentation.

Module KNOWLEDGE BASE

The knowledge stored in the knowledge base can be divided into a declarative part for representing static knowledge, and a procedural part for representing dynamic knowledge.
The declarative part describes how different types of objects, like the left ventricle and its four segments, motions and medical diagnoses are related with one another. A semantic network was chosen as

321

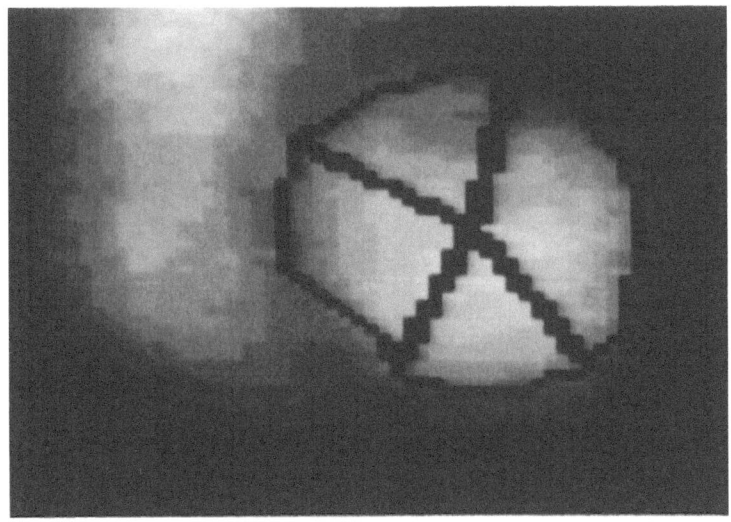

Fig. 4.7 Segmentation guided by medical criteria

representational formalism. The nodes are simply data structures, which consist of a name, attributes and attribute constraints. The latter form the relation to the procedural part of knowledge. The edges between the nodes are restricted to three relations: 'necessary-part-of', 'semantic-part-of' and 'specialization', and are implemented as links in a data structure. The difference between parts needed for calculating or structuring and parts describing hierarchies which result from the problem space [SAG 85]. Additionally, each node in the network has a procedure for calculating the certainty of an instance of it.

The principal structure of our network is shown in Fig. 4.8. There are 8 levels of abstraction. Level 0 provides the interface to the module METHODS by defining time parameters (the number of images in a sequence, for example), space parameters (angles of anatomical segments, for example), and giving the contours of our objects. Level 1 collects the information of level 0, level 2 descibes shape and size of the left ventricle and its segments. This knowledge is used to derive medical diagnoses of the kind "The left ventricle is enlarged". The levels 3 and 4 define the motion cycle of the heart in terms of contraction, expansion and stagnation. While at level 3 the changes of area from image i to i+1 of the sequence are observed, phases of motion are put together at level 4. An anatomical description of the heart cycle is given on level 5. There

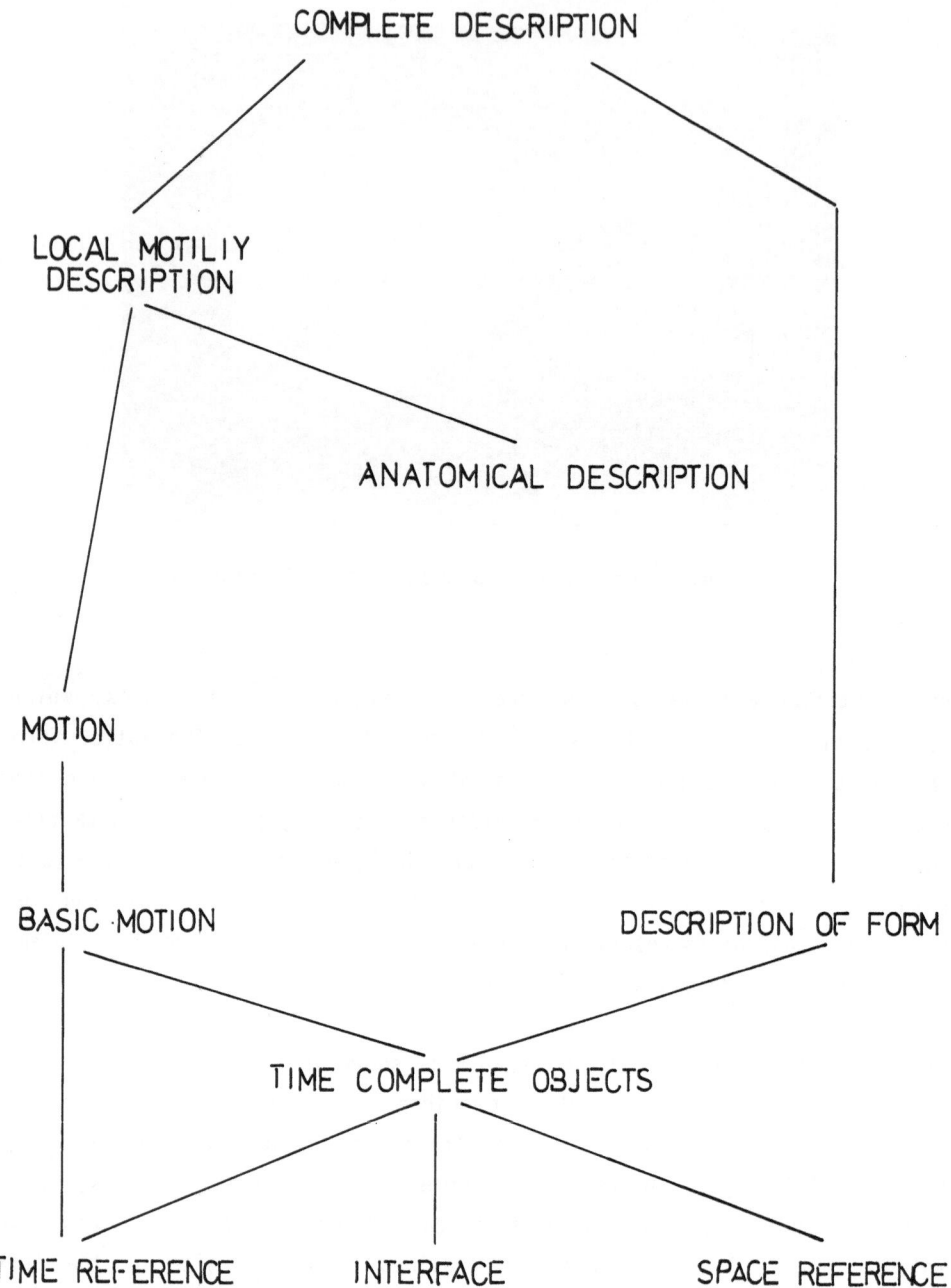

Fig. 4.8 Principal structure of the knowledge-base

are used medical terms like systole, pre-ejection-period, or diastole, for instance. The last two levels relate motion and shape of the left ventricle to medical evidence, level 6 on a local, level 7 on a global basis.

The procedures used are straight forward, in most of the cases. Angles, areas, the length of contours are to be calculated. But there are two classes of procedures, which utilise more sophisticated algorithms. It is the segmentation of volume curves into different motion phases and the inference of medical diagnoses.

The former problem is solved with the support of syntactic pattern recognition. Depending on the poor quality of the given input images a large degree of noise might be present. The decision whether a change in area from image i to i+1 should be interpreted as a contraction, expansion or stagnation, respectively, is made by assigning a certainty for each of those three possible ones. The medical basis for a motion description is given by 9 prototypes of heart cycles, each of which represents a normal or an abnormal behaviour of the left ventricle or his segments. The matching process between such a prototype and input data is performed by dynamic programming. Input data are a certainty matrix which assesses motion types as mentioned above. The optimal path is found in the resulting network of possible transitions. A detailed description of this approach is given in [BUNa84].

The latter problem attacks rule-based systems. Before describing it in more details, first, an introduction will be given to our scoring method. All assessment in this system is done with the help of fuzzy set algebra ([ZAH 65] or [ZAH 79]). The advantages against a stochastic approach are that it is possible to formulate statements using terms "a few", "weak", and so on without being forced to absolute decisions. The results are that there will be more possibilities which might be correct depending on a certain context. For example, the statements - "the left ventricle has weak motion" with certainty 0.8 and "the left ventricle is almost motionless" with certainty 0.4 - are allowed and express medical routine to a certain extent.

The method for infering medical diagnoses is similar to that in [SHO 76]. Presently the system knows about 45 diagnostic interpretations of the type "the inferioapical segment is nearly motionless". Each rule for inference is given a certainty for being correct.

```
    IF( A∧ B) ∨ (¬C)) THEN D
```

is transformed to

```
    CF(D) = max{min{CF(A),CF(B)},1-CF(C)}
```

with A, B, C and D names of nodes in the semantic network and CF certainty factor.

Those rules are used on level 6, 7 in the knowledge base. For example, consider the node 'KONGESTIVE CARDIOMYOPATHY' on level 7. A kongestive cardiomyopathy (KKMY) is characterized by an akinetic (AKIN) inferioapical (IA), posterolateral (PL), basal (B) and septal (S) segment combined with a nearly motionless (MLESS) as well as an enlarged (ENLAR) left ventricle (LV). Therefore the CF of diagnosing a kongestive cardiomyopathy is calculated by

```
    CF(KKMY)=min{CF(LV_ENLAR), CF(LV_MLESS),
                CF(IA_AKIN), CF(PL_AKIN), CF(B_AKIN), CF(S_AKIN)}
```

There are other rules, which are more complicated. All rules, however are based on the same mechanism as shown above.

Since those rules are embedded in the semantic net as attributes, as all procedural knowledge there is no need for an interpreter. The procedures attached to attributes are activated during an analysis by the module CONTROL.

Conclusion: In our approach there are combined some of the knowledge representation techniques mentioned in the last section. According to the information to be represented it is tried to find an optimal representation scheme. For integration of declarative knowledge describing relations, and procedural knowledge describing actions on those relations the chosen semantic network approach has given favourable results.

Module CONTROL

The knowledge base described above is used to guide the analysis of an input image sequence. The user starts analysis by labelling a goal node in the net which is interpreted as a request to create an instance of this node.

For each concept which is instantiated during the analysis, a certainty is calculated for the inference from input data. This assessment is done by means of fuzzy set logic as described in the paragraph above.

The strategy is determined by the assumption that the attributes of a node in the semantic network are only dependent on attributes of the immediate parts. Therefore, a node may be instantiated only if all of his parts are instantiated before. This yields for a top-down expansion of the network (depth-first search) along the 'part-of'-relation. So, nodes on level 0 of the net are instantiated first. Now the system traces back the specializations. This search path is fixed in the so-called EXPANDED-MODEL, which is some kind of notebook showing the amount of knowledge used already, and, therefore, indicating where additional knowledge is required.

Since there is the possibility for alternative decisions on various levels of the net by the procedural knowledge there will be competing instances to one node. For combinatorical reasons it is impossible to follow every alternative through all stages of the net. Therefore the control algorithm selects at each step only the best scoring instance for further computations. The protocol of each state of analysis (the knowledge expanded together with all instances created up to this moment) is fixed in the SEARCH TREE. Competing instances are kept in different nodes of the tree. Now, the algorithm chooses that leaf in the tree which contains the optimal search state given by the calculated certainty factor.

A detailed description of this algorithm is contained in [BUNb84]. In principle, the strategy is a modification of the heuristic graphsearch procedure: A^*-algorithm.

Module INSTANCES

The structure of the module INSTANCES is identical to that of the knowledge base with the exception that attributes contain the values computed from input data by activating the corresponding procedure.

Module OUTPUT DESCRIPTIONS

Since all results obtained during an analysis are contained in the module INSTANCES, it is possible to trace the path of decisions. A comfortable interface is implemented allowing for displaying all of these values to the user. Using the EXPANDED MODEL and SEARCH TREE created during the analysis the user may reconstruct the history of the analysis. Up to now no automatic justifications and explanations are possible. Consequently, the user has to know about the underlying structure of the system.

5. RESULTS AND FUTURE ASPECTS

The system described above has been completely implemented in RATFOR on a PDP11/34 residing on about 3M byte storage. Due to storage limitations, the whole system has been split into several processes communicating among each other by taking advantage of the RSX11-M operating system.

Two series of tests have been performed for validating the diagnostic interpretations obtained. Since all knowledge based analysis of the system is based on contours, a validation of the left ventricular contours is necessary and important. One of the standard techniques for image sequences, showing the human heart, is based on the medical parameter 'ejection fraction' EF [SIL 80]. For all image sequences in our database, that is 420 images, the contours have been determined manually by a physician, who - based on these ones - calculated the EF. The method for automatic left ventricular contour detection described in section 4 has been applied. Subjective visual inspection yielded 93% of the contours as found correctly. Calculating the EF, a correlation of r=0.93 was shown between contours outlined manually and automatically. That proved satisfactory.

A further test compared diagnoses infered by a phsician and those by the system. Fig. 4.9 shows some of the results obtained. Empty slots in the table are to be interpreted as a statement like "no decision" (certainty factor 0.0 for all hypotheses), or "no definite decision" (certainty factor almost identical for all hypotheses). Effects of abnormal behaviour on neighbouring segments can be observed comparing diagnoses. For instance, the akinesis of the inferioapical segment effects the neighbouring basal segment in image sequence S18.

For each sequence all medical interpretations contained in our knowledge base had been checked as hypotheses. The certainty factors for those

image sequence	physician	diagnoses system				
		LV	IA	PL	B	S
JAN04	ak(PL,S)	we 0.33		ak 0.5	nm 0.63 hy 0.4	hy 0.75
S07	hy	we 0.16	hy 0.32 ak 0.24	hy 0.14 ak 0.62		hy 0.23
S13	ak(IA,PL)		ak 1.0	ak 1.0		
S15	hy	nm 0.6 we 0.33	nm 0.09 hy 0.5	hy 0.75	nm 0.5	nm 0.18 hy 0.54
S17	we	we 0.37	hy 0.13	hy 0.13		ak 1.0
S18	ak(IA)		ak 0.73		ak 0.63	
S19	nm		nm 0.18	nm 0.33		nm 0.33
S21	we, ph	hy 0.49 ak 0.33 ph 0.25 we 0.33	hy 0.49 ak 0.49	nm 0.76 ph 0.25	nm 0.67	
S22	nm	nm 0.12				hy 0.01
S24	we, ph ph(IA)	we 0.33 mo 0.16	hy 0.33 ak 0.6 ph 0.3	nm 0.48 hy 0.5	nm 0.33 hy 0.25	hy 0.07 ak 0.4
J12	we(IA)		hy 0.48 ak 0.08			
J13	ak(IA)		hy 0.12 ak 0.45			
J14	ak(IA)		hy 0.1 ak 0.48			
S06	nm	nm 0.3	nm 0.7 hy 0.32			
S08	nm	nm 0.66 we 0.3				
S01	hy(IA,S)		hy 0.16			hy 0.17
S26	hy(S)					nm 0.11 hy 0.23

IA: inferioapical; PL: posterolateral; B: basal; S: septal LV: left ventricle

hy: hypokinesis; ak: akinesis; nm: normal behaviour; we: weak motion; ph: contrary phases

Fig. 4.9 Overview about results showing examples

instances having nodes in the semantic net corresponding to the diagnoses given by the physician had been significantly above the certainty of instances of other nodes. Three of 20 test sequences had been rejected as a whole. Besides numerical results the system produces a short report in medical terms describing the motional behaviour of the left ventricle, his shape and documenting some medical parameters like the EF, the length of some motion phases, and so on. The experiments performed so far have been very encouraging.

Besides the advantages of favourable cost and speed factors, the aim to free a physician from routine cases by installing knowledge based systems might be reached in near future. The results produced by such a system can be repeated whenever the neccessity occurs without affording a new examination of the patient. The knowledge contained in the knowledge base can be easily expanded and modified. The results of new research might be integrated in the course of time.

Having such a system there is the need for a comfortable user interface. If there occur results which are somewhat unexpected, a transparent system is neccessary for explaining und justifying the decisions made. Most of the existing systems do not offer such a possibility or the problem is not solved in a satisfying way because of narrow restrictions.

The years of research to come will bring interesting results.

Acknowledgements

The work presented in section 4 was performed in cooperation with the Institute of Nuclear Medicine of the University of Erlangen, and supported by the Deutsche Forschungsgemeinschaft DFG.

329

6. LITERATURE

[BAL 82] Ballard, D.; Brown, C.: Computer Vision, Prentice Hall, New York, 1982

[BAR 81] Barr, A.; Feigenbaum, F. (eds): The Handbook of Artificial Intelligence, Department of Computer Science, Stanford University, 1981

[BRA 83] Brachman, R. J.: On the Epistemological Status of Semantic Networks, in: Proc. IJCAI, 1983, Tutorial on Artificial Intelligence

[BRO 78] Brower, R.W.; Meester, G. T.: Spatial Resolution and Correlation between Segments in Regional Wall Motion Studies of the Left Ventricle, Computers in Cardiology, 1978, pp. 69-75

[BUC 84] Buchanan, B. G.; Shortliffe, E.: Rule-Based Expert Systems: The MYCIN Experiments of the Heuristic Programming Project, Addison-Wesley, 1984

[BUN 82] Bunke, H. et al: Smoothing, Thresholding and Contour Extraction in Images from Gated Blood Pool Studies, Proc. 1st IEEE Comp. Soc. Int. Symposium on Medical Imaging and Image Interpretation, Berlin, 1982, pp. 146-151

[BUNa84] Bunke, H.; Grebner, K.; Sagerer, G.: Syntactic analysis of noisy input strings with an application to the analysis of heart-volume curves, Proc. 7th ICPR, Montreal, 1984

[BUNb84] Bunke, H.; Sagerer, G.: Use and Representation of Knowledge in Image Understanding Based on Semantic Networks, Proc. 7th ICPR, Montreal, 1984

[CHA 79] Chandrasekaran, B.; Gomez, F.; Mittal, S.; Smith, J.: An approach to medical diagnosis based on conceptual structures, in: Proc. 6th IJCAI, Tokio, 1979, pp. 134-142

[DAT 81] Date, C. J.: An Introduction to Database Systems, Addison-Wesley, 1981

[DAV 80] Davis, R.; Lenat, D.: Knowledge-based systems in artificial intelligence, McGraw-Hill, New York, 1980

[DEL 79] Deliyanni, A.; Kowalski, R.: Logic and Semantic Networks, in: Communications of the ACM, march 1979, pp.184-192

[DUD 80] Duda, R. O.; Gaschnig, J. G.: Knowledge-based expert systems coming of age, in: BYTE 6, no. 9, 1981, pp. 238-281

[DUN 79] Dunn, R. A. (ed): The third annual symposium on computer application in medical care, Washington, october 14-17, 1979

[ENC 62] Encyclopaedia Britannica, Chicago, 1962

[ENG 79] Engelmore, R.; Terry, A.: Structure and Function of the CRYSALIS system, in: Proc. 6th IJCAI, Tokio, 1979, pp. 250-256

[ERM 80] Erman, L; Hayes-Roth, F.; Lesser, V.; Reddy, D.: The HEARSAY-II speech-understanding system: Integrating knowledge to resolve uncertainty, in: Computing Surveys 12, no. 2, 1980, pp. 213-253

[ERM 83] Erman, L.: Expert Systems Tutorial, in: Proc. IJCAI, 1983

[FEI 61] Feigenbaum, E.: The simulation of verbal learning behaviour, in: Proc. of the Western Joint Computer Confewrence, 19, 1961, pp. 121-132

[FIN 79] Findler, N. V. (ed): Associative Networks: Representation and Use of Knowledge by Computer, Academic Press, 1979

[GAS 80] Gaschnig, J.: An application of the PROSPECTOR system to the DOE's national uranium resource evaluation, in: AAAI 1, 1980, pp. 295-297

[GEO 82] Georgeff, M. P.: Procedural Control in Production Systems, in: Artificial Intelligence, vol. 18, 1982, pp. 175-201

[GER 81] Gerbramd, I. J.; Hoek, C.: Automated Left Ventricular Boundary Extraction from Technetium-99m Gated Blood Pool Scintigramm with Fixed or Moving Region of Interest, Proc. 2nd Int. Conf. on Visual Psychophysics and Medical Imaging, 1981, pp. 155-159

[GOL 77] Goldstein, I.; Roberts, R.: NUDGE: A Knowledge-Based Scheduling Program, in: Proc. IJCAI, 1977, Cambridge, Massachussetts

[GOR 81] Gorris, M. L.; McKluop, J. H.; Briandet, P. A.: A Fully Automated Determination of the Left Ventricular Region of Interest in Nuclear Angiocardiography, Cardiovasculor and Interventional Radiology 4, 1981, pp. 117-123

[GRE 79] Green, C.; Gabriel, R.; Kant, E.; Kedzierski, B.; McCune, B.; Philips, J.; Tappel, S.; Westfold, S. J.: Results in knowledge based program synthesis, in: Proc. 6th IJCAI, Tokio, 1979, pp. 342-344

[HAN 78] Hanson, A. R.; Riseman, E. M. (eds): Computer Vision Systems, Academic Press, New York, 1978

[HAR 68] Hartman, W.; Matthes, H.; Proeme, A.: Infpormation Systems Handbook, Philips Electrologica, Apeldoorn, 1968

[HAW 81] Hawman, E. G.: Digital Boundary Detection Techniques for the Analysis of Gated Cardiol Scintigrams, Optical Engineering 20, 1981, pp. 719-725

[HAY 83] Hayes-Roth, F.; Waterman, D. A.; Lenat, D. (eds): Building Expert Systems, Addison-Wesley, 1983

[HAY 84] Hayes-Roth, F.: The Knowledge-Based Expert System: A Tutorial, in: COMPUTER, Sep. 1984

[HEW 72] Hewitt, C.: Description and theoretical analysis (using schemata) of PLANNER, Rep. no. TR-258, AI Laboratory, MIT, 1972

[JOH 79] Johnson-Laird, P. N.; Wason, P. C. (eds): Thinking - Readings in Cognitive Science, Cambridge University Press, 1979

[KON 79] Konolige, K.: An inference net compiler for the PROSPECTOR rule-based consultation system, in: Proc. 6th IJCAI, Tokio, 1979

[KUN 78] Kunz, J. C.; Follat, R. J.; McCluny, D. H.; Osborn, J. J.; Votteri, R. A.; Nii, H. P.; Aikins, J. S.; Fagan, L. M.; Feigenbaum, E. A.: A physiological rule-based system for interpreting pulmonary function test results, Rept. HPP-78-19, Computer Science Department, Stanford University, 1978

[LIN 80] Lindsay, R. K.; Buchanan, B. G.; Feigenbaum, E. A.; Lederberg, J.: Applications of artificial intelligence for organic chemistry: the DENDRAL project, McGraw-Hill, New York, 1980

[LEV 78] Levine, M. D.: A Knowledge-based Computer Vision System, in: [HAN 78]

[LOC 78] Lockemann, P. C.; Mayr, H. C.: Rechnergestuetzte Informationssysteme, Springer-Verlag, 1978

[MAR 71] Martin, W. A.; Fateman, R. J.: The MACSYMA system, in: Proc. 2nd Symposium on Symbolic and Algebraic Manipulation, Los Angeles, 1971, pp. 59-75

[MCD 80] McDermott, J.: R1: An expert in the computer systems domain, in: AAAI 1, 1980, pp. 269-271

[MCG 82] McGraw-Hill Encyclopedia of Science & Technology, New York, 1982

[MEL 81] van Melle, W.; Shortliffe, E. H.; Buchanan, B. G.: EMYCIN: A domain independant system that aids in constructing knowledge-based consultation programs, in: Machine Intelligence, Infotech State of the Art Report 9, no. 3, 1981

[MIL 82] Miller, R. A.; Pople, H. E.; Myers, J. D.: INTERNIST-I, an experimental computer-based diagnostic consultant for general internal medicine, in: New England Journal of Medicine, August 19, 1982, pp. 468-476

[MIN 75] Minsky, M.: A Framework for Representing Knowledge, in: Winston, P. (ed): The Psychology of Computer Vision, McGraw-Hill, 1975

[MIT 79] Mittal, S.; Chandrasekaran, B.; Smith, J.: Overview of MDX - A system for medical diagnosis, in: [DUN 79]

[MIT 84] Mittal, S.; Chandrasekaran, B.; Sticklen, J.: Patrec: A knowledge-directed database for a diagnostic expert system, in: COMPUTER, Sep. 1984

[MYL 83] Mylopoulos, J.; Levesque, H.: An overview of knowledge representation, in: Brodie, M; Mylopoulos, J.; Schmidt, J. V.

(eds): On conceptual modelling: perspectives from artificial intelligence, databases, and programming languages, Springer-Verlag, New York, 1983

[NIE 81] Niemann, H.: Pattern Analysis, Springer-Verlag, 1981

[NIL 80] Nilsson, N. J.: Principles of Artificial Intelligence, Palo Alto, California, 1980

[POP 81] Pople, H. E.: Heuristic methods for imposing structure on ill-structured problems: The structuring of medical diagnostics, in: Szolovitz, P. (ed): Artificial intelligence in medicine, American Association for the Advancement of Science, 1981, pp. 119-185

[RAB 78] Rabiner, L. R.; Rosenberg, A. E. ; Levinson, S. E.: Considerations in dynamic time warping algorithms for discrete word recognition, in: IEEE Trans. ASSP-26, 1978, pp. 575-586

[SAG 85] Sagerer, G.: Darstellung und Nutzung von Expertenwissen fuer ein Bildanalysesystem, Dissertation, in preparation to appear 1985

[SHO 76] Shortliffe, E.: Computer-based Medical Consultations: MYCIN, American Elsevier, New York, 1976

[SIL 80] Silber, S. et al: Quantitative Beurteilung der linksventrikularen Funktion mit der Radionuklid-Ventrikulographie, Herz 5, 1980, pp. 146-158

[SIM 83] Simon, H. A.: Search and Reasoning in Problem Solving, in: Artificial Intelligence, vol. 21, 1983, pp. 7-29

[TSO 80] Tsotsos, J. K.: A Framework for Visual Motion Understanding, Technical Report, CSRG-114, June 1980

[WEI 81] Weiss, S. M.; Kulikowski, C. A.: Expert consultation systems: The EXPERT and CASNET projects, in: Machine Intelligence, Infotech State of the Art Report 9, no. 3, 1981

[WEY 80] Weyhrauch, R. W.: Prolegomena to a Theory of Mechanized Formal Reasoning, in: Artificial Intelligence 13, 1980, pp. 133-170

[WIL 80] Wilkins, D. E.: Using patterns and plans in chess, in: Artificial Intelligence 14, 1980, pp. 165-203

[WIN 77] Winston, P. H.: Artificial Intelligence, Reading Massachussetts, 1977

[WIN 83] Winograd, T.: Frame Representation and the Declarative/Procedural Controversy, in: Proc. IJCAI, 1983, Tutorial on Artificial Intelligence

[ZAH 65] Zadeh, L. A.: Fuzzy Sets, in: Information and Control 8, 1965, pp. 338-353

[ZAH 79] Zadeh, L. A.: A theory of approximate reasoning, in: Hayes, J.; Mitchie, D.; Mikulich, L. I. (eds): Machine Intelligence, vol 9, New York, 1979

SOFTWARE TOOLS FOR THE DEVELOPMENT OF
PICTORIAL INFORMATION SYSTEMS IN MEDICINE
-    THE ISQL EXPERIENCE   -

K. Aßmann, R. Venema, K.H. Höhne

Institut für Mathematik und Datenverarbeitung
in der Medizin (IMDM)
Universitätskrankenhaus Eppendorf
Martinistr. 52
2000 Hamburg 20
Federal Republic of Germany

Abstract

    While hardware components for pictorial information
systems gradually become available, software has not been
developed to a great extend. This paper presents a software
approach based on upgrading a conventional data base management
system by features for image management and the corresponding
language tool (ISQL). After a brief overview of existing image
data base software, requirements are defined. The concepts for
meeting these requirements are derived and their realization in
a prototype system is described.   The achieved properties
include integration into a department and hospital information
system, management of images, support of human-computer
interaction and adaptability to special user environments. It
is argued that the described tool ISQL with its properties of
high portability (implementation in PASCAL, use of SQL) and the
high level language approach (ISQL) can decisively facilitate
software development for pictorial information systems.

1. Objectives and Problems

    From the electronic storage and management of radiographic
images many advantages over the management with conventional
archives are expected. Some advantages of such systems, called
Picture Archiving and Communication Systems (PACS) are:

-   safe storage and easy retrieval of images in clinical
    work even in very large archives

-   arbitrary evaluation of the archive content

-   support of image analysis by image processing methods
    which are only possible with computers

NATO ASI Series, Vol. F19
Pictorial Information Systems in Medicine
Edited by K. H. Höhne
© Springer-Verlag Berlin Heidelberg 1986

- comparison of images from different modalities and simultaneous display

- linkage of textual and pictorial information in a uniform manner

- simultaneous access to images by several users

While the necessary hardware gradually becomes available, there are no general software solutions for such systems.

The problem of having a large amount of images can be solved quite well by means of optical storage devices available today. The display of images on a medical workstation is a problem not yet solved for all possible cases but there are some experimental prototypes of a workstation, e.g. described in (17). There are many projects which are concerned with the research and realization of the network connecting the components of a PACS (modalities, storage and workstations). Glass fiber technology will contribute substantially to solving the problem of the very high data rates for transmitting images. Publications on PACS are mainly concerned with hardware solutions. This situation is reflected by the topics of the two most important meetings (21,22) on the subject. The present situation with PACS resembles the situation of data base technology 15 years ago when the development of large mass storage devices was thought of as the solution for data management problems. Only the data base research, however, has provided general tools for the construction of operational information systems.

In the following a project is presented which studies a possible solution for the software problem. The concepts developed for the solution are mostly implemented in an experimental system. General ideas are discussed, a solution is outlined and practical results are presented with some examples.

There are typical differences between a conventional data base and an image data base (Fig. 1). In the design process of a conventional data base the part of the real world to be modelled is subject to perception and abstraction and is finally transformed into data base objects. In this process the level of abstraction is very high. For example a patient is represented by his name, sex, date of birth etc.

On the contrary, with an image data base the images themselves are the objects in the data base, their level of abstraction is low. So the main difference between a conventional data base and an image data base is due to the fact that the abstraction takes place on different levels. This fundamental difference results in two basic properties of an image data base management system:

- The target information, e.g. the diagnosis, is not obtained until the user has retrieved, displayed and interpreted an image by using additional knowledge (e.g. the experience of the radiologist). This requires means of <u>interaction</u>. After the interpretation the abstractions are added to the images by the user in an interactive manner. Therefore, the images are treated as unstructured objects the structures of which cannot be described formally. In the future a knowledge based system can be used for image interpretation. Related problems, however, are not even solved rudimentarily for the broad field of radiologic image interpretation.

- Due to the low degree of abstraction a variety of modes of storage, coding and display of images is possible. Nevertheless, an object oriented access is required, i.e. the user should be able to refer to the images irrespective of their physical properties.

The huge amount of data to be managed is not considered a software problem.

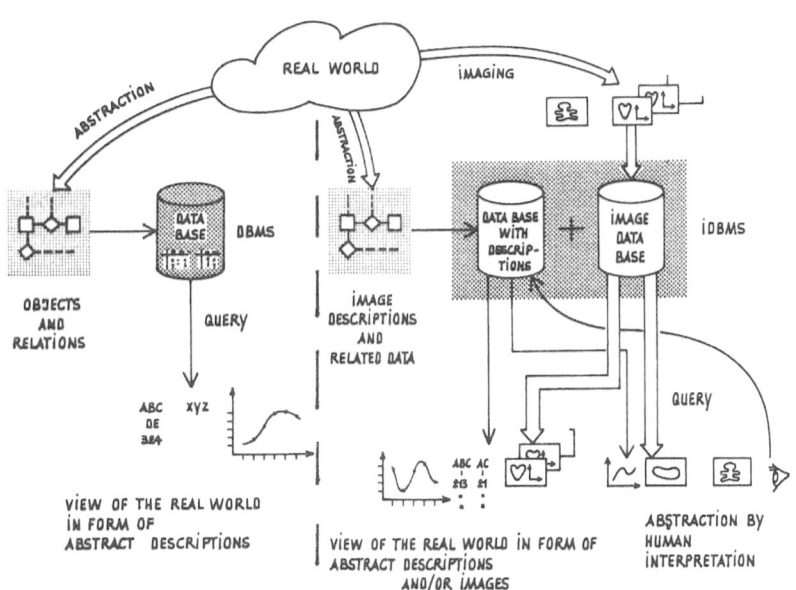

Fig.1   The function of a conventional data base system vs. the function of an image data base system

The tool for constructing conventional information systems is a data base management system (DBMS) which has various possibilities for application programming. In addition, there are utilities such as application generators, report generators and user interfaces. There are some corresponding approaches for pictorial information systems. The languages used in these approaches and the system background are briefly summarized below (Table 1). Details of the languages can be found in (6,8,9).

All DBMSs underlying the above mentioned languages are based on the relational data model. All applications originate from geography or related fields and have been developed on this background. All languages assume that the images are structured objects, i.e. there exists a structure which can be described formally. The languages are planned rather for image processing by a set of functions tailored to the specific needs of a particular application than for easy handling of images. Frequently, the implementation is an extension of APL. The only general approach seen is IDMS for which unfortunately no implementation has been published.

For the problem in question namely the construction of pictorial information systems in medicine the image data base languages mentioned are scarcely suitable. For the reasons mentioned above they are very specific and do not pay regard to special image types such as image sequences which are quite common in medicine (angiography). There are no concepts seen for easy handling of images and image display. At best the user interface is on a command level. The systems mentioned are described as image data bases but genuine data base operations are only designed in IDMS. It is worth mentioning that some language approaches have the idea of a set of APL procedures which can be augmented by user written procedures which allows an open and flexible system. A more recent approach to image data bases for satellite image evaluation is described as PICDMS (10).

There are three hardware oriented approaches for an image data base system: PICCOLO (26), the system designed by Yamamura (27), which shall also be used for medical applications and the system designed by Feng (11). The references describe hardware relevant aspects only so that one cannot judge on their suitability as a tool for the construction of pictorial information systems. There is an approach published for the management of higher descriptions for the analysis of image sequences which is based on the relational data model (5).

2. The ISQL-Approach

Having in mind the ideas described above the following approach for the development of tools for the construction of pictorial information systems in medicine has been made:

Using a scenario (4) of a department of radiology
PACS-functions are determined, then concepts for their
realization are proposed and tools for their implementation are
developed. Afterwards, applications are using these tools. At
that time results can be judged and problems still to be solved
can be recognized.

For the investigation of this approach a relational
conventional DBMS with the data base language SQL has been
used. It has been extended by data base functions for
structures and operations with images. The system has been
tested with various applications. A datatype ´Image´ has been
realized with an extended data dictionary (DD). The operations
are implemented by the extension of the data base language SQL
to the image data base language ISQL (Image-SQL). A
description of the language and first experiences can be found
in (2,3). The syntax of the language is described in the
appendix. Fig.2 shows the overall design of the system.

The prototype of the image data base system has been
implemented with a VAX 11/780 under VMS, the DBMS ORACLE and
its Host-Language-Interface (HLI) and PASCAL. At the moment
some 5000 physical and logical images are stored together with
their descriptions. The images are mostly computer tomograms.
About 1600 images are stored on on-line devices. As image
processing system a COMTAL VISION ONE/20 with two independent
work stations is used. Fig.3 shows the laboratory work station
for research on image data base management systems and
human-computer interaction.

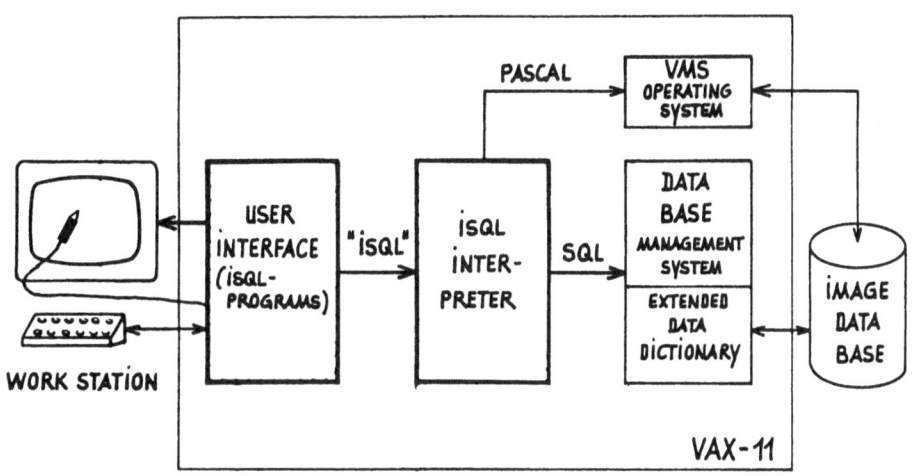

Fig.2   Overview of the software concept of the ISQL project

Feature/language	ADM	AQL	ARES	GADS	GRAIN	IDAMS	IDMS	IQ	QPE
Type of language	exten- sion of SEQUEL2 + com- mands	exten- sion of APL	N/A	retrie- val lan- guage	com- mands		exten- sion of SEQUEL2	very simple com- mands	2-dimen- sional
storage of images	?	?	?	?	?	?	matrix	files, tapes, 64 x 64 blocks	UNIX- files
Types of images	gray tone, sets(?) binary images	?	?	poly- gons	?	vectors	all conven- tional + PICTURE	satel- lite	
DB-operations	SEQUEL2 (?)	AQL	retrie- val only	EXTRACT only	GET only	retrie- val only	all, UPDATE is INSERT + DELETE	none	all
Implementation	partial- ly imple- mented	APL	asso- ciative proces- sor			APL	N/A	PDP 11/45 RATFOR FORTRAN	PDP 11/45 UNIX, C

Table 1   Approaches to picture query languages

Feature/language	ADM	AQL	ARES	GADS	GRAIN	IDAMS	IDMS	IQ	QPE
Display of images	?	?	?	?	SHOW command	graphics	concept of DEVICE	DISPLAY for 1 image	?
DBMS	SYSTEM/R	AQL (?)	N/A	GADS	GRAIN	?	?	IMDS	IMAID
Remarks	partially implemented images are aggregates	no display		decision support system for geographically related data		open system, collection of APL procedures	general approach some ideas on image coding	implemented in 1977	special operators for geographical operations
Reference	23	1	13	16	9	20	24	15	9

Legend :
?   cannot be clarified by means of the reference
(?)   uncertain
N/A   not applicable

Table 1 Approaches to picture query languages (contd.)

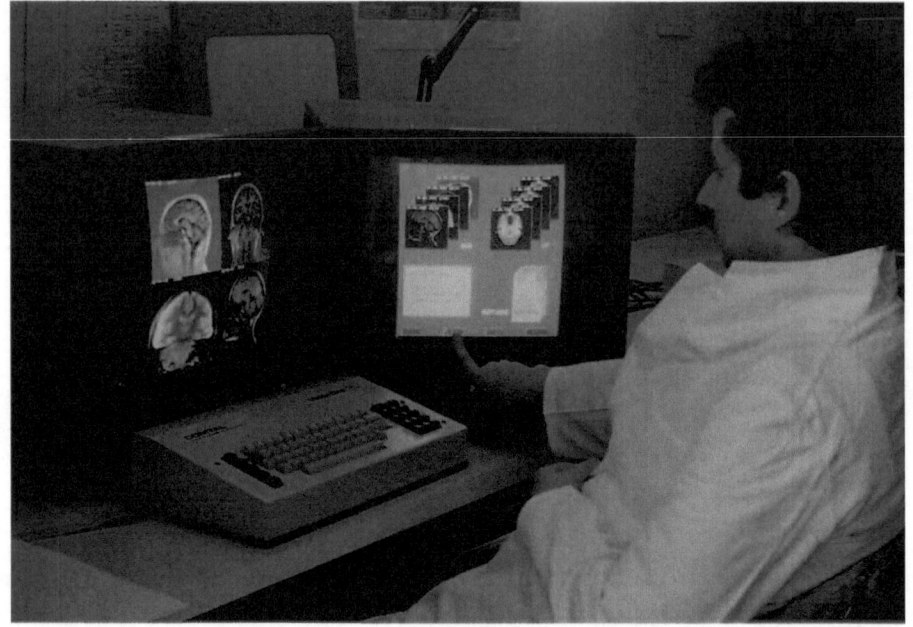

Fig.3   Laboratory work station

3. PACS-Functions, concepts and realization with the tool ISQL

In the following those functions of a PACS are described which can be realized by the concepts and tools of the described approach. For an overview the table below (Table 2) lists functions, concepts, realization in the prototype and remaining problems. The table also describes the present state of the prototype system.

3.1 Integration into the departmental or hospital organization

Certainly, an image data base in medicine makes only sense if it is integrated into a hospital organization. For this purpose a conventional DBMS is suitable. Using a DBMS, all functions which are not specific to images can be carried out. Already existing tools can be used such as a data base language, a transaction concept, data security and privacy and utilities. Therefore, the prototype has been realized with the DBMS ORACLE. Standard applications can be programmed very easily with its standard tools. The examples depicted in Fig.4 and 5 show a patient data entry screen and a histogram which have been generated by means of the application generator (IAF) and the user friendly interface (UFI), respectively, of ORACLE.

Required function	Concept	Realization	Example	Problem
Integration of the image data base into a departmental information system (conventioanl storage and retrieval)	use of a conventional DBMS with - data base language - transaction concept - data security and privacy - utilities	DBMS ORACLE	data entry screen, histogram Fig.4,5	performance
in addition: display and storage of images	extension of a conventional DBMS	extension of the data dictionary, operations: - SELECT IMAGE - DISPLAY - UPDATE IMAGE - INSERT IMAGE - DELETE IMAGE	display of MR-images, CT-image directory Fig.6,7	
userfriendly interaction	maintenance of a pictorial context	introduction of current objects	interaction via icons to produce a screen of MR-images and display of the image with full resolution Fig.8,9,10	
	various levels of communication	application generator SQL, ISQL, images used as icons		
easy adaption to a specific surrounding (hospital, radiologist)	generator for user environment	set of parameters, defaults constructs for user interfaces in language		

Table 2 Functions, concepts and realization with the tool ISQL

Required function	Concept	Realization	Example	Problem
optical display of non-image data	extension of the data dictionary by attributes	extension of ISQL for the display of optical attributes	not yet implemented	
image processing operations	implementation of image processing algorithms	extension of ISQL for image processing operations	not yet implemented	
storage of derived data	extension of the DBMS	- UPDATE IMAGE	under development	consistency
distribution of work-stations	distributed DBMS	no decision yet		
	multi user operations	ORACLE and ISQL are multi user systems		
reproduction of documents	interactive editing of images and textual data	document editor	not yet implemented	
retrieval by templates and feature extraction	complex data structures, pattern recognition	knowledge base is integrated into the data base	not to be started in the near future	
computer assisted diagnosis	knowledge based system	knowledge base is managed by ORACLE + AI-concepts	no decision yet	

Table 2   Functions, concepts and realization with the tool ISQL (contd.)

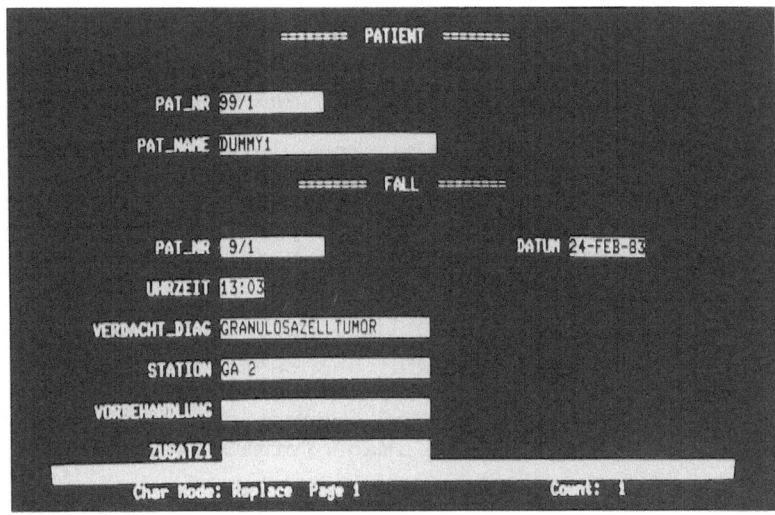

Fig.4   Patient data entry screen as designed with
         the standard tools of the DBMS

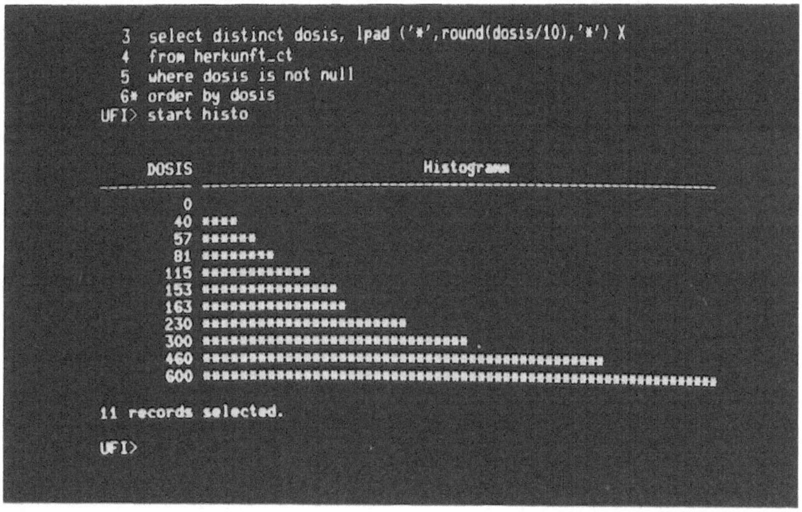

Fig.5   Query for a histogram of x-ray dosis in CT together
         with the result as an example of the use of
         standard DBMS tools

The underlying relational model of data has many advantages,
such as flexibility, well-understood mathematical background
and simplicity, which makes the software   part   of   the
construction of a departmental information system, with all due
respect, a relatively easy task.   There   are, however,
performance problems even with the conventional use of a
relational DBMS.

3.2   Management of Images

For the construction of a pictorial information system the conventional DBMS has to be extended for the additional functions of storage, management, retrieval and display of images.   Because of the low level of abstraction image presentation is one of the most prominent aspects.   To consider all possiblities sufficiently the display has to be independent of specific hardware devices and pecularities of images such as physical storage, coding and pixel depth.   To meet these requirements all descriptions are stored in relations with the management facilities of ORACLE.   The descriptions represent an extended data dictionary.   With the help of these relations a set of language constructs has been realized which represents the ISQL operations.   With these operations all image management can be performed independent of the pecularities of image display devices and image formats.   These operations are:

SELECT IMAGE ...   DISPLAY AS ...   WITH ...
This operation is used to retrieve images from the image data base and to display them in the desired appearance.   The images may be ordered or scaled down or displayed in any desired fashion as there is a display operation for each type of image (described in the extended DD).   Not only images may be retrieved but also text will be displayed with the images in a way the user can choose.   He simply indicates the text or the names of the columns the text is stored in.

Some images in the image data base have properties, e.g.   "no diagnosis", which should be understood at once when the image is displayed.   These properties can be assigned to the images as attributes and get displayed with the images as optical attributes.   For this purpose the data dictionary has to be extended for these attributes and ISQL must provide a display operation for this kind of display. This concept has been realized in the spatial data management project for alphanumerical data (12).

INSERT IMAGE
This operation is used to insert an image into the image data base.   This can be performed either by including the descriptions column by column or the descriptions may be read from a file with a set of descriptions.

DELETE IMAGE
This operation deletes an image from the image data base by deleting its descriptions.   An image may be deleted only by authorized personnel.

UPDATE IMAGE
At the moment an update on an image is considered to be the insertion of images which have been derived from the original image or textual data which is related to the image to be updated.   The original

image is the image which has been inserted in an earlier state of the image data base.

As an example of how to perform data base operations with ISQL fig.6 shows the retrieval and display of a set of MR-Images for a certain patient. This figure shows how images and text such as the patient's name and the date when the images were taken can be retrieved and displayed with one command.

Fig.7 shows how a directory of images of a certain day can be generated.

3.3 Human Computer Communication

Leaving traditional archives and working habits of the radiologist behind and going to electronic archives, computer assisted radiology could cause new problems for the user. This is due to the fact that in a fully electronic system more powerful and complex operations with images are possible. This complexity, however, must not be passed on to the user. Therefore, there have been implemented concepts for the easy access to the pictorial information system.

Choice of Communication Level. Experience with earlier projects (7) have shown that there should be a choice for the level of communication with the image data base to reflect the fact that there are different users with different levels of knowledge about how to use a computer. A language like ISQL is a suitable means.

The application programmer interacts with the image data base via SQL for non-image operations and uses ISQL for image operations or mixed operations (image and non-image).

The motivated user (e.g. a specially interested radiologist) may use (after some practice) ISQL as well. For operating in routine, however, ISQL itself is not suited.

For users in routine the application programmer can write programs in ISQL which create an interface to the image data base which is adapted to the needs and requests of the radiologist. For easy communication icons can be produced with ISQL. There is a special display operation (DISPLAY AS SOFTKEY) integrated in ISQL to perform the operation which is connected to the icon. The operation to be performed is a parameter of SOFTKEY. It is a special feature of ISQL that any image in the image data base may function as an icon. Fig.8 gives an example of how to program a user interface with icons by means of ISQL. Fig.9 shows an ISQL program which is activated by the icon. The shown ISQL program retrieves MRI images of a patient.

Context Management. When viewing images the radiologist has already built up a certain environment: the image actually under consideration, the sequence of images on the lightbox or the pile of jackets to be viewed later.

346

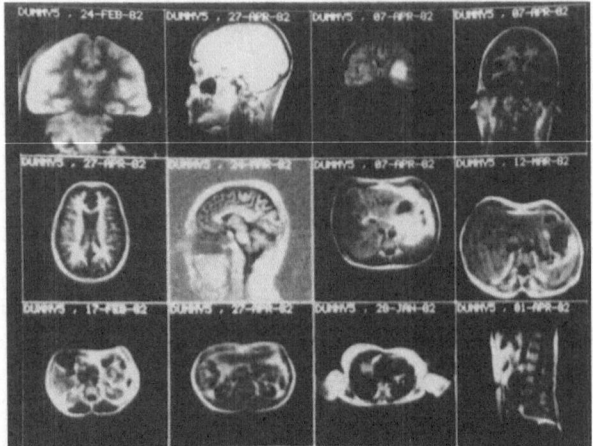

```
SELECT IMAGE pat_name,date
WHERE modality = 'NMR' AND
      pat_name = 'DUMMY5'
   DISPLAY WITH SCALE = 2;
```

Fig.6   Query for all MR images of a patient and the
         corresponding result screen

```
SELECT IMAGE
WHERE date = '07-SEP-82'
ORDER BY pat_name
   DISPLAY WITH SCALE = 4;
```

Fig.7   Request for a directory of all images of a
         specific day with the first 64 resulting images
         on screen. The image surrounded by a frame is the
         'current image'. The text to be displayed with
         the images is defined by default values.

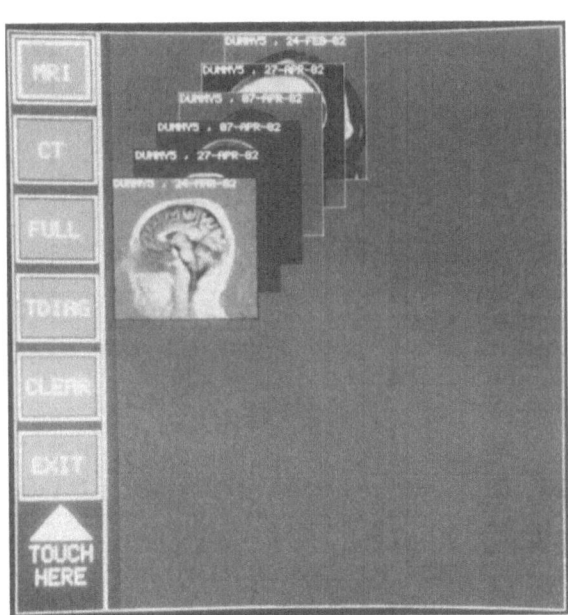

```
          :
          :
SET SCREENLEN 10
          :
SELECT STANDARD_ICON , 'MRI'
DISPLAY AS
    SOFTKEY (START mri_images)
WITH XPOSITION = 0,
YPOSITION = 0
          :
          :
```

Fig.8   Part of an ISQL program to producing the icon 'MRI'
         of the column of icons shown on the screen.
         The STANDARD_ICON is an image file.

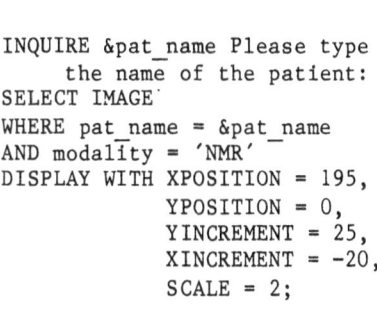

```
INQUIRE &pat_name Please type
      the name of the patient:
SELECT IMAGE
WHERE pat_name = &pat_name
AND modality = 'NMR'
DISPLAY WITH XPOSITION = 195,
             YPOSITION = 0,
             YINCREMENT = 25,
             XINCREMENT = -20,
             SCALE = 2;
```

Fig.9   Program activated by the icon 'MRI' (fig.8). It
         retrieves the MR images of a patient and displays
         them on the screen (right). The user is prompted
         for the name of the patient.

Operations that are trivial in this environment like "what is the tentative finding for the image on the lightbox" are very complex in an electronic system unless appropriate means to access the information about previous manipulations are provided. In the prototype this problems is solved by introducing "current objects". Presently, two are implemented: the current image and a set of current images, the current data base. By default the current image is the image most recently displayed on the screen. One can refer to that image, e.g. in order to get additional information concerning that particular image , by just making reference to the "current image". The image data base system then traces down to the required information with no other explicit description of how to access this information. Similarly, the current data base can be defined as the set of current images which have been retrieved by a previous query. In this way, the above mentioned pile of jackets may be modelled by the current data base. The concept of a current image data base in ISQL implements a local store for the images of a specific user but is restricted to retrieval and display. The current data base is a snapshot of the data base contents at a certain instant. Therefore it may be inconsistent on a global level. The concept of current objects is discussed in detail in (18). Fig.10 demonstrates how powerful the concept of current images may be used. To implement an operation like 'show this image in full resolution' only a one line ISQL-command has to be coded. It is not necessary to reference the image explicitly by some kind of identification.

It is a consequence of the relational approach that from the current image any item related to the image can be retrieved without the need of describing the access path. Thus also textual data such as the tentative diagnosis can be retrieved as shown in fig.11.

3.4   User environments

It is desirable to be able to adapt the pictorial information system to specific pecularities of the environment of an image data base, e.g. the hospital, the department or preferences of a radiologist. This can be accomplished by creating a user environment which may be generated according to the requirements. In ISQL, there are two tools for this purpose:

parameters for static properties such as colour, size and position of the text to be displayed with the images and ISQL procedures generating the user surface. Values for the parameters may be given either by specifying them explicitly with the SET command or they may take over predefined default values. These values are valid until specified otherwise.

More important is the possibility to create special user interfaces via ISQL procedures easily. As shown in figures 8 and 9, these are very simple and can be implemented rapidly. Mnemonic naming of commands and parameters makes ISQL procedures selfexplaining. This does not only facilitate the

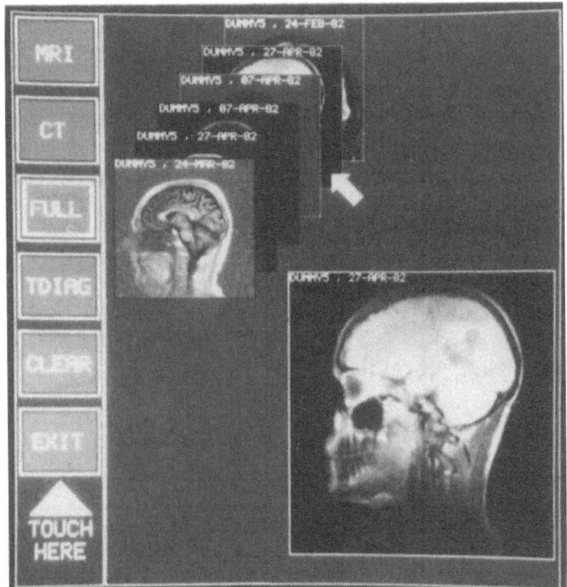

SELECT IMAGE
WHERE image_id = :CURRENT_ID;

Fig.10   Retrieval of a full scale version of the
          current image which has been marked by
          the user with the arrow.   The command shown
          above is performed by activating the icon 'FULL'.

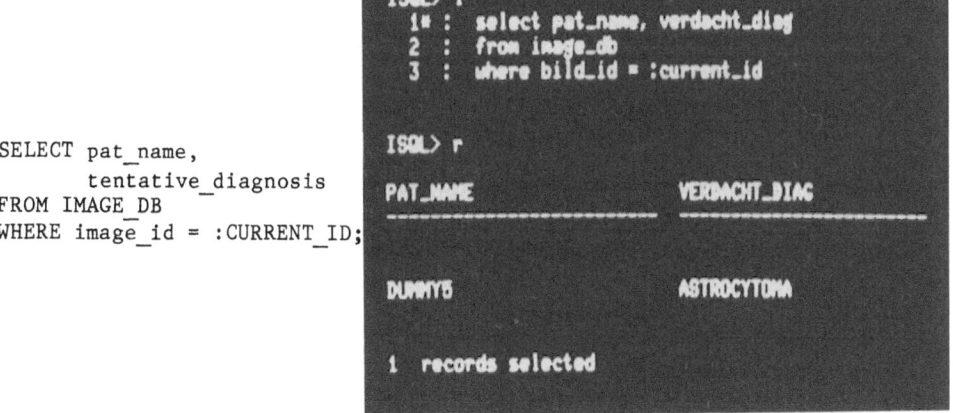

SELECT pat_name,
       tentative_diagnosis
FROM IMAGE_DB
WHERE image_id = :CURRENT_ID;

Fig.11   Example of the retrieval of non-image information
          related to the current image.   Here the
          tentative diagnosis is retrieved without specifying
          any explicit identification.

creation of the user interface but also supports maintenance and documentation. This is a decisive property if e.g. one system has to be adapted to many hospitals with different users.

It is a consequence of the approach that any image in the data base can be used as an icon. In the prototype images from the conventional film reading environment have been used as icons (see fig.12) in order to facilitate the access to the system for the radiologist. At the moment, this feature is being tested for its acceptance. Preliminary experience has shown that it helps the casual user to understand the system functions. It is not likely that it can decisively support routine utilisation. There might, however, be applications we have not thought of yet.

Fig.12   Initial screen of the image data base prototype
          showing real world images as icons
          for functions related to film reading.

3.5   Further developments

Integration of image processing operations. One of the new possibilities in computer assisted radiology is image processing. Image processing capabilities should be an integral part of a pictorial information system. One way of integrating these capabilities is to extend the image data base language by such operations. As a consequence different structures (e.g. pyramides and octrees (19)) according to the appropriate algorithms will have to be provided. For the management of these structures a relational DBMS seems to be suitable. These structures and operations have not been implemented yet in the prototype. Presently, the processing of images is performed separately. Subsequently the appropriate

data is included into the image data base by means of the
UPDATE IMAGE command. For the time being, the update on an
image is not well understood and will be subject to further
discussion and research, e.g. on transaction concepts and
concurrency.

Retrieval by feature extraction and templates. Presently,
image data base retrieval can be done only by making use of
descriptions already existing in the data base. Retrieval by
descriptions not yet existing in the image data base would
require the application of image understanding algorithms.
Such algorithms would be a necessary part of the search
strategies of the image data base management system. For the
near future operational solutions are not expected. Although
the retrieval by templates (e.g. "retrieve all images with
kidneys which look like the one in the example") seems to be an
easier task it still is an unsolved problem.

Assistance for diagnosis. Although an automatic interpretation
of radiological images, if ever possible, is in the far future,
knowledge based systems could be a substantial help.
Therefore, they should be taken into consideration during the
design process of a pictorial information system. In the
prototype the relational DBMS could be used as a tool for the
management of knowledge.

Support of a distributed system. Image workstations will be
spread over a radiological department or even a hospital. This
requires management of images at many sites in parallel.
Basically, the same problems occur as with conventional
distributed DBMS. There are no conventional distributed DBMSs
seen which are more than a prototype implementation. There are
conventional DBMS, however, where several local data bases may
communicate with a central data base. The problem of
distribution of data among different computer sites is even
more complicated with a distributed image data base as the
objects of a transaction are images. The special case of only
reading images by several sites seems to be a minor problem as
far as a subsequent update is not to be performed by more than
one site. In a PACS, images are distributed over many sites,
e.g. there is a local store in the modalities. The retrieval
of any image from an arbitrary site requires a global
management of a distributed image data base, which will not be
available in the near future. A multi-user system for a
central image data base is no major problem. An implementation
with ISQL will be available soon.

Document editing. In order to be able to communicate with
institutions outside the hospital and for documentation
purposes it is still necessary to provide means for the
production of documents. A database that integrates images and
text lends itself to the production of documents that combine
images and text. As a software solution one could think of a
document editing facility based on ISQL. No efforts have been
made into this direction yet.

3.6 Software properties of the tool ISQL

In order to keep the tool independent of any pecularities of hardware and software and thus make it a general tool, it should be portable. In ISQL, this is achieved by making use of standardized languages, i.e. the data base language SQL and the programming language PASCAL.

Pictorial information systems should be able to be constructed with little programming effort and uniform tools. A high level approach like ISQL is a suitable method.

A general requirement for pictorial information systems is the independence of any properties of special hardware, e.g. the display system. This independence can be achieved by managing all necessary descriptions with a DBMS.

There must be, of course, rapid access to any image in the image data base. This can be accomplished by making use of a storage hierarchy for the images: according to the response time required to retrieve the images are stored on storage means which reflect these requirements, i.e. images which have to be retrieved within a very short response time are stored on disks and others may be stored on an optical disk etc. No efforts have been made yet into this direction with ISQL.

The software properties of the tool ISQL are summarized in the table below:

Required property	I	Approach	I	Realization
portablity		use of standardized software tools		SQL, PASCAL
little programming effort, uniform tools		high level language approach		ISQL + utilities
hardware independence		descriptions are stored in the data base (DD)		ISQL structures
short response time with retrieval of images		hierarchy of storage devices		not yet implemented

4. Conclusion

In a protype system a software solution for the implementation of pictorial information systems has been described. It is based on the extension of a conventional relational DBMS. We have shown that most of the functions typical for a PACS can be implemented with the tool ISQL. These functions include besides conventional data base operations access to image data independent of the pecularities

of physical storage or display devices, support of
human-computer interaction and easy generation of user surfaces
and environments. It is demonstrated that a high level tool
not only speeds up PACS software development but also helps in
keeping up with the complexity of pictorial information
systems. The production version of ISQL will be portable as it
is implemented with the standardized languages SQL and PASCAL.

Some of the desirable functions still have to be
implemented. So image processing features are not yet included
in ISQL. Also the management of storage hierarchies is a
problem to be worked on. Finally it should be pointed out that
our approach is a good basis for research on advanced topics
such as knowledge based systems for image retrieval and
interpretation.

Acknowledgement

We would like to express our gratitude to F. Böcker,
U. Tiede and M. Riemer (Institut für Mathematik und
Datenverarbeitung in der Medizin) and G. Witte (Department of
Radiology) for many constructive discussions and for their help
with implementing ISQL.

5. References

1. Antonacci, F., Bartolo, L., Orco, P., Spadaveccia, N.:
   'AQL: A Relational Database Management System and its
   Geographical Applications', in (6), pp.569-599

2. Aßmann, K., Höhne, K.-H.: 'An Investigation of
   Structures and Operations for Medical Image Data
   Bases', Proc. 2nd Conf. on Picture Archiving and
   Communication Systems (PACS II), SPIE 418, 1983.

3. Aßmann, K., Venema, R., Riemer, M.,
   Höhne, K.H.: 'The ISQL-Language - a Uniform Tool for
   Managing Images and Non-image Data in an Image Data
   Base Management System', Proc. ISMII '84, IEEE
   International Symposium on Medical Images and Icons,
   Arlington, Virginia, July 24-27, 1984.

4. Aßmann, K.: 'Szenario für eine Medizinische
   Bilddatenbank', Internal report, Institut für
   Mathematik und Datenverarbeitung in der Medizin,
   Universität Hamburg, 1983

5. Benn, W.: 'Entwurf einer relationalen Datenbank zur
   Unterstützung der Analyse von Bildfolgen',
   Mitteilungen Nr. 116, Fachbereich Informatik,
   Universität Hamburg, Dezember 1983

6. Blaser, A.(ed.): 'Data Base Techniques for Pictorial
   Applications', Springer-Verlag, 1979

7. Böhm,M., Obermöller,U., Pfeiffer,G., Höhne,K.H.: 'Image Management in the System CA-1', Proc. 1st Int. Conf. on Picture Archiving and Communication Systems, Newport Beach, Proc. SPIE, 318 (1982), pp. 161-165

8. Bolc, L.(ed.): 'Natural Language Communication With Pictorial Information Systems', Symbolic Computation Series, Springer-Verlag, 1984

9. Chang, N.S., Fu, K.S.(eds.).: 'Pictorial Information Systems', Springer-Verlag, 1980

10. Chok, M., Cardenas, A.F., Klinger, A.: 'Database Structure and Manipulation Capabilities of a Picture Database Management System (PICDMS)', IEEE Transactions on Pattern Analysis and Machine Intelligence, Vol. PAMI-6, No. 4, July 1984

11. Feng,T.: 'A Very Large Data Base Computer', IEEE Workshop on Computer Architecture for Pattern Analysis and Database Management, Hot Springs, Virginia, Nov. 11-13, 1981, pp. 12-24

12. Herot,F.: 'Spatial Management of Data', ACM TODS, Vol.5,No. 4, 1980

13. Ichikawa, T., Kikuno, T., Hirakawa, M.: 'A Query Manipulation System for Image Data Retrieval by ARES', Proc. IEEE Workshop Picture Data Description and Management, Aug. 1980, pp.61-67

14. IEEE Special Issue on 'Pictorial Information Systems', Computer 14, Number 11, November 1981

15. Lien, Y.E., Harris, S.K.: 'Structured Implementation of an Image Query Language' in (9), pp.416-430

16. Mantey, P.E., Carlson, E.D.: 'Integrated Geographic Data Bases: The GADS Experience', in (6), pp.173-198

17. Meyer-Ebrecht, D., Wendler, Th,.: 'Concept of the Diagnostic Image Workstation', Proc. 2nd. Conf. on Picture Archiving and Communication Systems (PACS II), Newport Beach, SPIE 418, pp. 180-188, 1983

18. Pfeiffer, G.: 'Erzeugung interaktiver Bildverarbeitungssysteme im Dialog', Reihe Informatik Fachberichte Nr. 51, Springer Verlag, 1982

19. Rosenfeld, A. (ed.): 'Multiresolution Image Processing and Analysis', Springer, 1984

20. Schmutz, H.: 'The Integrated Data Analysis and Management System for Pictorial Applications', in (6), pp.475-493

21. SPIE Proc. 1st. Conf. on Picture Archiving and Communication Systems (PACS I), SPIE 318, 1982

22. SPIE Proc. 2nd. Conf. on Picture Archiving and Communication Systems (PACS II), SPIE 418, 1983

23. Takao, Y., Itoh, S., Iisak, J.: 'An Image-Oriented Database System', in (6), pp. 527-538

24. Tang,G.Y.: 'A Management System for an Integrated Database of Pictures and Alphanumeric Data', Computer Graphics an Image Processing 16, 1981, pp. 270-286

25. Venema, R,: 'Erweiterung der Datenbanksprache SQL zur Manipulation von Bildern (ISQL) - Entwurf und Implementation', Diplomarbeit, Fachbereich Informatik, Universtät Hamburg, in Vorbereitung

26. Yamaguchi,K., Kunii,T.L.: 'PICCOLO Logic for a Picture Database Computer and its Implementation', IEEE Transaction on Computers, Vol. C-31, No. 10, Oct. 1982, Special Issue on Computer Architecture for Pattern Analysis and Image Database Management, pp. 983-996

27. Yamamura,M., Kamibayashi,N., Ichikawa,T.: 'Organization of an Image Database Manipulation System', IEEE Workshop on Computer Architecture for Pattern Analysis and Database Management, Hot Springs, Virginia, Nov. 11-13, 1981, pp. 236-241

Appendix : Overview of the Syntax of ISQL

SELECT IMAGE [, colnam1, colnam2, ...]

[FROM CURRENT DB]

[WHERE where-clause]

[ORDER BY colnam]

    [DISPLAY [AS opnam[(par1, par2, ...)]]]   ***

        [WITH [DEV=devnam][,FRAME=n][,SCALE=n][,LOW=n]
        [,HIGH=n][,IMAGENO=n]

        [XPOSITION=n][,YPOSITION=n][,XINCREMENT=n][,YINCREMENT=n]]];

DELETE IMAGE WHERE BILD ID = {number | :CURRENT ID};

INSERT IMAGE FROM filnam;   or

INSERT IMAGE (colnam1,colnam2, ...)
        VALUES(val1,val2, ...);

UPDATE IMAGE SET colnam1=val1 [,colnam2=val2][,...]]
       [WHERE BILD ID = {number | :CURRENT ID}];

The extentions to SQL are underlined.

*** presently implemented operations:

    STANDARD   display of the images in the standard format

    MINI   display of miniaturized images

    FILM   display of images as a film

    SOFTKEY   display of an image as an icon

Please note, that there exists a display program for each image type described in the image data base and for every display device. The name of the display operation, however, is the same for all types and devices. ISQL automatically does the necessary mapping according to the descriptions in the extended data dictionary.

EXPERIENCE WITH A PROTOTYPE PACS SYSTEM

IN A CLINICAL ENVIRONMENT

D. F. Preston, S. J. Dwyer III, W. H. Anderson

K. S. Hensley, L. T. Cook, S. L. Fritz, Joy A. Johnson

Department of Diagnostic Radiology

University of Kansas Medical Center

Kansas City, Kansas, 66103, U.S.A.

In the Department of Diagnostic Radiology at the University of Kansas Medical Center we have designed and implemented an on-line prototype PACS system in a clinical environment which has permitted us to identify hardware, software and human interface problems associated with the PACS prototype. Currently, in Radiology, digital images are recorded on magnetic media, displayed on a cathode ray tube, placed on film by a multiformat video film recording system. The films from a patient are placed in a film jacket which is kept in the film file room. The film jacket is manually retrieved and is the only copy available. The radiologist's text report is also kept in the film jacket and is the only report of the films. Frequently films are mis-filed and lost, occasionally the patient's report folder is lost but rarely is the patient's entire film jacket lost. It is not uncommon, however, for there to be delays exceeding several hours when the film jacket and the report folder are unavailable.

We believe the PACS system will provide improved radiological diagnosis through the integration of imaging modalities and through better communication with the referring physician. We believe the PACS system will eliminate the contention for film by providing inexpensive multiple copies simultaneously at remote sites. We believe we can eliminate film costs for that approximate 25% of our radiology which is digital. This film saving in itself will make the PACS system cost-effective.

ON-LINE RADIOLOGY NETWORKING

The system is based on the Intel 8086 Microprocessor and Ethernet Network. The detectors consist of a nuclear medicine gamma camera, a CT scanner, 2-digital subtraction angiography units and an ultrasound sector scanner. Each imaging modality is interfaced to an acquisition node which connects with an Ethernet Communications Network.

NATO ASI Series, Vol. F19
Pictorial Information Systems in Medicine
Edited by K. H. Höhne
© Springer-Verlag Berlin Heidelberg 1986

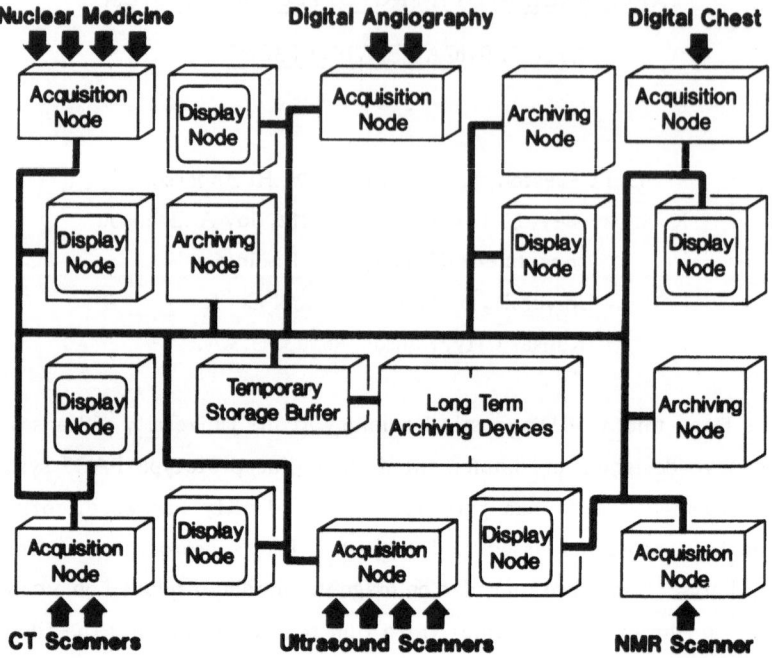

Figure 1. Schematic of Prototype PACS System.

The network in turn connects with display stations and archiving nodes. The system permits multiple copies of clinical images to be visualized with full dynamic range at remote sites simultaneously, thereby eliminating contention over the film jacket as the only source of patient films.

DAILY GENERATION OF DIGITAL DATA

Digital imaging in our department is approximately 23% of the workload and is increasing 4-7% per year. Our Department supports a 540-bed hospital. We are currently generating an average of 638 megabytes of digital information daily. (Table 1.) In the future we anticipate the digital chest area to examine twenty patients daily with two images per patient which will generate 84 megabytes daily, NMR of the head to examine ten patients daily with twenty images per patient (256 x 256 x 8) which will generate 13 megabytes daily with NMR body imaging to generate 33 megabytes daily for an additional 130 megabytes daily, or an expected total of 768 additional megabytes of digital information daily. The large image intensifier which we are adding to the system may increase that percentage to 35-40% of the total workload of the department as the image intensifier appears to be suitable for chest work, upper GI and barium enemas, and fluoroscopy.

DAILY REQUIREMENTS FOR DIGITAL TRANSMISSION

The digital data generated each day is a small part of the total transmission requirements. By measuring the activity in the film file room it is possible to estimate the amount of digital information which would have to be transmitted if we recorded and archived all of the digital information available. The average hospital stay for a patient at our hospital is 9 days. During the first 3 days of hospitalization, ten requests for the film file jacket are made. During the following 6 days there are 4 requests and for the remainder of the year on average there are 3 requests. The retrieval rate per day is 14 times the data generated per day. This is approximately ten gigabytes per day which must be transmitted to have a clinically useful PACS system. Assuming a twelve hour work day, the data transmission rate required is approximately 2 megabits per second.

Ethernet has a rated transmission speed of ten megabits per second which is well above the required minimum data transmission throughput of 2 megabits per second or even the desired throughput rate of 4-6 megabits per second. Ethernet, although it is slow, is fast enough to handle the calculated anticipated throughput. The slowest part of the data transfer system is in the bus of the microprocessor rather than Ethernet. It currently takes about 2 seconds to display a CT slice and this is thought to be slow. The general feeling of the radiologists who have used the system is that they would like to display a CT slice at a speed of one slice per second.

The data base management system is a B tree organization for the storing and retrieving of the digitally formatted patient data. Any patient can be accessed by either patient name or identification number. The specific image required can be obtained by then specifying the imaging modality and date.

Ethernet contains the logical protocol to permit collision detection and detection of errors in transmission. Using the 50-OHM coaxial cable and carefully tuning all parameters of the Ethernet network, we have been able to achieve throughput of only 850,000 bits per second of image data. This throughput is approximately one-third that of the anticipated minimum requirement of two megabits per second.

Extensive software was required. A total of 70,000 lines of code were written (mostly in Pascal) for the system operating under MP/M. The communication software was approximately 8,000 lines, the data base management software was approximately 9,000 lines and the interactive display station software was approximately 8,000 lines of code.

INTERACTIVE DIAGNOSIS DISPLAY STATION

Currently in CT body and CT head studies the final diagnosis is made from the multiformat video film recording rather than from the cathode ray tube. The final diagnosis, however, could be made from an interactive graphics display. In theory, the diagnosis made from the cathode ray tube should be superior to that made from the analog film from the multiformat video camera. The proper use of the interactive diagnosis display station requires display and multiple images from various modalities and control of the threshold and windowing functions. Manipulation of threshold and windowing functions is now understood by radiologists working with CT. To display multi-modality images on the same screen, the ability to handle the thresholding and windowing of each individual image is necessary. This requires multiple video lookup tables, each containing at least 12-bits.

It is necessary to have the capability to quantify values in the image within a region of interest to make distance/area/volume measurement and other objective numerical values from a region. Zoom capabilities, three-dimensional displays and functional images are also needed. Our system has zoom capabilities and the ability to quantify pixel values with any rectangular region of interest. Variable thresholding and windowing capability is also present and is always used when a multi-modality display is created on the cathode ray tube.

The display was an Advanced Electronic Design, AED-512, which permitted a 512 x 512 display. Because of the differences in dynamic range between CT, digital subtraction, ultrasound and nuclear medicine images, windowing and thresholding was performed on each image individually. This was quite inconvenient and time consuming. Automatic scaling each modality appropriately to the CRT would increase the throughput of the radiologist. For the relatively uninformed patient care physician, such a self-scaling capability would be mandatory.

The radiologist has unrealistic criteria for retrieval speed of images. Currently, in our well organized film file room, it takes at least five minutes to obtain films. The film file room is called by phone, a manual search is made in the film file room, the film jacket is brought to the reading room, the appropriate films are searched for by date, modality and specific image, the films are ranked in time sequence on multiple view boxes and the appropriate radiographic reports are read from the film report folder. Even simple cases where there are multimodalities and images to review, take another five minutes to sort and display the images. The radiologist accepts this ten minute wait without major complaint, but insists a PACS system should deliver images and text reports in seconds. We have not formally evaluated the radiologist's waiting times for films. Such an evaluation would be helpful to identify manual film retrieval system speed so that objective data would be available for comparison with PACS capabilities.

We believe a successful and acceptable PACS strategy would be to provide a que of patient names, images, modalities and dates requested of the archiving node. As the current patient images are being examined and interpreted, the next patient images and text report could be transferred to the display station memory from the archiving node. New images, when displayed, should be automatically grouped by modality and ranked in a time sequence and associated with the radiographic text report.

The ability to look at the radiographic text reports without the associated image may provide a strategy to decrease image traffic on the system. Frequently it is satisfactory just to know that the previous study was normal. The patient care physician will often be content with the radiological text report and not request the images if the report agrees with his preconceived notions of what the image should show.

In the last two years displays of 1024 x 1024 and 2048 x 2048 have reached the market. These displays would permit the display of a high quality chest radiographs or perhaps as many as 64 simultaneous, but relatively low resolution, nuclear medicine images.

HARD COPY RECORDING

Currently most digital radiographic images are placed on analog film. This is an expensive recording modality. During Fiscal '84, the cost of multiformat video film recording approached $1 million. In the last several years, electrostatic printers using paper media, and laser printers using film or paper media have been developed and judged satisfactory for nuclear medicine and ultrasound. We anticipate future improvements in the digital recording device will make the output satisfactory for the rest of digital radiology.

LONG-TERM ARCHIVING

Long-term archiving is thought to be a serious problem for PACS. We have investigated the frequency of retrieval from our 5-year analog film file room system. At our hospital, if a film jacket has been inactive for two years, it is placed in the long-term archiving file in the basement where it is kept for an additional five years. If the patient returns for another examination, the film jacket is moved to the current file. After inactivity of the jacket for two years, it is returned to the archives in the basement. We have in long-term storage 409,000 patient film jackets. Approximately 40 film jackets per day, or less than 2.4% of the film in long-term storage, will be retrieved in any given year. Additionally we have determined that approximately one-third of those films retrieved (approximately .8%) are not needed for immediate medical purposes. This means that approximately 1.5% of the films in long-term storage are retrieved in any given year for current medical purposes. Some consideration should be given to the strategy of eliminating films in the long-term film file room.

Several technologies are available for long-term archiving. Magnetic tapes and magnetic disks, optical tapes and optical disks, laser film archiving using film or optical tape storage media and paper archiving technology must all be considered.

During Fiscal Year '83, the cost of multiformat video film recording (without considering labor cost) was $186,000, whereas the estimated cost of linear optical tape was $75,000 and the estimated cost of linear film tape archiving was $55,000. The cost savings would be even greater if labor costs were considered, as the optical and linear tape systems are not labor intensive, whereas the multiformat video film recording system is labor intensive.

All of the digital, long-term archiving modalities permit complete spatial resolution and contrast range to be preserved. Data compression techniques may provide even greater savings and speed of retrieval. Techniques which permit noiseless compression are available with compression ratios of 5 to one. Techniques permitting 10-30 to one compressions have been implemented by others. Our system utilized no data compression.

PACS STANDARDS

Radiology is in need of PACS standards. Standards provide a defined performance for all PACS systems. They would permit the open interfacing of all manufacturers imaging equipment. There would be a reduction in the development time of new imaging equipment because the manufacturer had all the apriori knowledge of hardware requirements. Standards aid product procurement because of the wide spread availability of equipment built to recognized standards. Maintainability of PACS systems is enhanced due to the availability of test equipment built to these standard requirements. Standards will keep the designer in the mainstream of evolution and improve cost-effectiveness and thereby lower the cost of hardware and software development.

There are currently two efforts underway to develop PACS standards. The first is from the American College of Radiology and National Electrical Manufacturer's Association (ACR-NEMA) and The second is the IEEE Computational Medicine Technical Committee. The Digital Imaging and Communication Standard Committee is developing a medical image transfer interface standard. The Chairman is Gwilyn S. Lodwick, M.D., Massachusetts General Hospital, 55 Fruit Street, Boston, MA 02114.

SUMMARY

We have determined that in 1983, 23% of our departmental workload was digital and that this work load is increasing at an annual rate of 4-7% per year. The 23% workload amounts to 638 megabytes of digital information daily. Through measurement of activity in our film file room, we have determined that we must move 14 times that daily rate of information to support our 540 bed hospital. This means we must move 10.7 gigabytes per day. Assuming a 12-hour work day, this requires a transmission rate of 2 megabits per second to transmit 10.7 gigabytes per day.

A PACS system was designed, fabricated and programs developed. A multi-terminal system was developed which linked nuclear medicine, CT body, CT head, digital angiography, ultrasound, display stations and archiving nodes. Clinical images were acquired, archived, transmitted and analyzed. Under optimal circumstances we have achieved 850,000 bits per second of image throughput. This is approximately one-third the calculated required throughput. The slowest part of the system is the microprocessor bus rather than Ethernet, which specifications indicate could handle ten megabits per second.

We believe a properly designed PACS system will be both medically and economically effective. Medical effectiveness requires the ability to link text report with the image. An adequate number of images must be stored as digital studies so the radiologist anticipates images in storage which will be helpful. It is vital to be able to que requests for patient studies so transmission of images to display terminals can be made during times of low utilization. In that way, the physician will always have new patient data readily available for analysis. It is satisfactory to request patient data by hospital number, name, date, modality and image. Automated methods for initial thresholding and windowing of each modality should be developed. For each individual image, control of thresholding and windowing at the display station is manditory. The PACS system satisfactorily answers the contention problem for multiple copies simultaneously at remote sites.

Image perception studies comparing the CRT diagnosis with control of threshold and windowing with the diagnosis from film from a multiformat video film recording should be performed. The ability to look at the text report without the accompanying image should be developed as we believe most patient care physicians will be satisfied with the text report as long as it agrees with their preconceived clinical opinions. This capability may significantly decrease the need for massive image data transfers.

Economic justification is based on decreasing the patient's hospital stay by better management and distribution of radiological information as well as the replacement of film by either magnetic media storage, paper media or optical disk or tape technology.

OTHER SUGGESTED READING

Dwyer SJ III, Cook LT, Fritz SL, Lee KR, Preston DF, Batnitzky S, DeSmet AA: Medical Image Processing in Diagnostic Radiology. IEEE Trans Nuclear Science, Vol. NS-27, No. 3, pp. 1047-1055, June, 1980.

DeSmet AA, Tarlton MA, Cook LT, Fritz SL, Dwyer SJ III: A Radiographic Method for Three-Dimensional Analysis of Spinal Configuration. Radiology, 137:343-348, 1980.

Cook LT, Fritz SL, DeSmet AA, Dwyer SJ III: Applications of Quantitative Radiology. Proceedings of the 1980 International Conference on Cybernetics and Society, pp. 750-753, 1980.

Batnitzky S, Price HI, Cook PN, Cook LT, Dwyer SJ III: Three-Dimensional Computer Reconstruction from Surface Contours for Head CT Examinations. Journal of Computer Assisted Tomography, 5:60-67, 1981

Cook PN, Batnitzky S, Lee KR, Cook LT, Fritz SL, Dwyer SJ III, Charlson EJ: Three-Dimensional Reconstruction from Serial Sections for Medical Applications. Proceedings of the 13th Annual Hawaii International Conference on Systems Science, pp. 358-389, 1981.

Batnitzky S, Price HI, Cook PN, Dwyer SJ III: Three-Dimensional Computer Reconstruction in the Study of Brain Lesions. Automedica, 4:37-50, 1981.

Lee KR, Dwyer SJ III, Anderson WH, Betz DB, Faszold S, Preston DF, Robinson RG, Templeton AW: Continuous Image Recording Using Gray-Tone Dry Silver Paper. Radiology 139:493-496, 1981.

Dwyer SJ III, Anderson WH, Tarlton MA, Cook PN, Lee KR, Batnitzky S: Interactive Computer Graphics for Diagnostic Imaging. Computer Graphics World, Vol. 4, No. 11, pp. 46-53, November, 1981.

Duerinckx AJ, Dwyer SJ III, Pisa JE: Digital Image Archiving and Management. Diagnostic Imaging, pp. 62-63, November, 1981

Dwyer SJ III, Cook PN, Batnitzky S, Lee KR, Levine E, Price HI, Preston DF, Cook LT, Fritz SL, Anderson WH: Three-Dimensional Reconstruction from Serial Sections for Medical Applications. Proceedings SPIE Technical Symposium East '81, 3-D Machine Perception 182:98-105, 1981.

Dwyer SJ III, Templeton AW, Martin NL, Cook LT, Lee KR, Levine E, Batnitzky S, Preston, DF, Rosenthal SJ, Price HI, Anderson WH, Tarlton MA, Faszold S: Cost of Managing Digital Diagnostic Images to a 615 Bed Hospital. SPIE Vol. 318:308, 1982, Picture Archiving and Communication Systems (PACS) for Medical Applications.

Dwyer SJ III, Templeton AW, Anderson WH, Tarlton MA, Hensley KS, Lee KR, Preston DF, Batnitzky S, Levine E, Rosenthal SJ, Martin NL, Cook LT: Salient Characteristics for a Radiology Department, pp. 194-204, Ibid.

Dwyer SJ III: Computers in Radiology. Proceedings, IEEE Computer Society Fifth International Computer Software and Applications Conference, pp. 311-314, November 18-20, 1981.

Dwyer SJ III, Templeton AW, Martin NL, Lee KR, Levine E, Batnitzky S, Rosenthal SJ, Preston DF, Price HI, Faszold S, Anderson WH, Cook LT: The Cost of Managing Digital Diagnostic Image. Radiology, 144:313-318, July, 1982.

Templeton AW, Dwyer SJ III: Managing Digitally Formatted Diagnostic Image Data. AUR Research Symposium on Digital Radiolography, Williams & Wilkins, Publishers, Baltimore, MD (In Press).

Templeton AW, Dwyer SJ III, Rosenthal SJ, Hensley KS, Martin NL, Anderson WH, Robinson RG, Levine E, Batnitzky S, Lee KR: A Peripheralized Digital Image Management System: Prospectus. American Journal of Roentgenology 139:979-984, 1982.

Dwyer SJ III, Templeton AW, Anderson WH, Tarlton MA, Hensley KS, Betz D: A Diagnostic Digital Imaging Management System. Proceedings, IEEE Computer Society Conference on Pattern Recognition and Image Processing June 14-17, 1982. Los Alamitos, CA: The Computer Society Press, 294-299.

Templeton AW, Dwyer SJ III: Managing Digitally Formatted Diagnostic Image Data. Proceedings, AUR Automated Digital Information Conference, Baltimore: Williams and Wilkins, 1982.

Dwyer SJ III, Templeton AW, Martin NL, Preston DF, Anderson WH, Tarlton MA, Hensley KS, Betz D, Cook LT, Fritz SL, Wegst AV: A Distributed Diagnostic Imaging Management System. Proceedings, The Society of Nuclear Medicine, Digital Imaging of the Future, 1982.

Anderson WH, Tarlton MA, Hensley KS, Templeton AW, Dwyer SJ III: Implementation of a Diagnostic Display and Image Manipulation Node, SPIE, 418:225-232, 1983.

Dwyer SJ III, Templeton AW, Anderson WH, Tarlton MA, Hensley KS, Betz D, Lee KR, Levine E, Batnitzky S, Rosenthal SJ, Robinson RG, Preston DF, Price HI, Martin NL, Fritz SL: Archiving and Distribution Systems for Digitally Formatted Images. Proceedings, Advances in Digital Radiography, May 12-15, 1983.

Chang CHJ, Templeton AW, Dwyer SJ III: Management of Digitally Formatted Images. Medical Imaging Terminology, 15:36-45, 1983.

Dwyer SJ III, Templeton AW, Anderson WH, Tarlton MA, Hensley KS, Lee KR, Batnitzky S, Rosenthal SJ, Johnson JA, Preston DF: A Prototype Digital Image Management System. Proceedings, Seventh Annual Symposium on Computer Applications in Medical Care, IEEE Computer Society. October 23-26, 1983, pp 809-813.

366

Preston DF, Dwyer SJ III, Templeton AW, Anderson WH, Tarlton MA, Hensley KS, Betz D, Batnitzky S, Lee KR, Levine E, Rosenthal SJ, Martin NL, Robinson RG, Price HI, Johnson JA: A Digital Image Management System - Clinical Implementation of PACS. Proceedings, International Symposium on Electronic Imaging in Medicine, pp. 363-387, 1982.

Templeton AW, Dwyer SJ III, Johnson JA, Anderson WH, Hensley KS,.Rosenthal SJ, Lee KR, Preston DF, Batnitzky S, Price HI: An On-Line Digital Image Management System. Radiology 152:321-325, 1984.

Templeton AW, Johnson JA, Anderson WH, Cook LT, Dwyer SJ III, Preston DF, Lee KR, Rosenthal SJ, Batnitzky S, Levine E, Tarlton MA: Computer Graphics for Digitally Formatted Images. Radiology, 151:527-528, 1984.

Templeton AW, Dwyer SJ III, Johnson JA, Anderson WH, Hensley KS, Lee KR, Rosenthal SJ, Preston DF, Batnitzky S: Implementation of an On-Line and Long Term Digital Management System. RadioGraphics, January, 1985.

MODALITY	AVERAGE # PATIENTS DAILY	NUMBER OF IMAGES PER PATIENT	MEGABYTES PER DAY
CT BODY	16	20-40	264
CT HEAD	18	16-20	58
DIGITAL ANGIOGRAPHY	8	12-16	67
ULTRASOUND	20	30-42	242
STATIC AND DYNAMIC NUCLEAR MEDICINE	25	10	7
TOTAL MEGABYTES PER DAY			638

Table 1. DIGITAL WORKLOAD FOR 540 BED HOSPITAL

Recent Developments in Digital Radiology

H.K. Huang, D.Sc.
Professor and Chief
Medical Imaging Division
Department of Radiological Sciences
University of California, Los Angeles
CA 90024/USA

ABSTRACT

Digital radiology has experienced a tremendous growth in recent years, among the many advancements are the new developments in the area of image acquisition, picture archiving and communication systems, image compression, image viewing stations, and image processing. The involvement of our department related to these topics are reviewed in this paper.

I. INTRODUCTION

The arrival of high speed digital electronics and communication technologies in radiological sciences is gradually changing the method and procedure of acquiring, storing, viewing, and communicating diagnostic images. One natural development along this line is the emergence of a digital radiology department. We anticipate that some form of digital radiology will appear in a clinical environment in the next three to five years. A digital radiology department consists of two components: a radiology information management system and a digital imaging system. The radiology information system is concerned with the patient data management, which has been defined and well developed during the past few years[1], and is going to be discussed in other papers in this proceeding. Development of a digital radiology image system requires the technologies of image acquisition, archiving, communication, retrieval, display and processing. This paper describes the development on some of these topics in our department. Figure 1 shows the schematic of a digital radiology imaging system.

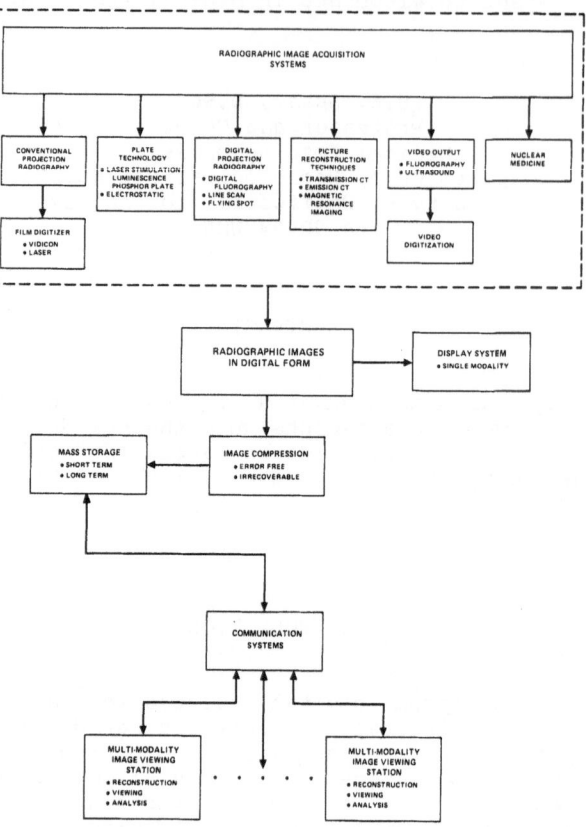

Figure 1. Schematic of a Digital Radiology Imaging System.

II. IMAGE ACQUISITION

Image acquisition is defined as the various radiographic image techniques used in acquiring images for diagnostic purposes. In this section we review some recently developed digital image acquisition techniques for diagnostic radiology.

2.1 Conventional Projection Radiography

Conventional projection radiographic techniques require the exposure of a screen/film cassette under the x-ray. The film is then developed and viewed from an illumnated light box and later stored in the film library for future use. Even though digital methods have been used extensively in diagnostic imaging, these conventional techniques are still used for about 80% of the examinations in a diagnostic radiology department (2). In order to establish a digital imaging system, these conventional techniques have to be modified. Two methods are being developed now: 1) Use a scanner to digitize conventional x-ray film into a digital form, and 2) Change the conventional screen/film cassette into a plate coated with specially designed phosphor, the plate is then scanned by a reader which converts the x-ray latent image on the plate into a digital image.

Film Scanner There are three types of scanners that are now being used to convert a conventional x-ray film into digital form. They are: the TV camera, the optical drum scanner, and the laser scanner. The advantage of the laser scanner is its high spatial and density resolutions. The scanning speed is about 10 to 20 seconds, and the set-up time is minimal. The disadvantage of the laser scanner is the price, it is much more expensive than the Vidicon camera and the drum scanner.

Both the vidicon camera and the drum scanner have been incorporated into the clinical setting in some hospitals and institutions. The laser scanner is still in the experimental stage and only a few hospitals are actually using it for digitizing. Since 80% of radiological examinations are still recorded on films, a high speed, reliable, and good quality scanner is essential to convert x-ray films into digital form. We anticipate the laser scanner will be used for high quality digitization, for instance, when details are required in a chest x-ray film. The vidicon camera will be used for films made for screening purposes. Figure 2 shows a laser scanner in our department used for film digitizing developed by Konishiroku Ltd., Japan[3].

Figure 2. A laser scanner used for digitizing a 14"x17" x-ray film to a high resolution image of up to 2048x2900x10.

Plate Technology. There are two different types of plate technology which are being investigated. The first type of plate technology utilized some specially selected phosphors coated onto a polyester support, for example, the system developed by Fuji[4] used BaFBr. This plate is placed inside a particularly designed cassette with the same dimensions as the conventional screen/film cassette used in conventional projection radiography. This cassette is then exposed to x-ray just like the conventional radiographic procedure, and after the exposure the plate is taken to a laser scanner for scanning. The phosphors, excited by the laser energy, emits light. The light is then converted to analog electric signals and then to digital signals to form an image. Preliminary studies indicated that this plate technology requires less x-ray dosage than the conventional screen/film method, and the digital image obtained from this plate preserves more diagnostic information than the conventional x-ray films. A few systems in Japan, U.S. and Europe are being used for clinical evaluation.

The other type of plate technology is the electrostatic imaging technique. In this case the conventional screen/film cassette is replaced by amorphorus selenium alloy on the flat aluminum substrate. After x-ray exposure the latent electrostatic image on the plate is scanned with multiple-electrometers forming a digital image. An experimental system called SELRAD is being evaluated clinically in our department[5]. Figure 3 shows the SELRAD system and a 1024x1024x12 image produced by the system.

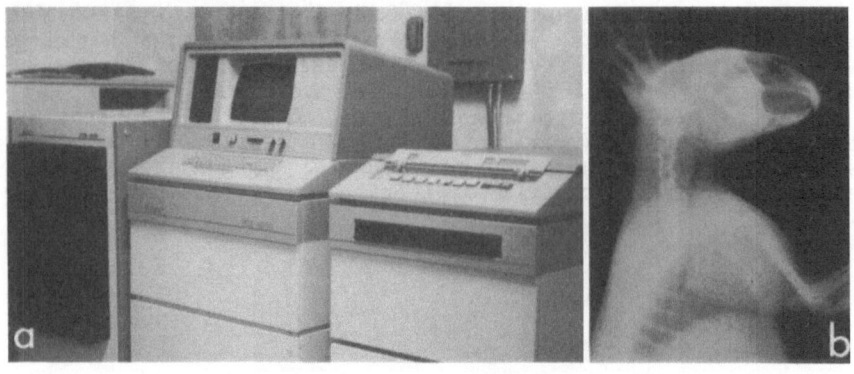

Figure 3. a. An experimental digital charged selenium plate projection x-ray (SELRAD) system which is a potential candidate to replace the conventional screen/film system in diagnostic radiology[5].
b. A 1024x1024x12 image of a rabbit produced by this system. (Contributed by P. Papin.)

The plate technology has great potential to replace conventional x-ray films. The result is a direct digital image which does not require the use of films as a storage medium. Since 80% of the radiographic examinations are still using the conventional projection radiographic techniques, the success of the plate technologies is essential for the future of digital radiology.

2.2 Digital Projection Radiography

This category is different than the plate technology in that some modifications have to be made on the image chain of a conventional projection radiography. There are three types of digital projection radiography that are now being used in a clinical environment or under clinical evaluation[6].

Digital fluorography (DF) is an extension of the conventional fluorographic system with a digital chain. A DF system consists of an x-ray tube, an image intensifier tube, a vidicon camera, an A/D, an image memory and processor, a digital storage, and an image display system. The DF has passed the stage of clinical trial and is being used in most major hospitals and clinics. This system is excellent for vascular radiographic examinations and the goal is to replace the invasive arteriographic technique.

The second type is the line-scan technique. In this case the x-ray is first collimated with a slit, it then transverses the patient and is detected by a linear detector array consisting of a phosphorus screen coupled to photodiodes. The output from the photodiodes are then converted into a digital image line-by-line.

The third type is the flying spot scanner. The x-ray is first collimated with a slit and further collimated into a pencil beam by a rotating disk with slits perpindicular to the beam. The pencil beam scans horizontally from left to right across the patient and is measured by a scintillator and a photomultiplier tube. Prototype systems of both types are being evaluated at clinical settings.

2.3 Utilization of Picture Reconstruction Technique

The utilization of picture reconstruction technique to form a diagnostic image has been discussed extensively in other literature[7]. In this case a cross-sectional image is reconstructed by many one-dimensional projections of the same cross section. The x-ray Computerized Tomography (CT) scanner which utilizes the transmission x-ray energy source is an example. This technology has caused many traditional diagnostic techniques to become obsolete in the past seven to eight years. The picture reconstruction technique can also be used for emission x-ray energy. There are the signal photon emission CT (SPEC) and the positron emission CT (PET). Both of them have been developed for a few years and are still under clinical evaluation and trial.

The magnetic resonance imaging (MRI) is a new technology in diagnostic radiology. It utilizes radio frequencies and strong magnetic fields as the energy sources. The description of this new imaging technique has been published in many radiological journals for the past two years[8] is going to be discussed in other papers in this proceeding.

III.   PICTURE ARCHIVING AND COMMUNICATION SYSTEMS (PACS)

The concepts of picture archiving and communicating system for diagnostic imaging originated from two workshops held in Newport Beach, California in January 1982 and Kansas City in 1983[9,10]. The PACS deals with four major components in a digital radiology department, namely image archiving, image compression, image communication, and image retrieval.

3.1 Storage Medium

The currently available technology for digital image storage is magnetic tapes and disks. A real-time digital magnetic disk which allows the storage of 800 512x512x8 images is being evaluated clinically in our department. However this storage capacity represents only a very small fraction required by a digital radiology department. A study shows that our department (685 beds and 140,000 procedures/year) uses about 400,000 x-ray films per year[11]. If each diagnostic image is assumed to be acquired in or digitized to a 1024x1024x8 bit it will require 400,000 megabytes to store the

information in one year. To accomodate this large capacity
requirement for diagnostic images, we are anxiously awaiting the
arrival of the digital optical disk. Without it a digital radiology
department cannot become a reality. We consider the storage problem
as still the major stumbling block in establishing a digital
radiology department.

3.2 Image Compression

A different approach to ease the storage problem is to consider image
compression, i.e., to compress the digital image into a more compact
form before storage and transmittal. There are two types of image
compression: error-free and irreversible. For radiological image,
error-free compression can achieve a compression ratio of about three
to one, whereas irreversible compression can achieve a much higher
compression ratio. However image reconstruction from irreversibly
compressed data introduces errors. The study of irreversible
compression which considers the relationship among the compression
ratio, error introduced, and speeds of compression and reconstruction
is an intensive topic of research in our laboratory. Figure 4 shows
some results on irreversible compression in our laboratory[12]. It
is seen that images reconstructed from compression ratio of up to
16:1 are very much acceptable for making clinical diagnosis.

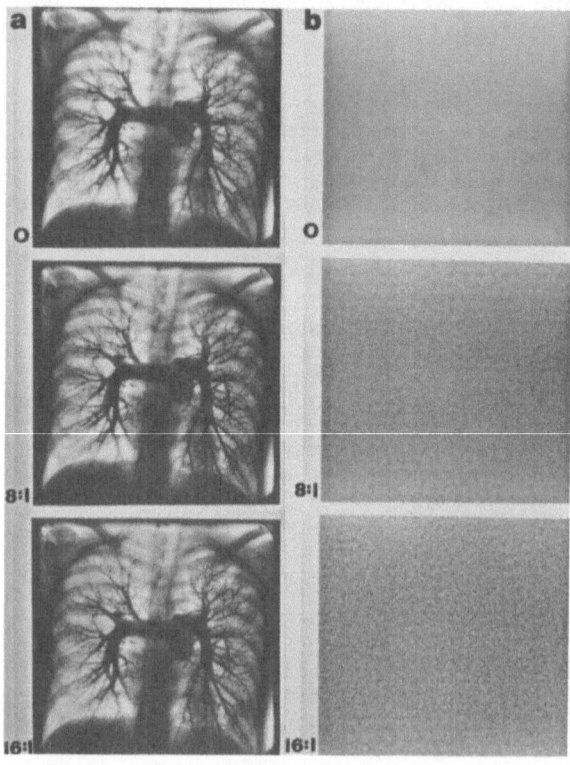

Figure 4. Irreversible
image compression on a
512x512x8 bit image.
a. Reconstructed images
from compression ratio of
8:1 and 16:1.
b. Difference images
between the original and
the reconstructed images.
(Contributed by C.S. Lo.)

3.3 Communication

We consider image communication within the same building in this context since our department and our hospital is within a building complex. The requirements of diagnostic image communication are different than that of the conventional computer terminal communication because in the former the emphasis is on large volume, high speed but a limited number of stations. At the present time magnetic tapes or diskettes, dedicated cabling, baseband network and broadband network are being used for image communication[13]. It requires about six minutes to send a 512x512x8 image through a 9600 BAUD terminal line, twenty seconds for a baseband network like the Ethernet system. Both of them are too slow for useful clinical application.

The broadband communication system can modulate the digital data (patient information) and video signals (images on TV monitors) and transmit them on a single cable instantaneously. Up to about twenty channels (or twenty different images) can be utilized at one time. The video components of this system are well standarized and can easily be interfaced into the diagnostic image equipment, and the modulation of the digital signal is also commercially available. This communication system is currently being implemented in our department. Figure 5 shows the connection of a three-channel system.

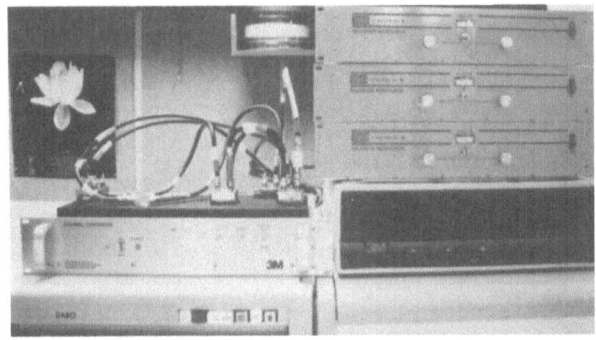

Figure 5. A working broadband communication system with three video channels and one digital terminal connected [13]. (Contributed by N. Mankovich.)

Other types of very high speed digital communication network are being researched and developed at this point. However none of them have been used for clinical trial as yet.

3.4 Retrieval

Diagnostic image retrieval requires the implementation of a data base structure which is related to image acquisition, archiving, and communication. There are short term and long term retrievals. The short term retrieval has two phases. Phase one is the first two weeks after the patient is admitted or after a procedure is performed. During this two week period the images will be requested very often by the referral physicians. Then there is a two month period of time when the images will remain active. After two months the images would be inactive and the long term retrieval can be from two months to seven years, and sometimes even longer. Therefore a specially designed data base structure geared to both short term and long term retrieval is essential for a digital radiology department to operate efficiently. Not too much work has been done along this line since the definition of optimum retrieval has yet been specified. In our department we are testing a data base package called MARS (Multi-modality Acquisition and Review Systems) developed by Gould/DeAnza based on the MUMPS system[14]. We are currently in the process of testing this system in a clinical environment as a short term retrieval system. Figure 6 shows the MARS system under clinical trial. The future image retrieval system has to be easy to use, allowing the maximum of storage, and a very fast retrieval speed.

Figure 6. The MARS system with three 1024 monitors and one touch sensitive screen. Left: (512) US, US, MRI, CT; Middle: Anatomical Section (512), MRI Sagittal (512), PET (128); Right (1024): Angiogram.

IV. IMAGE VIEWING AND PROCESSING

4.1 Multiple Viewing Station

Most radiological diagnoses require the examination of multiple images either from the same diagnostic modality or from multiple modalities. The current method of viewing these multiple modality images are from films. Many x-ray films are lined up by technicians on light boxes or an alternator prior to a film reading session. For digital image viewing TV monitors would be used in place of the conventional light boxes. In order for the physicians and radiologists to accept image viewing from TV monitors, the performance of the multiple image viewing display system has to be compatible with the conventional light box. A multiple image display system requires a large image memory, adequate video output control, and a powerful image processor.

The MARS system described in Section 3.3 has the capability of viewing three 1024x1024 images at the same time. Each 1024 image can be subdivided into smaller fields for multiple images within one 1024 TV monitor. A mobile multiple viewing station with six 512 monitors has been built in our department and is being tested in a clinical environment[15]. This system can display six different images at the same time, and each image can be manipulated individually. Figure 7 shows the viewing station displaying six different images.

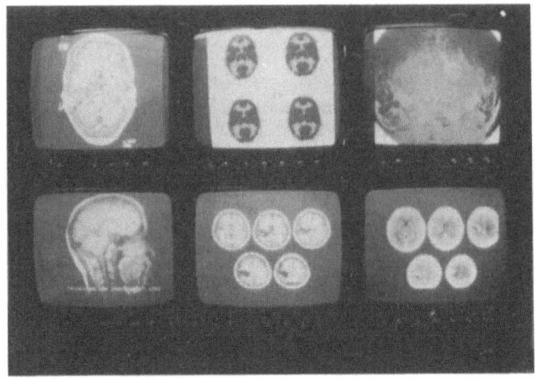

Figure 7. A multiple viewing station showing six different images simultaneously, upper left to right: anatomical section, PET, angiogram. lower: MRI sagittal, MRI transverse, CT.

4.2 Image Processing Laboratory

The picture archiving and communication systems (PACS) described in Section III and the image viewing station described here require the support of a general purpose image processing laboratory. This laboratory provides the service to all these components. Figure 8 shows the general purpose Image Processing Laboratory in our department.

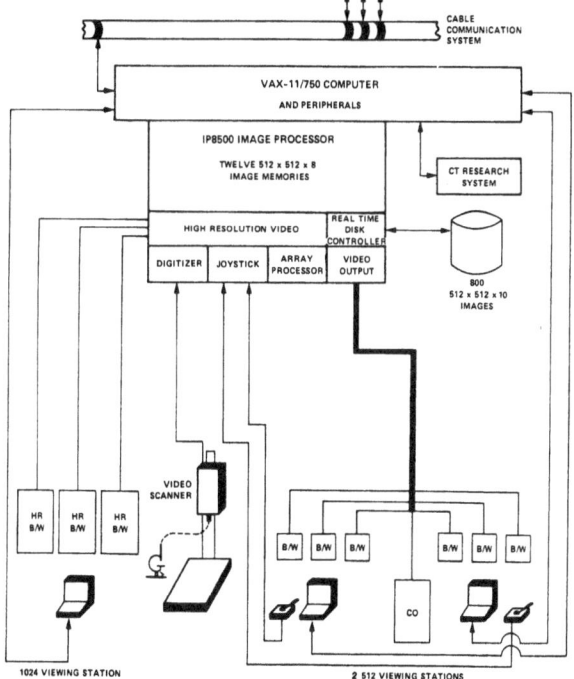

Figure 8. The Image Processing Laboratory at the Department of Radiological Sciences, UCLA.

V. SUMMARY

We have reviewed briefly the recent research and development related
to digital radiology in our department. Among these developments are
the image acquisition, picture archiving and communication systems,
image compression, image viewing stations, and image processing
laboratory. The image acquisition is the most advanced, with the
indroduction of the plate technology the chance of converting the
conventional projection radiographic techniques into a complete
digital form is very good. The image storage remains a potential
problem for developing a digital radiology department. To this end,
image compression technique has to be considered in order to ease the
storage problem. In image communication, the broadband communication
system has the potential of being accepted for communicating images
throughout the diagnostic radiology area. The acceptability of the
multiple viewing station by diagnostic radiologists and clincans is
the key to the success of establishing a digital radiology
department.

ACKNOWLEDGEMENT

The developments described in this paper are the result of generous
supports from Technicare Corporation, Gould/DeAnza, ADAC
Laboratories, Konishiroku Ltd., Light Signature, Inc. The
development of the multiple viewing station is supported by PHS Grant
Number 1RO1 CA 39063, awarded by the National Cancer Institute, DHHS.
The following staff from the image processing laboratory has
contributed significantly on various aspects of this development:
N. Mankovich, P. Papin, C. S. Lo, K.S. Chuang, R. Taira, and
B. Norton.

REFERENCES

1. R.A. Bauman (coordinator), "Computer Applications in Radiology",
   Proc. 7th Conf., Am. College Radiology, Boston, Mass., April
   25-28, 1982.

2. J.L. Johnson, D.L. Abernathy. "Diagnostic Imaging Procedure
   Volume in the United States", Radiology, Vol. 146, No. 3, 1983,
   pp. 851-854.

3. H.K. Huang, S.C. Lo, R.K. Taira, N.J. Mankovich, H. Takeuchi,
   "Preliminary Experience with a Laser Scanner and Laser Printer
   for Radiological Imaging," Presented at the 70th RSNA, November
   1984.

4. M. Sonoda, M. Takano, J. Miyahara, H. Kato, "Computed
   Radiography Utilizing Scanning Laser Stimulated Luminescence",
   Radiology, Vol. 148, No. 3, 833-838.

5. P.J. Papin, N.J. Mankovich, Z. Barbaric, H.K. Huang,
   "Preliminary experience with a charged selenium plate projection
   x-ray system", Proc. SPIE Conf. Medical Image Production,
   Processing, Display, and Archiving, San Diego, CA, Feb. 1984.

6. H.K. Huang, "Recent Developments in Medical Digital Radiography" Trans. Am. Nuclear Society, Vol. 45, TANSAO 45 1-884, 1983, 249-251.

7. H.K. Huang, "Biomedical Image Processing", CRC Critical Rev. in Bioengineering, Vol. 5, Issue 3, 1981.

8. I.L. Pykett, "NMR Imaging in Medicine", Scientific American, May, 1982, 78-88.

9. A.J. Duerinckx, "Picture Archiving and Communication Systems (PACS) for Medical Application", edited, Proc. SPIE: Picture Archiving and Communication Systems, Vol. 318, 1982.

10. S.J. Dwyer, III, "Picture Archiving and Communication Systems (PACS) for Medical Application", edited, Proc. SPIE: Picture Archiving and Communication Systems, Vol. 418, 1983.

11. H.K. Huang, Z. Barbaric, N.J. Mankovich, C. Moler, "Digital Radiology at the University of California, Los Angeles: A Feasibility Study.", Picture Archiving and Communication Systems (PACS) for Medical Application. edited, Proc. SPIE: Picture Archiving and Communication Systems, Vol. 418, 1983, 259-265.

12. S.C. Lo, H.K. Huang, "Radiological image compression using the full-frame bit allocation technique", to be published in Radiology 1985.

13. N.J. Mankovich, "Radiological Image Communication: Discussion and Preliminary Results", Proc. Computer Applications in Medicine, AAMSI Congress, San Francisco, CA, 1984.

14. H.G. Rutherford, M.J. Gray, "Digital light box, one of the integral pieces of PACS", Picture Archiving and Communication Systems (PACS) for Medical Application, edited, Proc. SPIE: Picture Archiving and Communication Systems, Vol. 418, 54-65, 1983.

15. H.K. Huang, N.J. Mankovich, Z. Barbaric, H. Kangarloo, C. Moler, "Design and Implementation of Multiple Digital Viewing Stations", Picture Archiving and Communication Systems (PACS) for Medical Application, edited, Proc. SPIE: Picture Archiving and Communication Systems, Vol. 418, 1983, 189-200.

THE FUTURE OF DIGITAL COMPUTERS IN MEDICAL IMAGING

by Roger A. Bauman, MD
The Massachusetts General Hospital
Boston, Massachusetts 02114/USA

Digital computers are now present in virtually all
Radiology departments. One of the first places they appeared
was in Nuclear Medicine in the late sixties. Quantifying
various physiological functions studied by nuclear techniques
was the main use; imaging was a secondary consideration.
Another early application of computers in Radiology departments
was to handle business functions. Information systems for
managing scheduling, film library functions and reporting
appeared in the late sixties and early seventies.

Computerized Tomographic scanning jolted Radiology a
decade ago. It was the first widely used imaging technique in
which the image data itself is collected in digital form
directly. CT scanning represented not only a startling
breakthrough, but in hindsight we realize it was the harbinger
of many changes to come.

Digital Fluoroscopy has become widely used; it is,
however, based on a clever integration and application of
previously existing technologies. More recently, Magnetic
Resonance Imaging scanning has emerged as the second major
unexpected modality.

There is a common thread associated with all of these
newer imaging modalities; the digital computer is an integral
part of each.

CT scanners and other digital imaging devices are now
well established. In the fully digital department of the
future such devices would be merely the source of digital
images. Computers would distribute, process, display, and
archive digital images. These functions have been lumped

under the unfortunate acronym PACS, Picture Archiving and
Communication Systems. This acronym is too vague; it doesn't
even include reference to the use of computers nor to the
single most important function, the interpretation of digital
images. I will use the term <u>fully</u> <u>digital</u> <u>department</u>
interchangeably with PACS.

THE FULLY DIGITAL DEPARTMENT

The fully digital department can be subdivided into six
functions:

1. Management
2. Image acquisition
3. Image distribution
4. Image interpretation
5. Image redisplay
6. Image archiving

The first function is control and management. Radiology
department information systems for management have been
evolving for more than a decade. <u>The integration of management
and imaging systems is essential to realizing the fully digital
department. The expansion of management computer systems to
provide an index and interface to image data makes full use of
the patient and exam data already in the system</u>. As these
changes go forward, these systems must be adapted for effective
use in small departments.

<u>The second function is image acquisition</u>. CT, DSA,
Nuclear Medicine gamma cameras with computers and MRI devices
acquire images in digital form. These digital modalities
comprise 25% of all imaging studies done today in some
departments. This percentage is likely to increase slowly as
these imaging modalities are used more frequently and more
widely.

Ultrasound presents a special case. Digital processing
is now used internally in many diagnostic Ultrasound units.

However, the output images from these devices are still in analog form. Some manufacturers now provide ports to access the digital data, including direct memory access (DMA). The future role of digital Ultrasound images is difficult to predict. There is a special major problem with real time Ultrasound; the amount of data from real time scanning is so immense that no good way exists to handle it in digital form.

The remaining 75% of imaging procedures include chest, bone and other plain radiographic studies, portable exams, fluoroscopic exams and a few miscellaneous categories. In the case of plain radiography there are several potentially exciting approaches to incorporating these studies into the digital imaging realm. The features are summarized in Table 1.

The third function is <u>image distribution</u>. Networks have the potential of making images available simultaneously in widely separated places. Exact digital copies overcome the one at a time user aspect of film based radiographs. Unfortunately, the immense volume of data in our images cannot be handled adequately by presently available local area networking technology; it is reasonable to expect industry to make progress in this area.

The fourth function of <u>image interpretation is the most important of all. It also is the greatest single problem blocking realization of the fully digital department</u>. I will say more about it later.

The fifth and sixth functions are <u>image redisplay and image archiving</u>. These functions together should result in multiple and easier access to imaging studies, in the potential for conferencing and perhaps at last in the elimination of lost studies. Various hardware approaches to handling the storage of images are available, but they are not yet optimal. Higher capacity units at still lower cost are needed.

<u>What are the benefits and the cost offsets of a fully digital department</u>? As we continue on the spiral of increased restrictions on health care costs we are forced to ask from where will come the money to pay for this technology?

Table 2 is a listing of the benefits that would derive from a fully digital department. These must be weighed against

the capital expense of computer hardware, the software costs
and the salary and fringe benefit costs for the people to
operate the computer equipment.

There are two major areas of expense which would be
sharply decreased in a fully digital department. These are the
cost of the film used at present and the salary expenses of a
large portion of the film library personnel. We believe with a
high level of confidence that cost savings from these two areas
will more than offset the expenses of implementing a fully
digital department.

There is a problem, however, in that we will go through
a transition period, perhaps several years long, in which we
will maintain a reduced film archive. We cannot leap directly
to the fully digital department nor to its cost savings.
Radiographic film or some other hard copy will be necessary
without question during this transition period. The savings in
film library personnel can not be achieved all at once.
Substantial savings in film costs can be realized by converting
from large size film formats to digital images recorded on
8x10-inch multiformat film.

MAJOR UNSOLVED PROBLEMS

The concept of a fully digital department is intuitively
attractive. After all, with 25% of our images already in
digital form, how can one resist the world of computers? In
fact, there are three major problems which must be solved
before a fully digitized department can be achieved. There
are, of course, an additional host of specific problems.

The first major problem is the establishment of industry
standards. Little attention has been paid to this area by
vendors until recently. Now almost all vendors recognize the
need for sharing of image data. Establishing standards is
vital for the fully digital department, yet the standards must
be done in a way that maintains the proprietary interests of

vendors. The joint effort of the American College of Radiology
and the National Electrical Manufacturers Association is well
underway.

The second major problem deals directly with the quality
of the digital images. Digital images will not be used if they
lack adequate information content. There is a myth that the
superb resolution of film based radiographs cannot be
reproduced in digital form. This is simply untrue. Dr. Lehr,
Dr. Doi and coworkers have studied image processing of chest
images with a monster sized matrix of more than 7,000 square
pixels. It is impossible to tell the original analog film
image from such a digital image. On the other hand, very large
matrices such as this are not practical.

The matrix and image size are fundamental to digital image
quality. Keep in mind that there is no one correct matrix
size. CT scanning, MRI and Nuclear Medicine computing have
relatively low resolution requirements; matrix sizes in common
use today in these modalities are generally recognized as
adequate. Imagers must be very careful not to request better
resolution only for the sake of prettier pictures; moving from
a 512 square to a 1024 square matrix quadruples the number of
pixels in the image, which impacts on the speed of handling of
the image, the memory requirements, the storage requirements
and the cost in a very direct way. Those departments with
newer CT scanners have already seen this; for example, magtapes
which could previously store 8 CT studies in 320 pixel format
now hold only 3 studies in a 512 matrix.

It is of great importance to determine proper matrix
sizes as soon as possible for the 75% of imaging studies which
demand higher resolution, the plain radiographs. The
teleradiology study by Gitlin and coworkers found that a 512
square matrix did not preserve information content adequately
for chest radiography. A further evaluation of a 1024 square
matrix is now underway. Will a larger image plane matrix be
needed? How many gray levels are needed? The usefulness of a
given matrix depends, of course, on the size of the image to
which it is applied.

To summarize, digital images must be adequate to allow

radiologists to achieve the same diagnostic accuracy achievable from analog images. Note that this does not require the same resolution as film per se. What must be assessed is the information content of the digital image. Unfortunately, this is not at all simple. If marketplace pressures force vendors to a larger matrix size unnecessarily, our capital dollars will pay for the luxury.

The third and most important problem which must be solved prior to achieving a fully digitized department is interpretation of digital images directly from digital displays. In spite of tracker balls and macro keys on physician display consoles, most CT and other digital images are interpreted from film based images displayed on a viewbox! This freezes the image in analog form at one preselected group of settings chosen either by the technologist or by the radiologist. The only digital images commonly interpreted directly in digital form today are in Nuclear Medicine, particularly the Nuclear Cardiology images where the movie like presentation of the data is important.

The current practice of interpreting digital images from film at the viewbox rather than directly from a digital display represents a failure of the digital display system. The fully digital department will not be possible unless digital interpretation stations can be improved so that radiologists elect to use them.

The most important reason for this state of affairs is the burden placed on the radiologist by the display system. The radiologist becomes a typist or wears down his or her finger making choices from a menu. The interaction is cumbersome and time consuming. A good test of equipment utility is whether or not it is actually used. Even the most modern digital interpretation stations fail this test. In fact, they have in common a high degree of user unfriendliness.

Several extinct Radiology reporting systems which required direct entry of information by the radiologists were discarded because they required too much effort and time. It is not at all unreasonable for the radiologist to require that interpretation of digital imaging studies from a display console be faster than can be done from analog radiographs on a viewbox. Indeed, nothing less is apt to succeed.

Imagers and allied scientists must investigate, understand and make known what is important in extracting diagnostic information from digital images. What aspects of data presentation, data manipulation, display of comparison images, reconstructed or reformatted images, and/or paging through image sequences are important? These are complex questions which need careful study if future digital interpretation stations are to meet the need. How many display screens are needed? Will comparison studies be shown side by side on one screen or on adjacent screens? What human factors are important? Are cathode ray tube displays adequate? How should multimodality displays be handled? What digital filtering, gray scale mapping, zoom, region of interest, et cetera, functions are needed?

CONCLUSION

How soon will the fully digital department be a reality? Most of the pieces of hardware required can be purchased in one form or another today. Nevertheless, the fully digital department is some time away because no adequate digital interpretation station exists which radiologists will use in preference to the viewbox, and that is a fundamental requirement which must occur before the benefits of full digital archiving and communications can be realized. Digital archiving is useless if the archived images will not be used. The evolution to the fully digital department will require more than four years, perhaps as much as nine or more years to accomplish in an exemplary manner. In many ways, the hardware is the smaller problem; the system design and the programs needed to make it work constitute a far larger task. The most difficult challenge is the digital interpretation station.

It is clear that entirely new strategies must be devised

and programmed. The digital interpretation station should
present one case after another to the radiologist in an
automatic way requiring almost no interaction except for
initial identification of the radiologist and the group of
studies to be reviewed. Simple, human engineered interactions
should permit intervention and further manipulation of data
when desired. The various images, possibly including some
preselected processed views such as reconstructions, should be
displayed to the radiologists in a manner which allows
extraction of the diagnostic information with a minimum of
effort in a minimum of time.

Digital imaging has profoundly affected Radiology.
Computers will be an even more important tool in the future.
Adequate digital interpretation stations will make possible the
fully digital department. I believe that through the combined
efforts of many we will meet the challenge of harnessing
computer technology to enable imaging to more effectively serve
patients.

Table 1

DIRECT DIGITAL ACQUISITION
OF PLAIN RADIOGRAPHIC STUDIES

Scanner Type	Patient Dose*	Use Present Equipment	Flexibility
Point Scanner	low	no	good
Line Scanner	standard	no	chest only
Area Scanner	standard	yes	excellent

*Doses vary and are application specific.

Table 2

BENEFITS OF A FULLY DIGITAL DEPARTMENT

1. Images can be retrieved in a short time.
2. Images can be transmitted to areas where needed.
3. Images can be displayed in many geographical areas.
4. Coequal images can be used simultaneously in multiple areas.
5. Studies are available to authorized viewers immediately after acquisition.
6. Even very recent studies and reports on a patient are available.
7. Exam sequencing, exam tailoring and integration of diagnostic data would be possible.

Looking Back at PACS Attempts -

What Has Happened Since PACS I

Gerald Q. Maguire Jr.
Department of Computer Science
Columbia University in the City of New York
New York, NY 10027 USA

1. Introduction

My collaborators and I originally came to this PACS effort, bringing the expertise of Physics, Computer Science and Radiology to bear on the subject. Since the time of our original PACS I paper [7] many years ago, we have been able to influence a number of PACS efforts: namely, a large nuclear medicine system at Middlesex University Hospital in New Brunswick, NJ [8], a broad band radiology system incorporating commercial viewing stations overseen by Y.Bizais in Nantes, Brittany (France), and a broad band video system linking the many different floors on which the radiology department is located overseen by P. Cahill at New York Hospital, Cornell Medical Center in New York City and reported on at PACS II. [2] Now, the PACS effort has moved passed the development stage and into the commercial exploitation stage. However, a number of interesting questions concerning the implementation of PACS in medical applications, particularly with respect to Radiology remain. (Unfortunately, due to lack of support for PACS activities, some of our computer science colleagues (Baxter, Schimpf and Zeleznik) have moved into research in other areas.)

This paper will attempt to focus on these questions.

2. Original Needs

We first noticed that things were less than ideal, when one of us, James H. Schimpf, was asked to do a comparison of CT scanners for the purpose of assessing image quality and stability. This necessitated taking a standard phantom from site to site. It was discovered that some CT scanners had magnetic tape for long term storage and image exchange while others only had floppy diskettes. Then of course, there was the task of reading each of these media into an existing and fortunately friendly computer system for intercomparison. The lack of image format standards led to many hours of programming simply to read the data. This prompted the call for at least a standard for image exchange and lead eventually to the publication of the AAPM Report "A Standard Format for Digital Image Exchange" [1]. The lack of acceptance for even this most rudimentary document has been disappointing.

Secondly, we as many others, were faced with the fact that our CT scanners were quickly becoming old and obsolete technologically, and hence being replaced. The decommissioning of the early scanners, accompanied by the resulting inability to view the

NATO ASI Series, Vol. F19
Pictorial Information Systems in Medicine
Edited by K. H. Höhne
© Springer-Verlag Berlin Heidelberg 1986

data collected on them, led us to the idea of creating a universal viewing console. This would allow any number of images collected from different sources, be they CT, Nuclear Medicine, or Ultrasound, to be viewed digitally, on the same viewing console. This, of course, led us back to the standard image format.

While working on this common image viewing and processing console, we were approached by a group of psychiatrists interested in comparing positron emission tomography (PET) scans with matched CT scans. We found that we could easily support this effort in the context of our wider goals. Additionally, a plastic surgeon interested in cranial/facial reconstruction, as well as a radiologist interested in viewing and analyzing spinal disorders from CT scans, were interested in using our system.

3. Guidelines

We saw early on that our only hope of success was to set up some guidelines which would contain the problem and make it possible to accomplish. We tried to develop the *minimum* in new concepts and equipment. We took seriously the adage that "a good engineer is one who can recognize a good idea (even if it is not his own) and use it - with the serial numbers appropriately filed off".[1] We therefore set out to use existing concepts and equipment wherever possible and to exploit each of these to the fullest.

One key to success was the decision to partition the problem. In doing so, we left the field open so that some problems could be solved by others (perhaps with some encouragement). We used existing research and hence were able to let someone else bear some of the cost burden. We were able to hide the details and reduce the complexity. By going slowly we were able to take an evolutionary path which permitted us to gain experience and take advantage of feedback from the people using our system. By getting the users addicted to being able to do things simply, we allowed our system to grow in popularity, thus providing more opportunities for feedback. We tried above all to avoid unnecessary work and development. We worried less about early performance, in the spirit of Don Knuth, and tried to remember that if we didn't have the time to do it right the first time, we might never have the time to fix it.

4. Hardware

Meanwhile, processors continued to improve in performance and drop in price, making it more and more feasible to off-load central control onto small stations placed throughout a given department. Table 4-1 shows the remarkable improvement in the speed of processors.

In addition, the development of Local Area Networks (LANs) utilizing broadband RF technology (which takes advantage of the cable TV industry), allowed a very fast and flexible network to be configured with available, relatively inexpensive and reliable parts. The real beauty of broadband systems lies in their adaptibilty to changing needs. One of the common recurrent occurences is that our needs and/or desires outstrip our capabilities.

---

[1]Theodore Sturgeon

Processors	MIPS
DG Eclipse S/200	0.7
Prime 750	0.75
VAX 11/750	0.75
Motorola 68000 10MHz	0.75
DG Eclipse MV4000	1.0
Vax 11/780	1.1
Motorola 68000 16Mhz	1.270
IBM 4341 Group 2	1.270
Ridge 32	1.5
Dec 2060	1.5
Gould 32/97	4.67
AMDAHL 5860	10.0
Gould Power Node 6000	26.67
Cray 1	80.0-100.0
Cyber 205	120.0
Hitachi and others	400-500
Fifth Generation Machines (1990)	$>10^4$

Table 4-1:   Table showing processor capabilities in Millions of Instructions per Second (MIPS)

With broadband, it is possible to extend the network by inserting another set of frame buffers and moving down one level in the cable tree.   This is possible due to the use of frequency selective splitters and taps in mid-split cable systems.   Figure 4-1 gives a representation of this by showing the headend of the system.   When (as in case 1) only one framebuffer is located at the root of the network and transmitting on a particular frequency, everyone receives the same signal (**A**).   However, (as shown in case 2) when two framebuffers in separate subtrees are both broadcasting on the same frequency, then two different signals (a and b) can be sent.   Displays in one part of the tree will receive a (the lefthand side in the figure) and those in the other part of the tree will receive b. Recall that in a mid-split broadband system, the high frequencies are outbound on the network, while the low frequencies are inbound.   This results in being able to send information to the headend from both the a and b trees and being confident that the information, when it is outbound traffic, is confined to a given subtree.

5. Data Volume

It is necessary to consider the amount of data that will have to be both transferred over the network and available to it.   The formatted data volume consisting of a minimal set of attributes per exam which may be defined as follows (from the New York University radiology database):

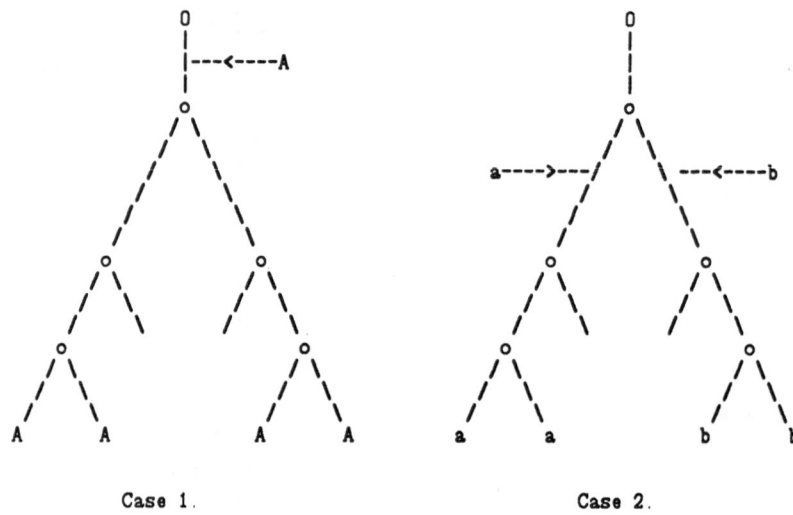

Case 1.   Case 2.

Figure 4-1:   Broadband Tree

patient indentification number(7 digits)
name(30 characters)
birth date(6 digits)
date of exam(6 digits)
type of exam(5 characters)

Assuming ASCII storage, this amounts to 54 bytes per exam.

The additional text data volume consisting of the average diagnostic report which contains about 200 words adds an additional 1200 characters. If we assume that the remaining text data (composed of histories, preliminary reports, annotations, etc.) produces again this amount, and that there are at least four types of documents per exam, then we can estimate about 4800 text characters per exam.

In terms of image data volume, we estimated [7] on the basis of figures from NYU, $2.75 \cdot 10^9$ pixels (2 bytes per pixel) per day from all modalities. This included estimates assuming digitization of all radiographs. NYU is a 700 bed hospital. Dwyer et al. [4] at the University of Kansas Medical Center, estimated 2.5-$5.0 \cdot 10^8$ bytes per day, solely from digital systems. The University of Kansas has 614 beds.

These estimates match those for the University of Utah Medical Center, when normalized for the institution size difference, and also match the data presented by D. Meyer-Ebrecht. [3]

Thus assuming that there are 250 working days per year, and allowing for additional images resulting from image processing, the total digital storage requirements for image data over a ten year period was estimated to be 10^{14} to 10^{15} bits to 10^{14} bytes. Handling this amount of data is not unreasonable. For example, the Lawrence-Livermore computing facility estimates that data is accumulated at the rate of 10^{13} bits every six months. A lot of the data produced is never looked at, but the Livermore philosophy is that it is easier to keep the data than rerun the program. Although this represents a large amount of data, it is not unique to PACS and thus commercial manufactures are developing storage systems to provide for these data volumes.

This brings us to the question of archival storage. There are several computer manufacturers (Tandem, Stratus, Synapse ...) who build extremely reliable hardware. Thus as storage media are developed - 6250 bpi tape, streaming tape drives, optical disks, these manufacturers offer a soultion to archive system reliability and flexibility. However, they do not have the staff to build specialized network interfaces. But, their price is dropping and they provide isolation from storage technology. Hence, the use of available, plug compatible network hardware then becomes a necessity.

6. Survey of Medical Image Source and Computer Suppliers

In preparation for this talk, I attempted to survey about one hundred and fifty manufacturers who sell either image source equipment (CT scanners, Digital X-ray Equipment, etc.), or computer systems for use with medical equipment (such as nuclear medicine computer systems). The questions asked each were as follows:

1. What are your plans for an image standard? Will you participate in or follow the American College of Radiology/National Electrical Manufacturers Association (ACR/NEMA) recommendations when and if they are promulgated and adopted?

2. What electrical interface do you intend to use? For example, will you use the IEEE 802 standards?

3. What software protocols will you support? For example, International Organization for Standardization (ISO), Open Systems Interconnect (OSI), European Computer Manufacturers Association-72 (ECMA72) or National Bureau of Standards (NBS) TP or Department of Defense (DOD) TCP? What protocols do you plan to use for layers 5,6, and 7?

4. What functions do you intend to provide? For example,

   a. transfer of images
   b. transfer of directory lists
   c. remote procedure invocation
   d. additional functions

One manufacturer[2] sent the following reply:

1. The electrical interface conforms to:

   - IEEE 802.3
   - Ethernet 1.0
   - ECMA 80 and ECMA 81

2. Software protocol:

   - Layer 2:

---

[2]Thomson-CSF

    * IEEE 802.2 Class 1 and IEEE 802.3
    * Ethernet 1.0
    * ECMA 82 and ECMA TR14

- Layer 3:

    * ECMA TR14

- Layer 4:

    * ECMA 72 and ECMA TR 14
    * ISO DIS 8072 and DIS 8073 Class 4

- Layer 5:

    * ECMA 75 Basic subset
    * ISO DIS 8326 and DIS 8327 Basic subset

- Layer 6:

    * empty layer

- Layer 7:

    * A home-made simplified subset of the ECMA 85 Basic subset or the corresponding ISO technical reports.
    * Only a file transfer protocol has been considered.

This was the most complete response received, several other firms indicated that they were following or participating in the ACR/NEMA activities. Quite a large number of firms indicated that they simply sold equipment to OEMs who integrated it into systems for final sale.

7. PACS Goals

There seems to be a difference in goals regarding the many PACS efforts. Is the goal to get images out onto a network, or is the goal to provide a standard plug? This latter will provide a longer term solution. What is the time interval for actually putting imaging nodes on a network? For PACS standards, what are the goals of such efforts as ACR/NEMA?

For instance one major goal of the ACR/NEMA effort seems to be to provide a physical piece of hardware (plug) between the network interface and whatever is connected the interface (media), and the different vendors computer hardware. The intention of this plug is to decouple the internal computer backplane (bus) from the media specific network interface. This is useful if there are a lot of different backplanes. But, how many are there really? Unibus, Q-bus, Multibus, Versabus, VMEbus, DG-Eclipse come to mind, but networking manufacturers already handle these. It would seem then that this effort (the plug) is unnecessary.

However, the other part of the ACR/NEMA activity, specifying the data format, is very important. Once these file formats are described, existing software can be used to transfer these files from one system to another.

In general, what is needed is the appropriate decomposition of the problem, to decouple the independent decisions - to allow them to be made indepentently. This might be viewed as building a firewall between the different phases of the problem. On one side of one firewall, we might place the device manufacturers, i.e., those involved in imaging technology, storage technology or display technology. These people should have a well defined interface to the computer manufacturer for the devices they produce. The computer manufacturers on the other side of this first firewall, should take the information from the device manufacturers' equipment, and/or pass it to this equipment in a well specified format. In addition, the computer manufacturer should be able to reformat the information obtained from these devices into the network specific format, or to convert from that format into that necessary for information to be passed to other peripheral devices. This is manifestly a software (or a special purpose VLSI chip) task. In addition, there should then be a second firewall built between the computer manufacturers and the network manufacturers. Using hardware standards such as those provided by ISO, the network manufacturers will be able to accept the standard interface that the computer manufacturers are providing to the network interface. In this way, each of the manufacturers can concentrate on what they do best, i.e., display manufacturers do not need to worry about the details of imaging technology and neither need worry about archive technology.

An important issue then, is to specify the protocols that will be used to access the network. The first generation of network equipment, such as ethernet and broadband CATV, implemented ISO levels one and two in hardware and expected the processor following them to implement levels three to seven. The second generation of networking equipment, such as that built by Communication Machinery Corp., implements levels one through four in hardware and thus the host processor must only implement levels five through seven. This results in a great savings in processor overhead. As of yet, ACR/NEMA has not specified what protocols it's plug will implement. Thus the plug between the computer backplane and the network is only a partial solution of the problem.

8. Successes and Failures

Our initial plans for NYU have largely gone unfulfilled because the interest there is mainly in a viewing station. The physicians there are not yet willing to accept a full fledged network. Our efforts were mostly unfunded and bootlegged from other activities, which resulted in little time actually being spent on the project. In addition, the chairman was largely unsupportive, there were a very small number of people involved and these few were politically isolated, as no senior physician was involved. In our early enthusiasm we did too much for too little and failed to account for warring fiefdoms both within and between departments.

On the bright side, both New York Hospital (Cornell Medical Center) in New York City and Henry Ford Hospital in Detroit and recently, UCLA Medical Center [5, 6] have all installed broadband systems along the outlines proposed in PACS 1 [7]. In addition, in Nantes, Brittany (France), Y. Bizais is installing a series of VICOM display systems all running UNIX[3] in conjuction with his networking effort. We were instrumental in

---

[3]UNIX is a trademark of Bell Laboratories

influencing the shape of this installation.

9. Future Efforts

It is hoped that future efforts in **PACS** for medical applications will implement the following points:

1. broad band technology (with fiber optics incorporated where appropriate and as it becomes available)
2. tree-structured architectures
3. frequency agile modems, or at least the use of multiple channels
4. interfaces supporting transport services (**ISO** level four)
5. simple file transfer
6. local disk as a cache (for example, at viewing stations, image source devices, etc.)
7. a move toward multi-media electronic mail standards and ideas

References

[1]   Baxter, B, S., Hitchner, L. E. and Maguire Jr., G. Q.
A Standard Format for Digital Image Exchange.
American Association of Physicists in Medicine Series, Report number 10. N.Y., N.Y., 1982.
11 pages.

[2]   Cahil, P.T., McCarthy, R.H., Kaplan, P. and Hunt, W.
ICDBM: an Image Communications and Data Base Management System for Radiological Imaging.
In *Picture Archiving and Communication Systems (PACSII) for Medical Applications*, pages 134. Society of Photo-Optical Engineers, May, 1983.

[3]   Meyer-Ebrecht, D., Bohring, D., Grewer, R., Monnich, K.-J., Schimdt, J. and Wendler, Th.
Hierarchical Approach to Distributed Picture Information Systems.
In *First International Conference on Picture Archiving and Communication Systems (PACS) for Medical Applications*, pages 112-116. Society of Photo-Optical Engineers, January, 1982.

[4]   Dwyer III, S.J., Templeton, A.W., Martin, N.L., Cook, L.T., Lee, K.R., Levine, E., Batnitzky, S., Preston, D.F., Rosenthal, S.J., Price, H.I., Anderson, W.H., Tarlton, M.A. and Faszold, S.
Cost of Managing Digital Diagnostic Images for a 614 Bed Hospital.
In *First International Conference on Picture Archiving and Communication Systems (PACS) for Medical Applications*. Society of Photo-Optical Engineers, January, 1982

[5]   Huang, H.K., Mankovich, N.J., Barbaric, Z., Kangarloo, H. and Moler, C.
Design and Implementation of Multiple Viewing Stations.
In *Picture Archiving and Communication Systems (PACSII) for Medical Applications*, pages 189-198. Society of Photo-Optical Engineers, May, 1983.

[6]   Huang, H.K., Barbaric, Z., Mankovich, N.J., and Moler, C.
Digital Radiology at the University of California, Los Angelos.
In *Picture Archiving and Communication Systems (PACSII) for Medical Applications*, pages 259-265. Society of Photo-Optical Engineers, May, 1983.

[7]   Maguire Jr., G. Q., Zeleznik, M. P., Horii, S. C., Schimpf, J. H. and Noz, M. E.
Image processing requirements in hospitals and an integrated systems approach.
In *First International Conference on Picture Archiving and Communication Systems (PACS) for Medical Applications*, pages 206-213. Society of Photo-Optical Engineers, January, 1982.

[8]   Noz, M. E., Erdman, W. A., Maguire Jr., G. Q., Stahl, T. J., Tokarz, R. J., Menken, K. L. and Salviani, J. A.
Modus Operandi for a Picture Archiving and Communication System (PACS).
Radiology 152(1):221-223, July, 1984.

DESIGN CONSIDERATIONS FOR MULTI MODALITY
MEDICAL IMAGE WORKSTATIONS

Th. Wendler, R. Grewer, K.J. Mönnich, H. Svensson

Philips GmbH, Forschungslaboratorium Hamburg
Vogt-Kölln Str. 30, 2000 Hamburg 54, FRG

Abstract

The introduction of fully digital pictorial information systems in hospitals will result in a number of changes in the clinical work practice. The acceptance of new technologies will decisively depend on the design of medical image workstations. They have to fit into the organization of the diagnostic imaging department, adapt to the attitudes and the behavior of cliniciants and obey psychovisual and ergonomic rules. Technical complexity has to be made completely transparent for the clinical user by appropriate human interface design.

The paper desribes an image workstation approach aiming at these properties. Fundamental conflicts and problems in routine work with images are identified and problem solving strategies and functions are proposed. Most critical functions for convenient image handling and presentation, implemented in an experimental image workstation, are discussed in detail. User interaction principles that avoid the use of keyboards have been introduced that are based on simple rules, easy to understand and consistent for all work procedures. A new touch screen device has been developed for this purpose, combining minimum image degradation and high touch resolution.

Introduction

In the near future we expect electronic pictorial information systems to be introduced in hospitals, replacing today's photographic film based image archives in diagnostic medicine. Strategies for the introduction of picture archiving and communication systems (PACS) in the medical field have been intensively discussed [1], [2], [3]. The fully digital diagnostic imaging department will support filmless image aquisition, distribution, archiving and interpretation. As a result, the traditional way of medical image evaluation by means of film/lightbox systems will be replaced by new work procedures using electronic image displays. These changes in the user's work practice is one of the particularly important problems we have to pay attention to.

New electronic display systems have to be designed in a way that appropriately supports the diagnostic process. They will be, very unlike lightboxes, for more than simple tools only capable of presenting a number of images on a collection of monitor screens. Additionally, they have to provide facilities for an easy and convenient filmless image handling, for userfriendly communication with pictorial- and alphanumeric data bases, for appropriate high quality presentation of all kinds of images and related information on high resolution screens, for selected problem solving functions in the fields of image modification and image analysis. We refer to these devices as 'medical image workstations' [4]. They are, from the technical point of view, complex interfaces between pictorial information systems and their clinical users.

According to the organization of digital imaging departments, to different types of work procedures and to different kinds of clinical users, there will be a variety of different workstation types. The 'multi modality medical image workstation' is the most general category of devices we can think of in this area, capable of simultaneously handling, presenting and processing images from different imaging modalities. This means the absense of restrictions concerning spacial resolution, dynamic range and relevant operations to be applied to images. In particular, 'multi modality' stations have to include operations with large format, high resolution radiographs, which are and will be the majority of images we have to deal with.

The medical image workstation will be one of the principal radiologist's working places, therefore the key component to access all the available resources that are needed for the diagnostic process. So workstation design appears to be one of the most important and most critical issues among all the PACS problems, and the solutions in this field will decisively influence the overall acceptance of pictorial information systems in hospitals.

The user's situation

For the introduction of new technologies in the field of medical diagnostics a number of severe problems must be solved which all have to do with the fact that pictorial information systems have to be utilized by human operators (different kinds of medical professionals), who are interested in specific questions in a specific situation of routine work. A workstation concept will only be accepted by the medical community, if it functionally fits into the clinical work procedures and if its man-machine interface design is adequat for technically non-trained users. A high priority has to be given to all measures that make the technical complexity of PACS completely transparent to human operators.

A number of conflict areas can be identified that arise from a mismatching between the user's wishes (originating from the traditional work routine with photographic films) and the advanced technical solutions of electronic display systems. If 'softcopy' viewing instead of 'hardcopy' image handling is introduced, the result will be a number of changes in the user's work practice, which seem to be in dispensable due to some fundamental differences between film/lightbox and display systems. There are a number of different technical capabilities and limitations, different procedures of image handling and image presentation, different perceptual properties and image quality requirements. A number of workstation functions for the improvement of the diagnostic process will be desired which are not even known today. The introduction of novelties that differ from the traditional work procedure will not be rejected by cliniciants, if the image workstation design is oriented towards the goals of radiology and if technology driven approaches are avoided.

Workstation Design Considerations

Acceptance problems in the introduction phase of pictorial information systems can be solved, if medical workstation design is based on a thorough understanding of
- the attitudes and the behavior of clinical users
- psychovisual and ergonomic factors
- the capabilities and limitations of digital image processing techniques.

The user's attitudes and behaviors in the future are of course difficult to predict. They have partially to be derived from today's work practice for systems with new and different capabilities. What we can expect, regardless of how the work procedures will change, is that cliniciants want to do their work at least with the same convenience, accuracy and efficiency as today. A close cooperation between people from technical and medical disciplines will be necessary to arive at these properties. This will be a process with several iterative loops.

The functional workstation design objectives will to some extent depend on the organization of digital imaging departments and on the habits of individual radiologists, whereas a number of design considerations, especially in the field of human interface design, should be common in all workstation approaches. In fully electronic systems the man-machine interface becomes a most important factor, and a number of conflicts between technical solutions and human properties can be identified, if psychovisual and ergonomic rules are violated. If or ifnot workstation functions are accepted for the practical work, will depend on how they adapt to human perception, recognition and interaction.

It should be well understood that medical image workstation design is a very sensitive field. We should think of it as an iterative and interdisciplinary process, involving experts from medicine, psychology, human engineering, physics and electrical engineering.

The workstation approach described in this paper is based on 4 design steps (fig. 1):

1. Identify conflicts and problems that will appear with the introduction of filmless pictorial information systems.
2. Define appropriate user functions to avoid conflicts and solve problems in routine work.
3. Define man-machine interaction stategies to enable non-technical users to utilize the workstation functions.
4. Technical implementation (hardware, software).

These steps should be taken in this order.

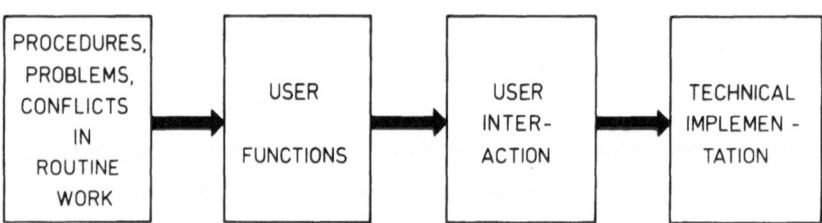

Fig. 1: Medical Image Workstation design steps

Conflicts in the filmless imaging department

Medical image workstations have to provide a number of user functions with images which may be grouped in 4 categories

- image handling (the process of accessing pictorial information in a distributed storage system)
- image presentation (the process of visualizing the accessed information)
- image modification (the process of manipulating images for the benefits of improved information extraction)
- image analysis (the process of extracting non-pictorial information from an image)

The first 2 categories, handling and presentation of images, play a basic role in every type of workstation and requires attention. The paper will therefore focus on these fundamental problems.

The handling of electronic images seems to be one of the most critical functions, as differences to the use of photographic film are obvious. The user's 'geographical' orientation, the knowledge about the physical location of the images he is interested in, is lost. Images to be accessed are resident in 'anonymous' storage devices somewhere in a distributed technical system. A high priority has to be given to the design of user functions that help to handle digital images with the same convenience as photographic films that are located in the patient's film jacket. Even more, it should be possible to create image handling procedures which are better adapted to the clinical work procedure, easy to learn and understand, and that operate much faster than today. This has to be supported by appropriate image storage and management strategies (as e.g. described in [5]).

The fundamental differences in the display capabilities of 'hard' images and 'soft' images result in some conflicts in the field of image presentation:

- spatial resolution conflict
- contrast resolution conflict
- confict with desired standard presentations
- conflict between multiple images to be presented on a few display screens

- image information conflicting the display of overlaid non-image information

A simple set of functional tools should be available in every workstation that allows to modify a given display presentation according to the user's wishes in the various work situations. A tool set for most frequently needed image presentation functions is proposed in the next paragraphs.

User interface design

The majority of existing digital image viewing stations is not really used as a diagnostic tool. The evaluation of images, even from electronic ima- ging modalities (like CT), is still based on photographic film, a result of the high degree of user unfriendliness caused by inadequat human interface design. The fully digital diagnostic imaging department will never exist, if this situation continues. Much more emphasis has to be given to the principles of man-machine communication.

Human interface design involves a number of aspects which are not independent of each other:
- perception: Design of display systems that support the mechanism of human vision (resolution, phosphor, flicker, surrounding light conditions etc.)
- recognition: Problem fitting presentation of information on display screens that support the process of recognizing relevant information based on the user's experience (desired image orientation, size, standard presentation etc.)
- orientation: User guidance by appropriate presentation of information that allows the user to understand the status and capabi- lities of the system he works with at any point of time.
- interaction: Design of interaction strategies which are based on a set of simple rules, easy to learn, easy to understand, con- sistent for all work procedures. Design of interactive elements which meet the requirements of ergonomy. Overall system design that gives high responsiveness in all work situations.

Not all of these aspects can be discussed in this paper. In the following a subset of workstation functions is described which, from the interaction point of view, are based on:

- a touch panel in front of a high resolution image screen (1024^2) as the principle interactive element (fig. 2....fig. 37)
- an interaction strategy which needs a single rule ("Whatever problem you have, you can solve it by touching the screen")
- a menue oriented activation of functions without keyboard use
- immediate response for all described functions.

Selected workstation functions for image handling and presentation

INDEX:

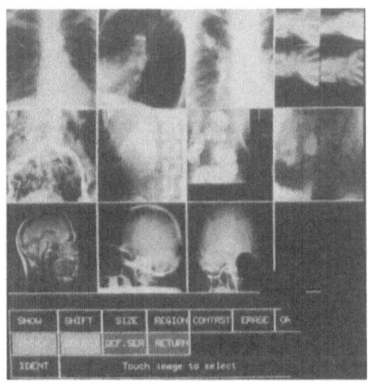

Fig. 2: INDEX (few images)

In any work situation the user can receive the presentation of a pictorial directory from which it is possible to select a number of images for detailed viewing. This INDEX function is a normalized survey of a set of images (fig. 2). According to different work procedures, the presented images may be:

- all available images from one patient ('patient's film jacket'). This type of directory is useful to select images to be preloaded into the workstation's local workfile for fast access during a later diagnistic work session
- a subset of the patient's images which are relevant for the current investigation (e.g. organ specific) and used during the diagnostic work session

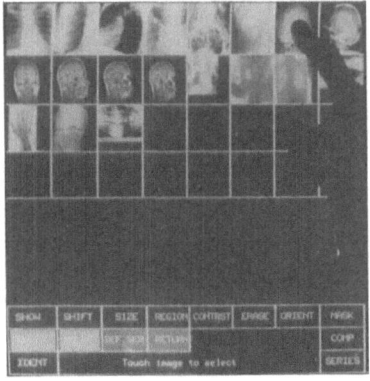

Fig. 3: INDEX (more images)

- a set of images collected from different patients, prepared for the purpose

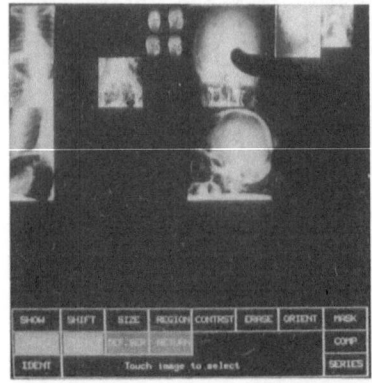

Fig. 4: INDEX (memory survey)

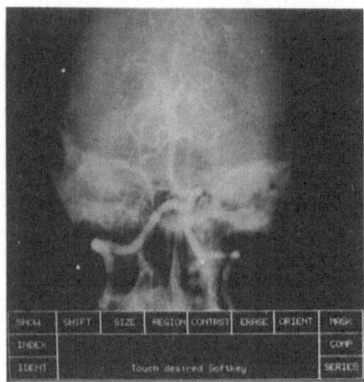

Fig. 5: OVERVIEW

of consultation (presentation to medical colleagues), teaching or research.

For scheduled work sessions, the images presented to the radiologist with 'INDEX' will be already resident in the workstations local workfile or semiconductor memory for very fast access. The preparation of this situation (preloading) is the typical work procedure for the clinical technicians. For non-scheduled retrievals (e.g. emergency), the 'INDEX' presentation is generated on the screen by directly accessing the central archive. In this case it will be a desirable feature to be able to access coarse surveys of images which can be retrieved ultimately fast. The S-transform picture coding scheme [6] is one of the most elegant ways to arrive at this situation.

Images may be selected from the 'INDEX' presentation for full resolution viewing by touching them. All touch-selected

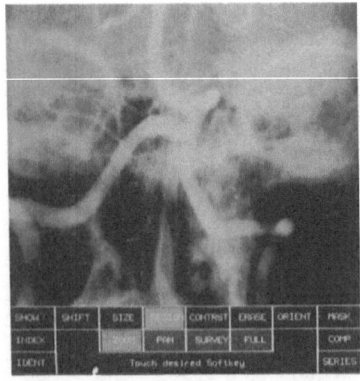

Fig. 6: REGION/ZOOM (1)

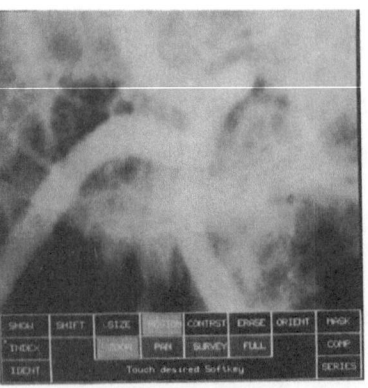

Fig. 7: REGION/ZOOM (2)

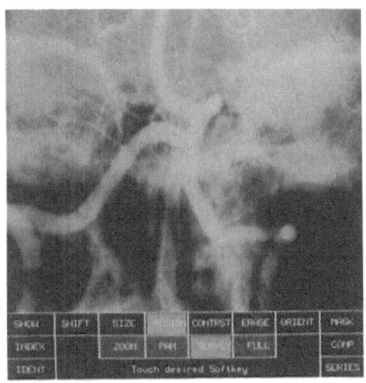

Fig. 8: REGION/SURVEY

images will be put on a stack, the top which will be visible when returning from the INDEX-mode. Fig. 3 gives the situation if more images have to be presented for a directory (the normalizing factor is selected automatically). In fig. 4 an alternative INDEX presentation is shown that gives a survey of the complete physical semiconductor image memory of the workstation.

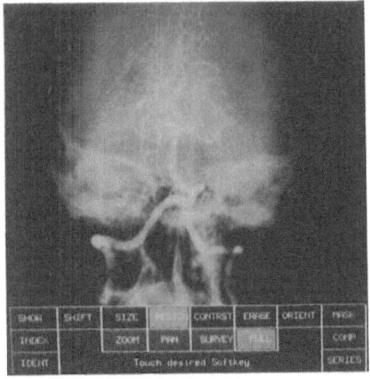

Fig. 9: REGION/FULL

OVERVIEW:

Returning from the directory after selecting images will result in the overview presentation of the most recently touched image.This is the presentation with the highest possible resolution that shows the document as a whole. The example in fig. 5 shows a 2048^2 pixel radiograph, now presented with 1024^2 pixel on the screen (overview factor 2).

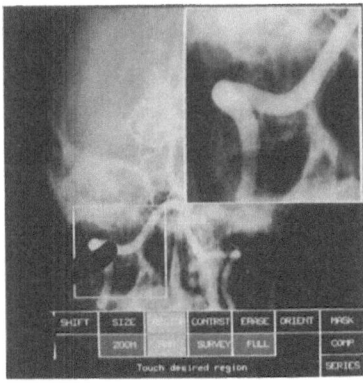

Fig. 10: REGION/PAN (I)

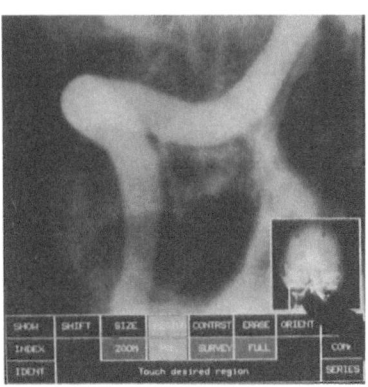

Fig. 11: REGION/PAN (II)

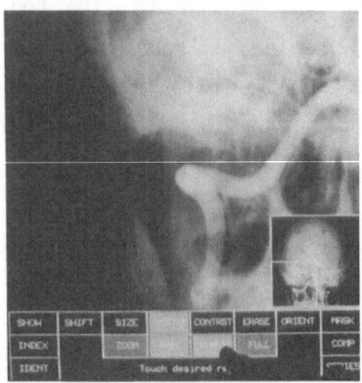

Fig. 12: REGION/PAN/SURVEY

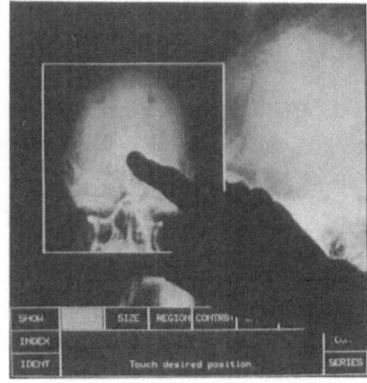

Fig. 13: SHIFT to position

REGION:

As the full image resolution is available for the user at any point of time, spatial resolution conflicts can be avoided by optimizing the presentation of interesting regions. By touching the appropriate softkey, the center region of the radiograph in fig. 5 will be magnified in different steps (REGION/ZOOM) without any time delay (fig. 6, fig. 7). REGION/SURVEY will be the similar function in the inverse direction (fig. 8). REGION/FULL will return to the overview presentation of the image as a whole, a desirable function that takes the human visual perception as a global process into account.

Previously defined regions may be moved across the image using any interactive element that generates X/Y coordinates (REGION/PAN). There are different possibilites and styles to implement the PAN-

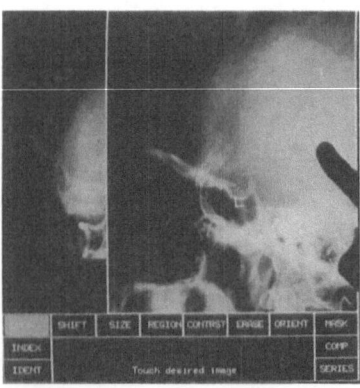

Fig. 14: SHOW (Image 1)

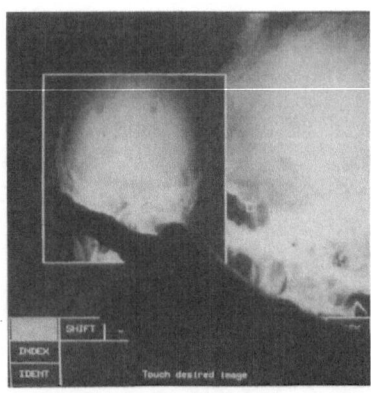

Fig. 15: SHOW (Image 2)

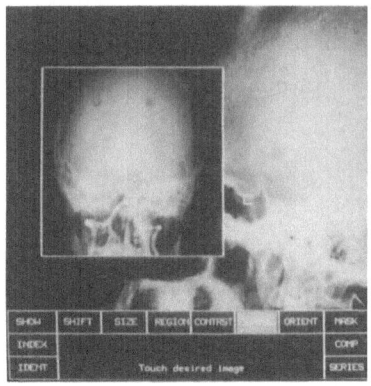

Fig. 16: ERASE (function call)

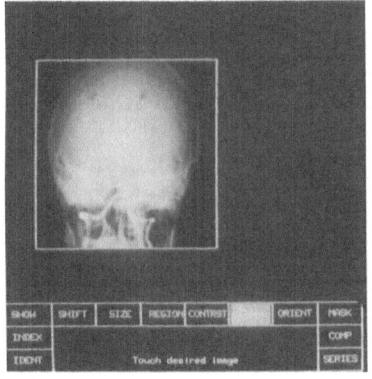

Fig. 17: ERASE (execution)

function. The region defined in fig. 7 may be moved directly (e.g. by trackball, joystick etc.). The modes shown in fig. 10 and fig. 11 offer the capability to move regions by using the touch panel if this is desired. Particularly the use of the small survey presentation of the entire image shown in fig. 11 (PAN II) seems to be an adequat human interface design for this function, as it supports global viewing, helps the user to keep an orientation of where he is panning and allows to use the overview area as a touch field to determine the window's position. Any combination of REGION sub-functions is possible (e.g. PAN II/SURVEY in fig. 12).

SHIFT:

If more than one image is present on the screen, the operator may wish to change the position of these images. The SHIFT

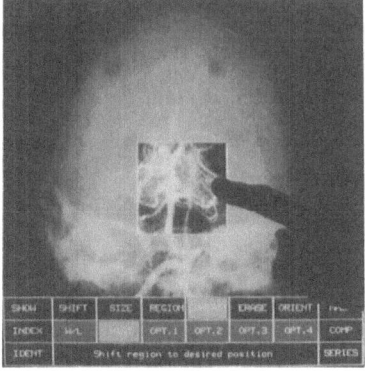

Fig. 18: CONTRAST

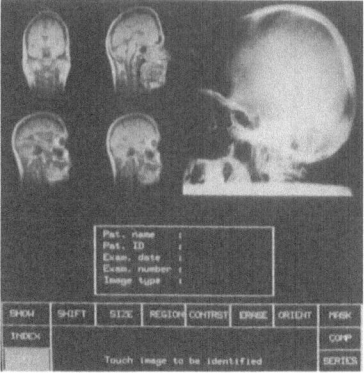

Fig. 19: IDENTIFY (function call)

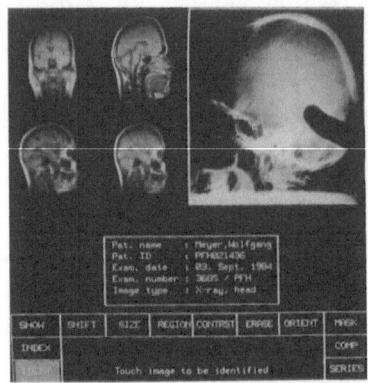

Fig. 20: IDENTIFY (execution)

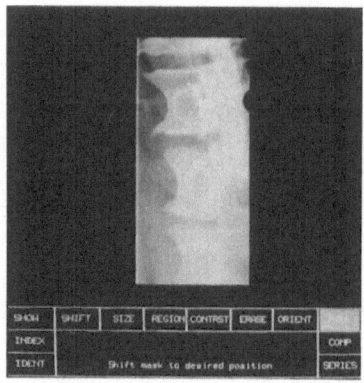

Fig. 21: MASK

function (fig. 13) in combination with the touchpanel offers an easy way to compose the screen layout.

SHOW:

Very unlike transparent X-ray pictures, digital images may be regarded to be opaque.Consequently, the presentation of multiple images might result in overlapping. The function SHOW (fig. 14 and fig. 15) in combination with the touch panel makes it possible to assign the highest display priority to a desired image in a simple and convenient way.

ERASE:

Images that are no longer needed on the display may be erased immediately from the screen (fig. 16 and fig. 17). They are, of course, still physically present

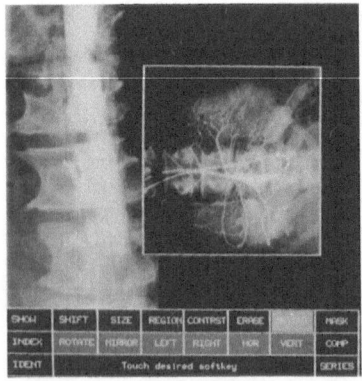

Fig. 22: ORIENTATION
(function call)

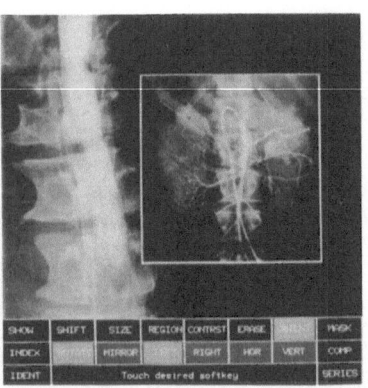

Fig. 23: ORIENTATION (execution)

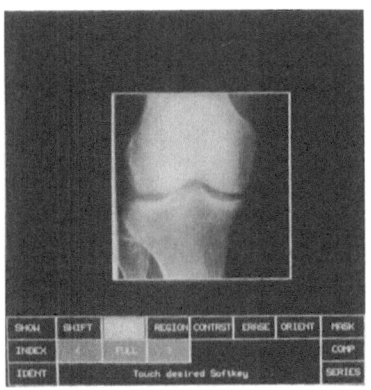

Fig. 24: SIZE (function call)

in the workstation and might be rapidly visualized again by using INDEX/SELECT.

CONTRAST:

A number of techniques that optimize the contrast range of a presented image are well known. Fig. 18 gives, as an example, the implementation of a real-time histogram equalization using interactively defined regions of interest.

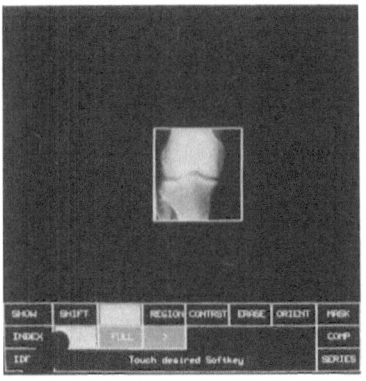

Fig. 25: SIZE (REDUCE)

IDENTIFY:

For all presented images related alphanumeric data sets are available. Part of this information will be made visible for the selected (touched) image by using the IDENTIFY function (fig. 19 and fig. 20).

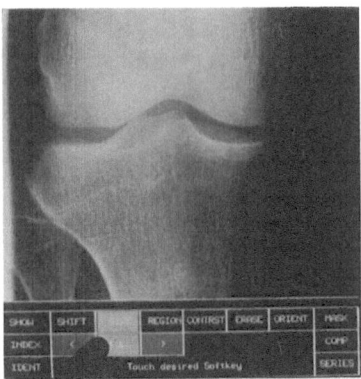

Fig. 26: SIZE/FULL

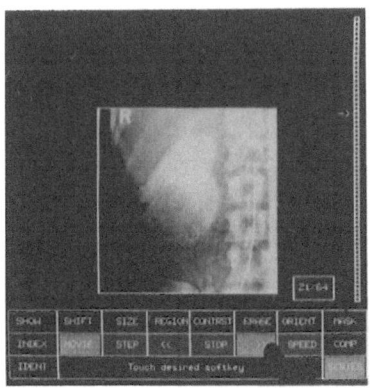

Fig. 27: SERIES/MOVIE

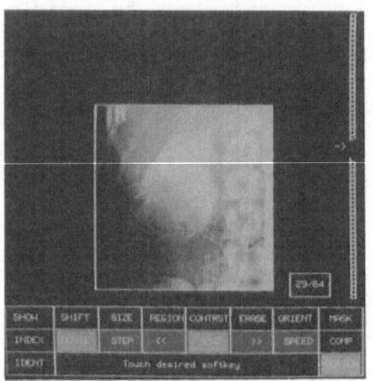

Fig. 28: SERIES/MOVIE (stopped)

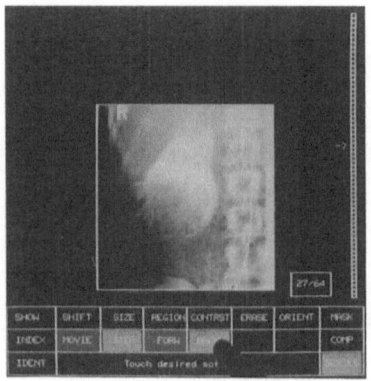

Fig. 29: SERIES/STEP

MASK:

At the lightbox, the contrast of a specific region may be optimized by using
shutters which cover a part of the image
area. A similar function is applied here.
Background greylevel can be drawn across
the screen from the 4 edges of the monitor (MASK, fig. 21).

ORIENTATION:

To arrive at a desired standard presentation it should be possible to change the
orientation of images (left/right, top/
bottom, angle). Fig. 22 and fig. 23 give
the example of a 90 rotation to the
left.

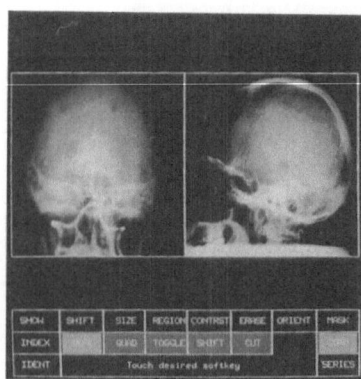

Fig. 30: COMPARE/DUAL

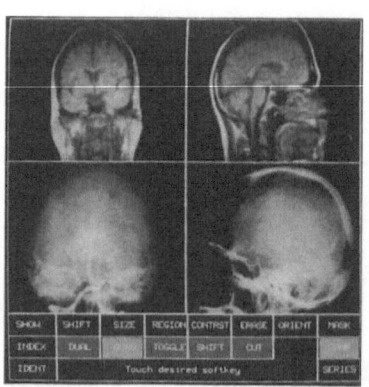

Fig. 31: COMPARE/QUAD

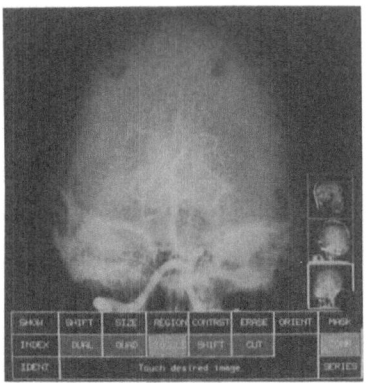

Fig. 32: COMPARE/TOGGLE

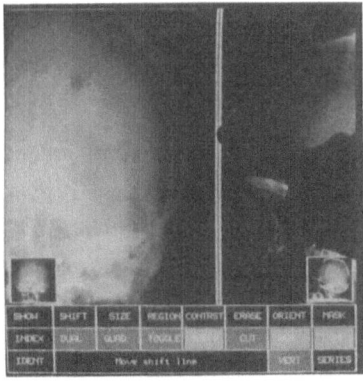

Fig. 33: COMPARE/SHIFT (hor.)

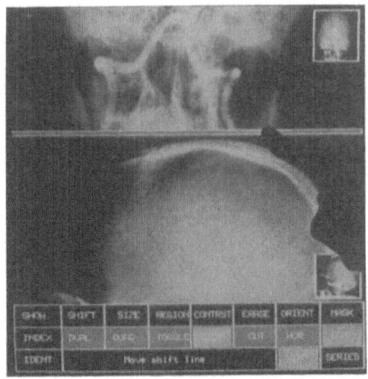

Fig. 34: COMPARE/SHIFT (vert.)

SIZE:

The presented actual size of electronic images will depend on their resolution and the resolution and physical dimensions of the display screen. In some cases a modification of the presented image size seems to be useful (e.g. life-size presentation). Fig. 24 and fig. 25 show the result of the SIZE operation (enlarge/reduce) applied to the X-ray image of a knee. The screen size is about 25×25 cm. The presentation of an image that covers the entire screen may be obtained by SIZE/FULL (fig. 26).

SERIES:

For dynamic viewing of image series (time sequences or slices) a number of presentation functions should be provided. In fig. 27 and fig. 28 the MOVIE presentation of a time sequence is demonstrated

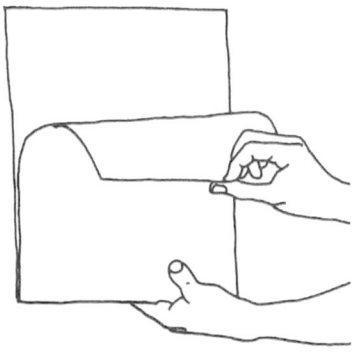

Fig. 35: COMPARE/CUT (principle)

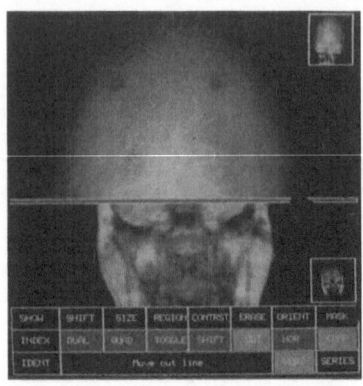

Fig. 36: COMPARE/CUT (vert.)

which operates very much like a video cassette recorder (foreward-stop-backward) with adjustable motion speed. In the stop position any image of that sequence may be selected by touching the vertical bar at the right part of the display which is also used as an indicator what part of the serie is looked at. The STEP mode (fig. 29) can be used to precisely ajust on a particularly interesting image (e.g. arrival time if contrast agent, maximum filling etc.).

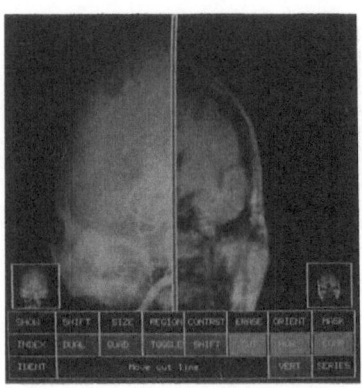

Fig. 37: COMPARE/CUT (hor.)

COMPARE:

As the comparison of two or more images is a frequently needed function in diagnostic medicine, a number of compare modes are proposed which support this process. All images to be compared will be touch-selected as described with the INDEX function.

The COMPARE/DUAL mode will present the two images resident on top of the select stack in the normalized way shown in fig. 30. To avoid the limitations in spatial resolution, this function could be utilized with the same type of interaction but on two seperated high resolution image screens. The function COMPARE/QUAD (fig. 31) works similarly with 4 images. The installation of 4 monitor screens appears, for psychovisual and ergonomic reasons, suspect.

A number of functions have been introduced that allow image comparisons with high resolution on a single screen. The function COMPARE/TOGGLE will switch between the selected images as soon as the according survey presentations in fig. 32 are touched. COMPARE/SHIFT presents two images in parallel and allows the user to see both alternatively by shifting the border line between them horizontally (fig. 33) or vertically (fig. 34). This function may be applied, if images of the same object from different views (e.g. frontal, lateral) are to be looked at. The principle of COMPARE/CUT is demonstrated in fig. 35. One image is put on top of the other and can be partially removed by moving the 'cut'-line in vertical (fig. 36) and horizontal (fig. 37) direction. This mode seems to be adequat for comparisons, where differences between images showing the same object (e.g. before and after treatment) are of interest. The 'shift'- and 'cut'-lines may be moved by touchpanel or any other interactive element.

Most of the described functions use a specific kind of information mapping from the image memory to the display. For clarity, symbolic graphical presentations of these mapping mechanisms are shown in fig. 38.

Implementation

The described workstation approach is being realized in the laboratory on the basis of an image processor building block system described in [7]. As high responsiveness is imperative for a userfriendly workstation, most functions concerning images are implemented with special hardware modules (e.g. all described image presentation functions will be realized with an intelligent address generator in real-time). Software structures for the flexible realization of user interface are being investigated.

Special attention has been paid to the design of the touch-panel. As can be seen from the image presentation functions, high touch resolution and minimum image degradations are essential requirements. This problem is being solved with a glass plane in front of the CRT tube that is attached to 4 strain gauges (expected resolution: 256^2).

Summary and Conclusion

A number user conflicts and acceptance problems in future fully digital pictorial information systems in hospitals have been identified and some problem solving workstation functions and interaction strategies have been proposed for image handling and presentation. Although necessarily incomp-

418

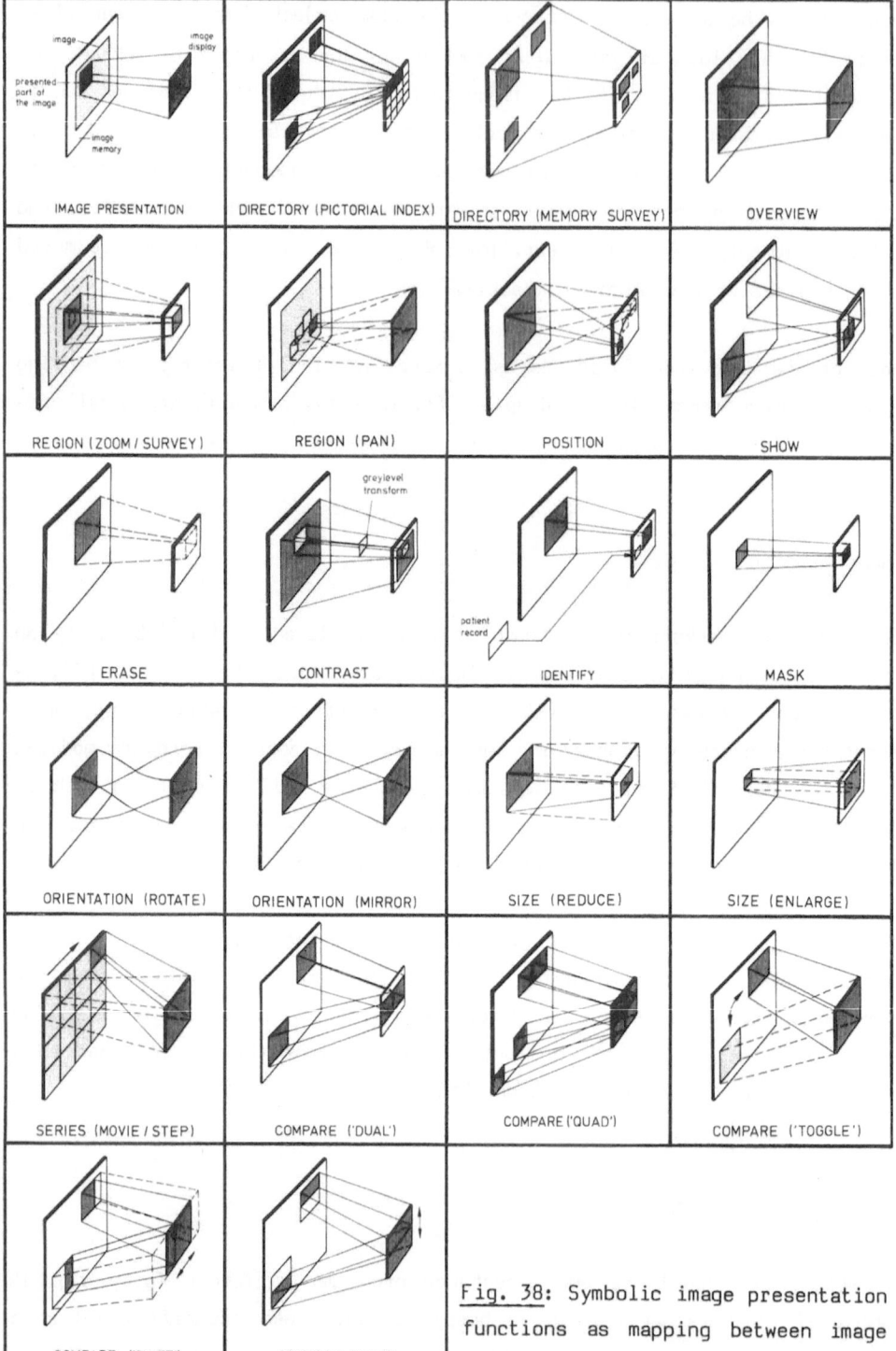

IMAGE PRESENTATION   DIRECTORY (PICTORIAL INDEX)   DIRECTORY (MEMORY SURVEY)   OVERVIEW

REGION (ZOOM / SURVEY)   REGION (PAN)   POSITION   SHOW

ERASE   CONTRAST   IDENTIFY   MASK

ORIENTATION (ROTATE)   ORIENTATION (MIRROR)   SIZE (REDUCE)   SIZE (ENLARGE)

SERIES (MOVIE / STEP)   COMPARE ('DUAL')   COMPARE ('QUAD')   COMPARE ('TOGGLE')

COMPARE ('SHIFT')   COMPARE ('CUT')

Fig. 38: Symbolic image presentation functions as mapping between image memory and display.

necessarily incomplete, this proposal might be a good basis to start with application studies in hospitals to gain feedback from the medical community.

During the investigation of user functions it became obvious, that a touch-panel is one of the most general among the interactive devices, supporting a very direct, comprehensive and consistent interaction strategy. However, we should avoid building up a religion around it. A workstation approach might benefit from additional more 'remote' interactive elements that the user can handle leaning back in his chair. Further studies are required in this field.

The discussion of image presentation functions adapted to the clinical work routine has shown, that already lots of things can be done conveniently and fast with a single high resolution screen. A second monitor will add some benefits in special work situations. We should therefore start to reject the workstation 'cockpit'-approaches that still dominate the discussion with a lost of CRT monitors and keyboard interaction, which are all techno-logy driven and lack appropriate human interface design.

Acknowledgement

The authors are greatly obliged to Prof. Dr. D. Meyer-Ebrecht (RWTH Aachen) for his valuable contributions in the initial phase of the workstation pro-ject and would like to thank all the colleagues from the Philips Medical Systems Devision who supported our work with helpful comments and discus-sions.

The described work was funded by the German Ministry of Research and Technology (BMFT) under grant No. 08R8301 4. Only the authors are respon-sible for the contents of the publication.

References

[1] A.J. DUERINCKX (editor):
SPIE Vol. 318, Proceedings of the 1st Int. Conf. on PACS for Medical Applications, Newport Beach, Jan. 1982

[2] S.J. DWYER III (editor):
SPIE Vol. 418, Proceedings of the 2nd Int. Conf. on PACS for Medical Applications, Kansas City, May 1983

[3] D. MEYER-EBRECHT, Th. WENDLER:
An Architectural Route through PACS
IEEE Computer, Vol. 16, No. 8, Aug. 1983

[4] D. MEYER-EBRECHT, Th. WENDLER:
Concept of the Diagnostic Image Workstation
Proc. SPIE Vol. 418, 'PACS II', Kansas City, May 1983

[5] K.J. MÖNNICH, Th. WENDLER:
Stategies for Storage and Distribution of Medical Images and their Realization in a Picture Base Prototype.
1984 Int. Joint Alpine Symposium, Medical Computer Graphics and Image Communication & Clinical Advances in Neuro CT/NMR, Innsbruck, Feb. 1984

[6] Th. WENDLER, D. MEYER-EBRECHT:
Proposed Standard for Variable Format Picture Processing and a Codec Approach to match Diverse Imaging Devices.
Proc. SPIE Vol. 318, 'PACS I', Newport Beach, Jan. 1982

[7] Th. WENDLER, D. BÖHRING, D. MEYER-EBRECHT, J. SCHMIDT, H. SVENSSON:
Modular Multiprocessor Picture Computer Architecture for Distributed Picture Information Systems.
Proc. SPIE Vol. 318, 'PACS I', Newport Beach, Jan. 1982.

MEDICAL WORK STATIONS IN RADIOLOGY

H.U.Lemke,   Technical University of Berlin

Abstract

Radiological and related services have been augmented by several
"computer technology driven" diagnostic imaging tools, such as
X-ray Computer Tomography, Nuclear Magnetic Resonance and Digital
Subtraction Radiography etc. With PACS, these services will
experience a further computer technology driven change in the
management and archiving of images.

More fundamental changes in the working profile of radiologists
and related medical disciplines, however, will come with the
introduction of graphic oriented Medical Work Stations (MWSs).
The user requirement space for such MWSs allows for the
specification of a multitude of different types of work stations
for diagnostic, therapeutic and teaching purposes.

The computer science disciplines applicable for the realisation
of these work stations are described. As a paper discussing concepts,
emphasis will be given to a general systems approach.

NATO ASI Series, Vol. F 19
Pictorial Information Systems in Medicine
Edited by K. H. Höhne
© Springer-Verlag Berlin Heidelberg 1986

1 INTRODUCTION

1.1 Radiological Services

It is widely expected that there will be an information science
derived evolution of techniques which will assist in diagnosis
and therapy planning for radiology and related departments.
This development will transfer geographically, organisationally
and/or mentally isolated imaging activities towards fully integrated
multi imaging modality diagnostic departments.
In the main this implies:

a) a transition from analog film systems to digital image
    generation systems,

b) integration of digital imaging modalities through Picture
    Archiving and Communication Systems (PACS) and

c) the graduated employment of Medical Work Stations (MWS).

The advantages which are expected from this development are:

a) increased accessibility of image information etc.,

b) improved reliability,

c) ease of use,

d) composite imaging,

e) retention of dynamic diagnostic information,

f) transmission and display of images to multiple geographical
    areas,

g) ease of interaction between medical subspecialists,
    e.g. radiologists with referring physicians,

h) expertise in subspecialities of diagnostic imaging can be
    widely disseminated,

i) studies are available to authorised viewers immediately
    after image acquisition (1),

j) very recent studies and reports on a patient are available,

k) exam sequencing, exam tailoring and integration of diagnostic
    data would be possible, and

l) cost savings.

According to a World Health Organisation (WHO) definition, a well-
structured diagnostic x-ray service at the country level should
form a pyramid consisting of three levels of sophistication:

1) Basic Radiological Service (BRS),
    the broad base of the pyramid, and available to the mass
    of the population requiring uncomplicated radiographic examinations.

2) General Purpose Radiological Service (GPRS),
    at the intermediate level, functioning as a back-up service
    for the BRS facility and a filter station for the sophisticated
    department at the top. An example is Digital Subtraction
    Radiography (DSR).

3) Specialised Radiological Service (SRS),
    performing specialised radiodiagnostic procedures, and
    undertaking research and training. Examples are x-ray CT
    and NMR.

Within this structure, there is an increasing dependence on computer
methods towards the upper or third level of the radiological services.
The advantages pointed out above are therefore somewhat restricted
to the upper levels.

Although many of the imaging modalities are in a stricter sense not
"radiological", as for example, ultrasound and those derived from
nuclear medicine, in the following they will be discussed under the
general heading of radiology.

1.2 Computer Assisted Radiology

In principle, computer assistance can be given to any of the three
levels as defined by the WHO. In practice, however, for a variety
of reasons, computer assistance has been mainly applied to the
upper or third level of the radiological services. Computer assistance
is here applied to

a) image generation,

b) storing and transferring of images, and

c) viewing, analysing and interpreting of images.

The application of computers to these activities, (activities which
characterise radiological departments), may be defined as Computer
Assisted Radiology (CAR).

Up to the present time, CAR has been applied with more or less
success to a variety of imaging modalities. These have been in the
main special purpose "stand alone" solutions to image processing
problems. Economic and health care considerations, however, show
that an integrated approach is necessary in the long term.

Concepts to this end have been developed in the United States and
Japan, and a number of prototype systems are already installed in
some university hospitals. These may be referred to as integrated
CAR systems. A Japanese concept for integrated CAR is shown in
Fig.1.1 (2). The basic concept behind a well advanced implementation
of an integrated CAR system in the USA is shown in Fig.1.2 (3).
Medical Work Stations are an essential part of these concepts.

2 MEDICAL WORK STATIONS

In the process of medical diagnosis and therapy, information is usually presented by means of the written word, pictures, graphics and the spoken word. For a particular patient, the sum-total of this information may be labelled the Medical Record (MR). In the interest of a patient oriented health care system, there are a number of important, if not vital, requirements on how the information in the MR should be organised and used, e.g. there should be:

a) access to the information in the MR at the right place, in the right time, by the right people,

b) maximum utilisation of information for diagnostic and therapeutic and teaching purposes,

c) reliable linkage of all patient specific information into one MR.

In addition, there are some desirable features of data representation and processing for the medical practioner, e.g. there should be:

d) uniform, structured and easy to understand data representation of MRs,

e) easily extendable MRs,

f) safe, protected and easily accessable MRs,

g) speedy statistical data gathering facilities on MRs,

and most important of all

h) flexible conferencing and consulting mode facilities, using MRs and all modes of communication (i.e.word, picture and voice communication).

Each of the above requirements for information management and evaluation can be satisfied by using medical work stations (MWSs) in a local network. A number of such MWSs, satisfying differing user requirements, have been developed, and are being used in various clinical settings.

2.1 User Requirement Space

The user requirement space for MWSs may be derived from the

a) imaging modality,

b) clinical speciality and

c) representation dimensionally.

These specify a 3-dimensional user requirement space as shown in Fig.2.1. Within this user requirement space, MWSs may occupy positions ranging from simple 2-dimensional diagnostic MWSs to sophisticated 4-dimensional teaching MWSs.

They may be single or multiple imaging modality oriented, and may be non-specific in their clinical or organic system or designed towards one particular clinical speciality (e.g.neurology or cardiology).

The type of MWS required has a considerable effect on the design techniques (and therefore the costs) to be employed for it's realisation.

The single most important issue to be solved is to allow for compatability between the MWSs in hardware and software. The aim should be to allow for upward compatability from 2-dimensional diagnostic work stations to 4-dimensional teaching work stations.

2.2 MWS Design Techniques

The technical disciplines which are emerging for the design of MWSs are largely derived from the area of Technical Informatics. They are(Fig.2.2):

- a) Computer Vision

- b) Computer Graphics

- c) Modelling

- d) Man-Computer-Interaction

- e) Application Programming.

Computer Vision

Computer Vision algorithms may be applied to the preprocessing, segmentation or analysis of images. They should take account of the psycho-visual behaviour of the human observer, and should provide quantitative information on processed images.
A very important emerging aspect is the a priori knowledge representation and model driven analysis of images. With this, concepts of machine intelligence are increasingly being employed in Computer Vision.

Computer Graphics

The most important Computer Graphic techniques available for 2-dimensional and 3-dimensional display are:

- a) rotational, translational and scaling transformations,

- b) parallel and perspective projections,

- c) kinetic depths cues,

- d) stereoscopic viewing,

e) hidden line and hidden surface removal,

f) illumination models,

g) "true" volumetric 3-dimensional imaging by means of varyfocal
    mirror displays or helical surface laser displays.

"Natural" representations of organic, physiologic and pathologic models,
with special surface texture algorithms, represent a particular challenge
to Computer Graphics.

Modelling

Although 2-dimensional and 3-dimensional modelling of patient specific
data from processed images represents a key function of a MWS, little
attention has been given to this important topic.
The single most significant aspect of modelling is the computer
representation of geometry. Current techniques evolve around polygones,
cuberilles, B-splines and cubic splines representations. For composite
medical imaging, composite modelling with the above and other techniques,
will become increasingly important. Extraction of 2-dimensional and
3-dimensional metric information must be possible, particularly for
diagnostic purposes. Model shaping operations should be provided for
therapeutic and teaching medical work stations.

Man-Computer-Interaction

This will imply appropriate interface design for the handling of alphanumeric, picture, graphic, voice and tactile information. As yet, there
are no generally accepted "human engineering" concepts on how to design
this interface. Current Man-Computer-Interaction systems for medical
imaging are ad hoc designs.

Application Programming

These are generally application specific modules which partially may be
executed on the MWS. Typical examples are modules for volume calculations,
surface area estimation, stereotactic frame data generation, isodose
surface generation, computer-aided diagnosis, statistics, surgical
planning and control, and prosthesis planning and manufacturing.

When these Technical Informatic disciplines are mapped against the axes
of imaging modalities (see Table 2.1) a matrix results. Each matrix
element may be looked at from the medical or technical point of view
(Table 2.2) showing problem areas which are still the subject of research.

According to the medical usage, MWSs may be classified into

a) diagnostic MWSs,

b) therapeutic MWSs and

c) teaching MWSs.

2.3 Diagnostic Work Station

At present, by the far the greatest demand exists for diagnostic MWSs. These are stations, which typically will provide the user with a number of functions for the display, manipulation, archiving and retrieval of images.

A diagnostic medical work station may be enhanced further with functions supporting Computer Assisted Decision Making (CADM). Several techniques have been developed and are known to work in particular clinical settings. Acceptability issues, however, remain to be solved. Formal knowledge structures, such as symbolic reasoning techniques may be one way to improve acceptability (4), but there seem to be still a long way to go.

2.4 Therapeutic Work Stations

The common requirement for therapeutic work stations is 3-dimensional modelling and representation. Although the 3rd dimension is also of some use in diagnostic work stations, as for example, for seperation of adjacent structures or volume and distance calculations, in therapy planning it is far more important. Typical usage for 3-dimensional modelling, display and manipulation is in the area of:

a) radiotherapy,

b) stereotactic surgery,

c) biopsy and general surgical planning,

d) orthopedics,

e) reconstructive plastic surgery,

f) prothesis design.

The Technical Informatic components necessary for 3-dimensional therapeutic work stations are shown in Fig.2.2.

2.5 Teaching Work Stations

Diagnostic or therapeutic work stations may be augmented to teaching work stations by providing special teaching files and adding interactive manipulation and correction software. At present, such work stations are being developed in a research environment; none appear to be in active use in either preclinical or clinical medical education.

In radiology, they are of particular interest for training purposes because of the independence on film and viewing boxes.

3 SUMMARY AND CONCLUSION

PACS is well on the way to find its place in medical imaging and
associated departments. Because of teleconferencing and teleradiology,
it can serve as a communication bridge within the subdivisions of
radiology and to other clinical specialities.
It is likely to bring with it a complete reorganisation of the communication
flow in health care systems. There is some painful work ahead to integrate
PACS with existing hospital information systems. Standards, such as those
discussed by ACR-NEMA, are a vital prerequisite to successful integration.

Although in the short and medium term, PACS will lead the way in the
introduction of advanced technology into radiology, it is the MWS which
will have in the long term the greatest effect on the work profile of
the radiologist.
One to two diagnostic work stations per department will eventually give
way to the trend of one work station per radiologist. The real number
of work stations needed may still be higher when internists, orthopedists
and urologists (which together outnumber radiologists several times)
also use these stations.

The advantages of the CAR components PACS and MWS are many and of
considerable significance. Their acceptance, however, is not only
a matter of costs but also a matter of the awareness of their potential,
and not least, the willingness of the medical community to adapt to changes
in their working profile.
There are voices, from the medical community, which predict, that the
traditional general radiologist will increasingly become a "creature" of
the past, and that the departments of radiology will become confederations
of imaging specialists.
In addition to physics and the traditional medical subjects, future
radiologists will need to be trained in computer science topics which are
relevant to imaging. This may also help the often made call for the
"algorithmic" (or some other computer oriented paradigm) approach to diagnosis.
There is also a need for a new radiology or imaging consultant, who is
capable of deciding on the type and sequence of diagnostic studies
appropriate for a particular patient. She or he must be familiar with
efficacy, costs, and applicability of various imaging modalities, but may
not be an expert in all the technical details. Taken together, this suggests
a team approach of radiologists, physisists and computer scientists in
Computer Assisted Radiology.

REFERENCES

(1) R. A. Bauman et al   "The Digital Computer in Medical Imaging:
                            A Critical Review", Radiology, Vol.153
                            Oct.1984, pp 73-75.
(2) S. Hashimoto   "Integrated Medical Imaging", AJR, Vol.142
                            Jan.1984, pp 192-193.
(3) H. W. Templeton   "An On-Line Digital Image Management System",
     et al,   Radiology, Vol.152, Aug.1984, pp 321-325.
(4) J. C. Kunz et al   "Computer-Assisted Decision Making in
                            Medicine", The Journal of Medicine and
                            Philosophy, Vol.9, May 1984, pp 135-160.

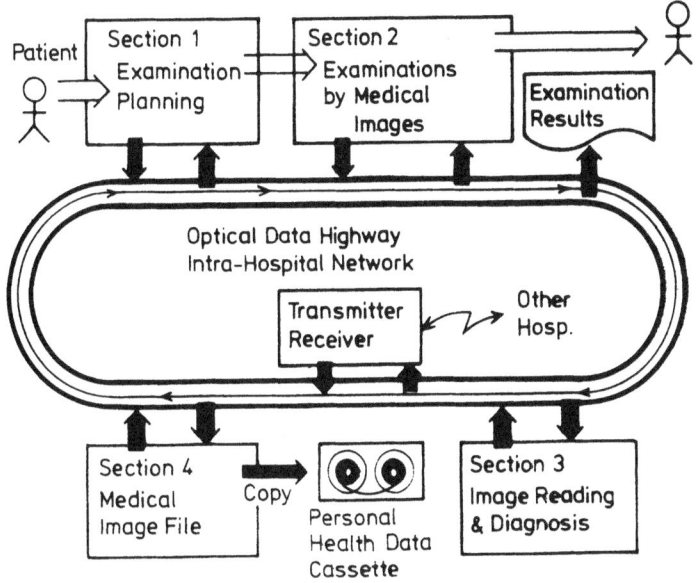

Fig. 1.1 A concept of integrated CAR in Japan (2)

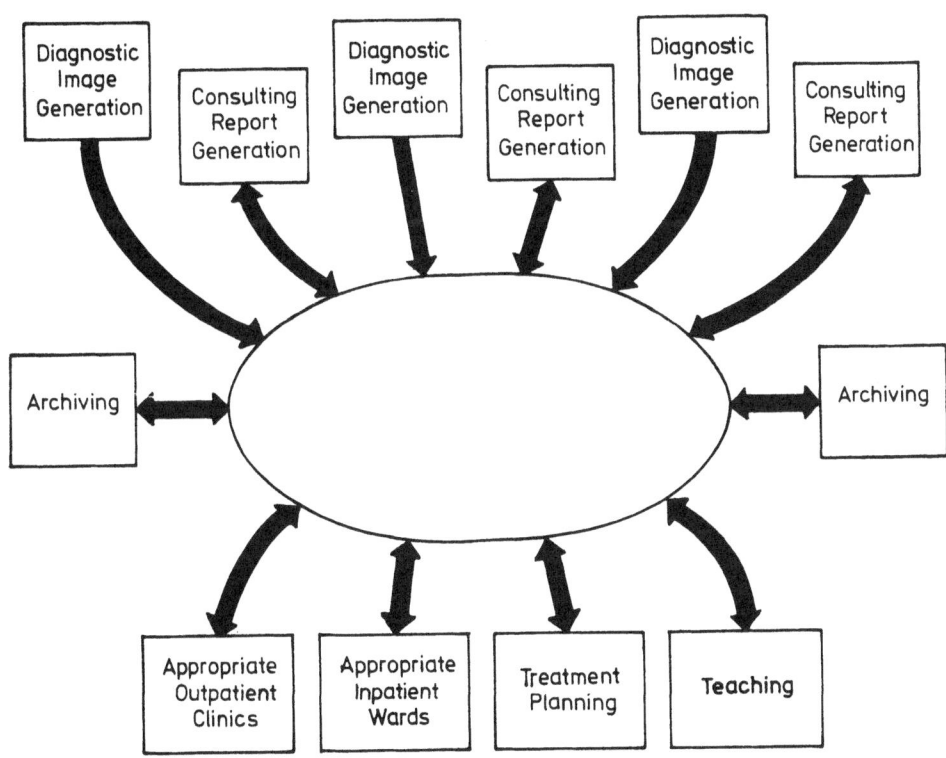

Fig. 1.2 A concept of integrated CAR in the USA (3)

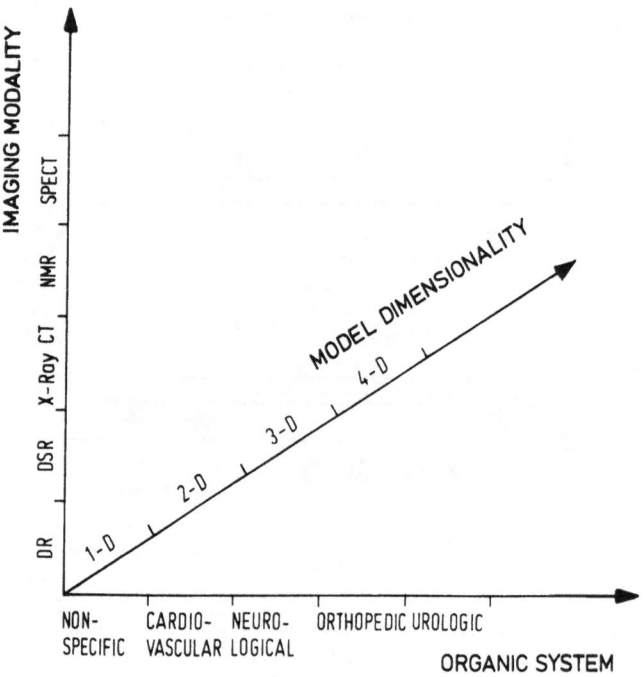

Fig. 2.1 User requirement space

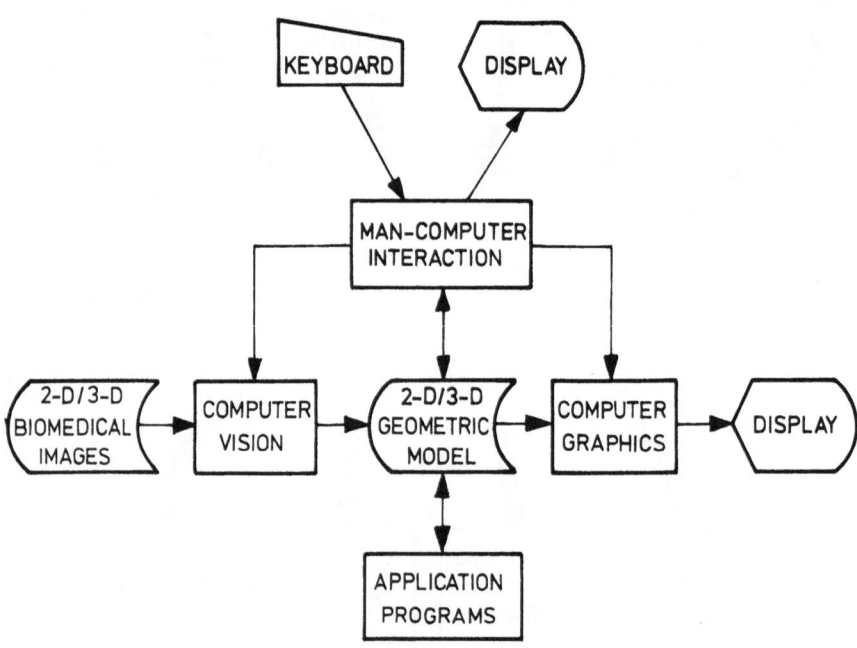

Fig.2.2 Information processing components of MWS

431

Computing functional capability \ Medical Imaging	COMPUTER TOMOGRAPHY				DIGITAL SUBTRACTION RADIOGRAPHY DSR	DIGITAL RADIOGRAPHY DR
	X-RAY CT	NMR	PET	US		
DIGITAL IMAGE GENERATION : DIG						
APPLICATION PROGRAMS : AP						
COMPUTER VISION : CV						
COMPUTER GRAPHICS : CG						computat. methods & technology \ Medical problems & req
MODELLING : MOD						
MAN-COMPUTER-INTERACTION : MCI						
PICTURE ARCHIV. & COMMUNICAT. : PACS SYSTEMS						

Table 2.1   Imaging modalities and information processing disciplines

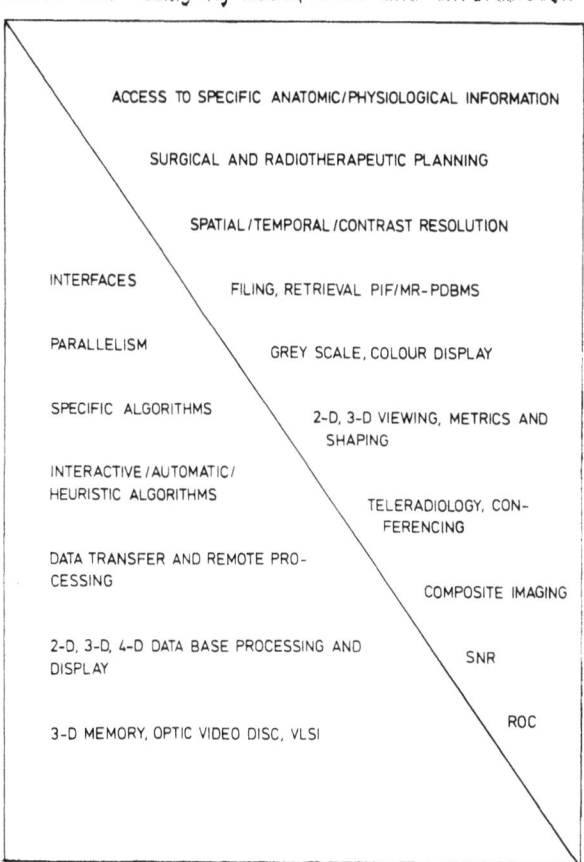

Table 2.2

Examples of medical and computational aspects

ACCESS TO SPECIFIC ANATOMIC/PHYSIOLOGICAL INFORMATION

SURGICAL AND RADIOTHERAPEUTIC PLANNING

SPATIAL/TEMPORAL/CONTRAST RESOLUTION

INTERFACES

FILING, RETRIEVAL PIF/MR-PDBMS

PARALLELISM

GREY SCALE, COLOUR DISPLAY

SPECIFIC ALGORITHMS

2-D, 3-D VIEWING, METRICS AND SHAPING

INTERACTIVE/AUTOMATIC/HEURISTIC ALGORITHMS

TELERADIOLOGY, CON-FERENCING

DATA TRANSFER AND REMOTE PRO-CESSING

COMPOSITE IMAGING

2-D, 3-D, 4-D DATA BASE PROCESSING AND DISPLAY

SNR

3-D MEMORY, OPTIC VIDEO DISC, VLSI

ROC

THE DIMI SYSTEM PHILOSOPHY AND STATE OF DEVELOPMENT

Y. BIZAIS, M. BABA-AMI, E. BOISSINOT, S. ROY
The DIMI Project, Centre Hospitalier Régional
Hôpital G. & R. LAENNEC
44035 NANTES CEDEX - FRANCE

1 - INTRODUCTION

Since the early 70's, use of computers in Medical Imaging increased tremendously. Digital images can be processed for optimal display, extraction of quantitative information or improvement of signal characteristics, as it is the case in Nuclear Medicine, Radiology (X - CT, DSA), Ultrasounds and now MRI. Systems used for such tasks are called Image Processing Systems (IPS). In addition to image processing, images and text information connected to images may be archived and distributed in digital form. Systems performing image management are known as Picture Archiving and Communication Systems (PACS) or Image Management Systems (IMS). Several such systems are being developed in North America and Europe (1).

It is clear that just as IPS has changed the role of Medical Imaging, IMS will change the way Medical Imaging Departments (MID) are functioning. Furthermore, because digital images represent a very large amount of information as compared to text, development of PACS requires the use of very powerful hardware devices in terms of transfer rate and storage capacity. Therefore prototype PACS are now being developed to test whether systems accepted by radiologists can be built from currently available technology. This paper describes how the IPS/IMS system being realized at the University Hospital of Nantes, France, was designed in order to involve only commercially available technology and to fulfill users' requirements. In this process, most of the key issues concerning PACS functionalities and design will be reviewed.

NATO ASI Series, Vol. F19
Pictorial Information Systems in Medicine
Edited by K. H. Höhne
© Springer-Verlag Berlin Heidelberg 1986

2 - THE DIMI SYSTEM ENVIRONMENT

Two specific points of the DIMI project must be described :

- First, this project was started in Oct 82 and was conceived during the building of a new hospital specialized in thoracic and neurological pathologies. Therefore choices concerning the MID building, the acquisition devices and the PAC system were made simultaneously and consistently.

- Secondly, the system functionalities were defined by a group of computer scientists and radiologists. The system was designed, so to speak, interactively. It had to be operational quickly and to allow extensions at the same time.

Because of these two points no constraint originating in preexisting equipment or lack of collaboration with radiologists was imposed on the system architecture. However the retained solution was by no means specific to the environment in order to build a true prototype. For the same reason, private companies were associated to the project very early in the design.

3 - IMAGE MANAGEMENT FUNCTIONALITIES

PACS are concerned by two aspects of the functioning of Medical Imaging Departments :

The first aspect is communication (2). It is true that PACS will improve communication within and potentially outside Medical Imaging Departments . It is possible to achieve effective transfer rates of 3 Mb/s today, which correspond to about 1.5 512 x 512 digital images every second. Furthermore copies of images rather than originals are transferred so that fear of losing images crucial to patient care vanishes. Finally the same image can be displayed at several sites simultaneously, facilitating team diagnosis. Nevertheless it should be noticed that generalized distribution of digital images requires transfer rates as high as 15 Mb/s which is beyond today's technological capabilities.

The second aspect is archiving. PACS, to be successful, have to provide a fast-access archiving system. Any study can be retrieved and shipped to a workstation in less than 10 minutes using fast tape drive, and in less than 30 seconds using a DOR juke box (3). This is much faster and more efficient than current manual methods. PACS must also provide better structured archives. To improve the efficiency of MIDs, it must be possible to easily perform retrospective as well as prospective studies. Therefore PACS must allow the selection of images on the basis of complex criteria, through the use of a data base query langage.

PACS will improve the short term as well as the long term efficiency of MIDs by improving the accessibility to image data and the understanding of diagnostic value of imaging procedures. The improvement will be even greater when the image data base is integrated into a Hospital Information System.

4 - IMAGE PROCESSING FUNCTIONALITIES

Within a PACS, a radiologist has two types of activities :

- First, he is a radiologist : he acquires images, analyzes images using conventional light boxes or a dedicated computer depending on which imaging modality he practices, and he dictates his report.

- Secondly, he wants to review images, to compare images or to perform quantitative analysis on images produced by himself or one of his colleagues.

In most PACS described in the literature, both of the above functions are completely disconnected because the former is performed on preexisting equipments while the later involves general-purpose image workstations. This results in very poor ergonomic systems and for this reason an alternative solution has to be proposed if successful PACS are to be built. A second reason to merge IP and IM issues into one system is efficiency in software development.

5 - DIGITIZATION OF MEDICAL IMAGES

Two types of imaging procedures can be defined. The first class is made up of complex studies such as heart catherization or X-CT scans. Usually few studies of this sort are performed every week by radiologists used to technological changes and eager to improve the quality of their investigations. Interpretation is not only made from visual inspection of images but also from quantitative criteria. Spatial resolution is not as critical as for simple studies and 512 x 512 digital images usually suffice.

The second class consists of simple studies such as chest or skull X-rays. Numerous studies of this type are performed routinely in very simple environments. Diagnosis is essentially visual such that very high resolution matrices (e.g. 1024 x 1024 or 2048 x 2048) are needed.

It follows that integration of complex procedures into PACS is at the same time less technologically challenging and more important than digitization of simple studies. This suggests that prototype PACS should involve complex studies only, and that acceptance of PACS should be tested on this subset of imaging procedures. By the time field tests are completed, it will be technologically possible to include simple studies and to generalize the concept of PACS.

6 - PACS ARCHITECTURE

Most PACS decribed in the literature (4) can be described as follows (Fig 1) :

- On one side of the LAN are connected image acquisition devices (X rays, Nuclear Medicine, Ultrasounds, MRI). They may be equipped with their own dedicated computer system and images may be processed locally. In any case, at one point, images are digitized such that they can be shipped through the network to the archiving mode.

- On the other side of the LAN images can be retrieved, displayed and processed in a number of workstations of different kinds.

This architecture parallels the duality of the radiologist's activity described in section 4.

PACS - CONVENTIONAL ARCHITECTURE

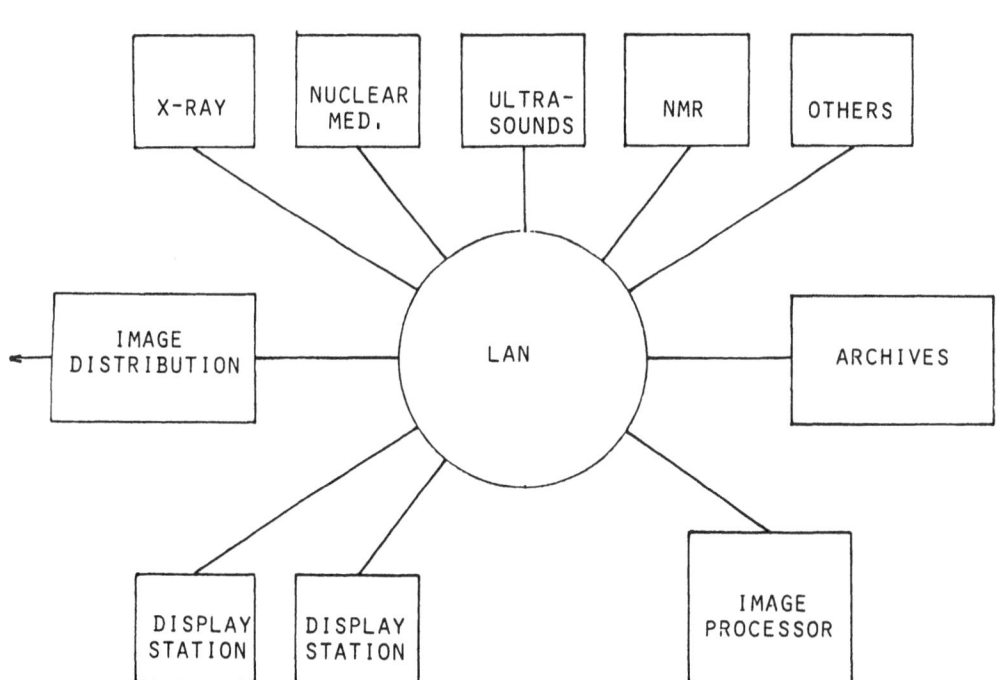

Architecture of the DIMI system can be seen in Fig 2 and summarized as follows :

- Acquisition devices sharing some kind of common properties are organized around an image processor (workstation). For many imaging modalities, images are digitized and preprocessed before entering the workstation. The radiologist may review or display images wherever they come from in the same way. There can exist workstations to which no acquisiton device is connected, which makes no conceptual difference.

This architecture integrates both functions performed by the radiologist, thus resulting in an ergonomic system.

Furthermore, if nodes are software compatible, software development costs are minimized and upgrades can be done very easily.

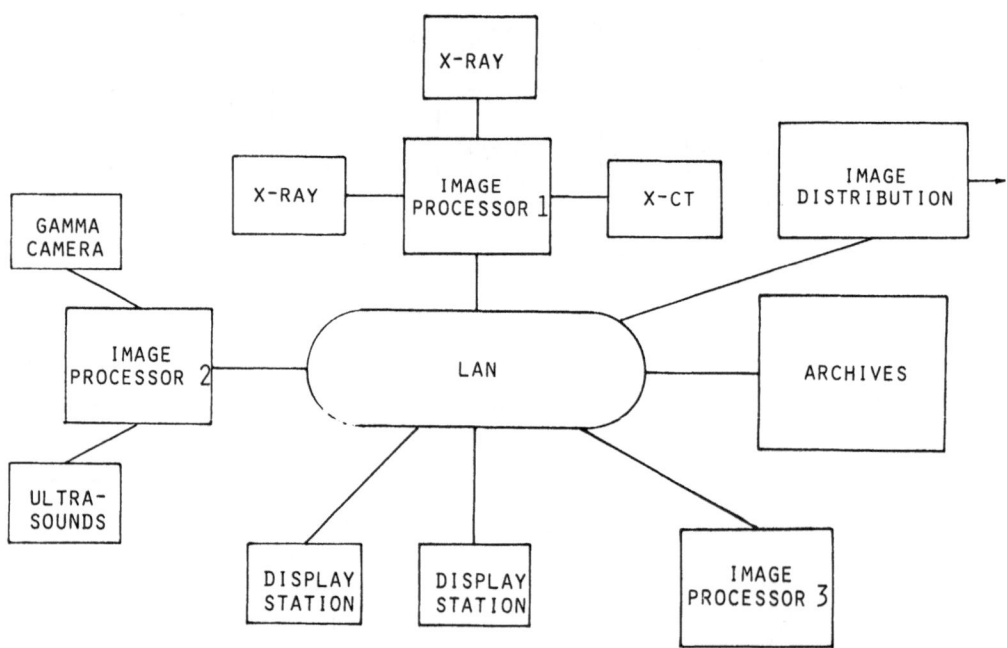

7 - THE DIMI SYSTEM

The medical imaging community at the University Hospital of NANTES was so convinced that a PACS system would be extremely useful, a group of radiologists and computer scientists, in conjunction with the efforts of private companies, started to design a IPS/IMS system to be built for the new hospital. Developing the concepts of imaging modality selection and of IP/IM fusion resulted in the system described in Fig. 3 and 4.

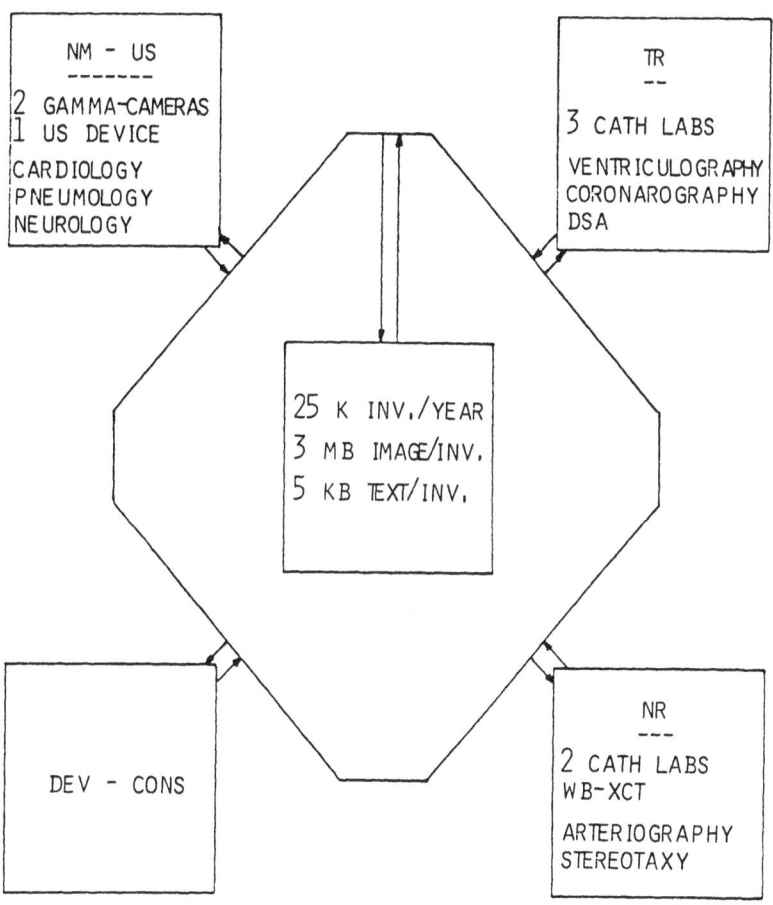

Fig. 3 details the imaging modalities involved in the DIMI system and Fig 4 gives the amount of image information generated by each type of study. It should be noticed that 3 MB of image data corresponds to what radiologists consider as useful for diagnostic purposes only. Data selection before long-term archiving permits the storage of only 75 GB every year, a figure which could be further reduced to 20 GB if a simple compression algorithm was used. DOR should quickly provide an efficient means to store such an amount of data. For each study, an estimated 5 KB of text data will be created. This data is administrative and medical information, will be stored permanently on-line and accessible through the data base management system.

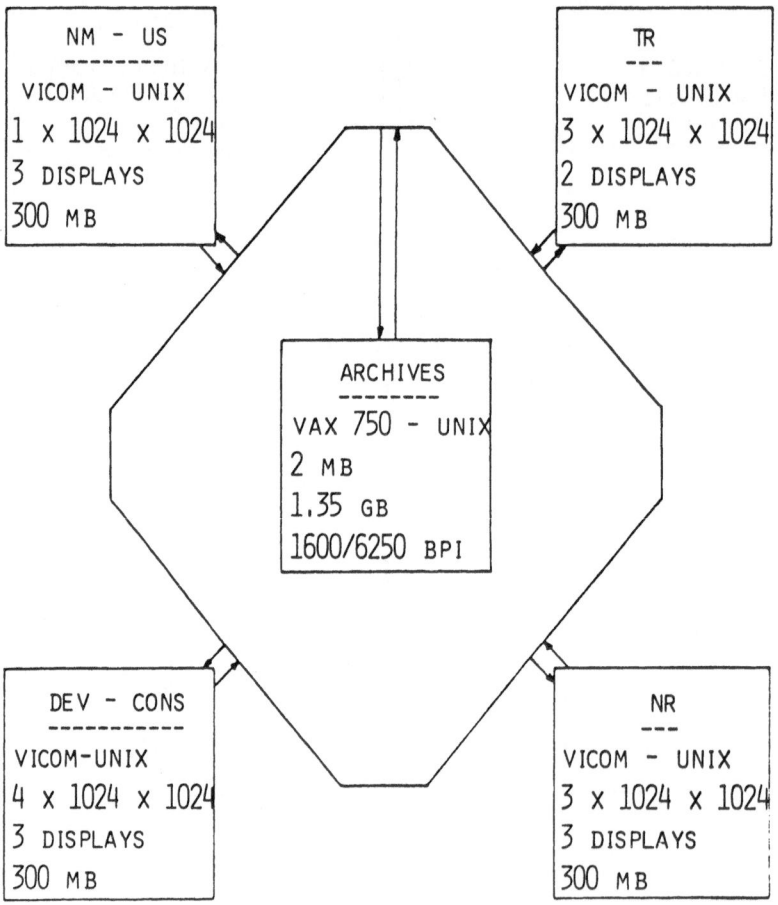

Fig. 5 describes the hardware involved in the system. Each image processing station uses a VICOM system running under UNIX which implies software developed for any station can be run in any other station. The archiving node is based on a VAX 750 computer running under UNIX.

N.M	EXAMENS		IMAGE	IMAGE/ STUDY	VOLUME/ STUDY	STUDY/ WEEK	LOAD/ WEEK
	HEART	thallium	128^2x16b.	4	0.15	25	4
		tomography	64^2x16b.	120	1	25	25
		angiography	64^2x16b.	80	0.65	25	16
	NEURO	scintigraphy	128^2x16b.	4	0.15	20	3
		angiography	64^2x16b.	130	1	20	20
		tomography	64^2x16b.	120	1	5	5
						120	100
N.R.	arteriography		1024^2x8b.	10	10	15	150
	stereotaxy		1024^2x8b.	10	10	5	50
	x-ET		512^2x8b.	20	5	100	500
						120	700
U.S.	abdominal		256^2x4b.	8	0.25	150	40
	heart		256^2x8b.	24	1.5	40	60
	vascular		64^2x8b.	8	0.03	100	3
						290	100
T.R.	ventriculography		512^2x8b.	20	5	25	100
	coronarography		1024^2x8b.	8	8	25	200
	lung angiography		512^2x8b.	5	1.3	6	8
						60	310
						590	1210

User-interface is similar in every node for image processing and image management tasks. Therefore, the system will be very easy to utilize as soon as the user knows how to run one task in one node. Furthermore, software will be accessible from multilevel menus for occasional users and at the command level for expert users.

Concerning data flow and data storage, the DIMI system involves a three-layer structure :

First during the acquisition, processing and report of a study, data is stored locally and is considered as private to the node. In other words, it can be accessed only with creator's permission.

Secondly data is transferred to the archiving node, where it is stored on mag disks for 10 days.

Thirdly, image data is archived on mag tapes, while text data is kept in the data base. As soon as data is in the archiving mode, it is considered as public and can be accessed according to privileges built in the DBMS.

It should be noted that transfers for either archiving purposes or for review, which can be planned in advance, will be performed at night. Thus the LAN will be used during daytime only for emergency cases and data base updating.

8 - REALIZATION AND EXTENSIONS

By mid-1985, an IPS/IMS will have been built by three groups (an in-hospital group and two private companies). This system will offer basic functionalities in image processing (Digital Radiology, Nuclear Medicine and Ultrasounds) and in image management (communication and archiving). The set of functionalities will be wide enough and the system user-friendly enough for routine use by the Medical Imaging Department.

During the second phase of the project, the system is intended to be used in two complementary directions : first, as an MID facility which will feed the data base and secondly as an applied research tool to develop image processing and image understanding software, and to evaluate investigation performances as well as the acceptance of the system by the medical community.

Several extensions are planned :

- The generalization of digitization to modalities not included in Phase 1 (i.e. chest X-rays, MRI)

- The generalization of image distribution to medical and surgical departments.
- The connection to other data bases, especially the Hospital Information System and computerized medical records. (Today, the various systems share a common key)

9 - CONCLUSION

The system presently being developed at the University Hospital of NANTES can be considered as both a distributed IP and IM system. Key features include integration of IP and IM tasks, selection of imaging modalities and user-friendliness.

Because of its in-hospital implementation, the DIMI system will allow one to evaluate the acceptance of PACS by the medical imaging community. Moreover, the data base fed by the MID activity will represent an ideal input for testing the software to be developed.

REFERENCES

(1) DUERINCKX A.J., DWYER S.J., PREWITT J.M. : Digital Picture
   Archiving and Communication Systems in Medicine.
   Computer Mag. 16, 8 p 14 (1983)

(2) COX J.R., BLAINE G.J., HILL R.L.,JOST R.G. : Study of a
   Distributed Picture Archiving and Communication System for Radiology.
   Proc PACS I, p 133 (1982)

(3) COLBY R.L. BARTUSKA A.J., HERZOG D.G. : Optical Disk Data
   Storage.
   Proc PACS I, p 36 (1982)

(4) DWYER S.J., TREMPLETON A.W., ANDERSON W.H. et al : Salient
   Characteristics of a Distributed Diagnostic Imaging Management System
   for a Radiology Department.
   Proc PACS I, p 194 (1982)

INTERACTIVE DISPLAY OF 3D MEDICAL OBJECTS

Jayaram K. Udupa, Hsui-Mei Hung, and Lih-Shyang Chen
Medical Image Processing Group
Department of Radiology
University of Pennsylvania
3701 Chestnut Street
Philadelphia, Pennsylvania 19104

ABSTRACT

Interactive display of internal human organs has recently received much attention because of its potential applications. Capabilities for visualization, manipulation and quantitation are the common requirements of most medical applications. For interactive display and manipulation to be feasible, the medium of interaction should effectively communicate 3D information to the user, and the result of a manipulative action by the user should follow immediately after initiating the action. The paper reports on our attempts to answer this challenge in a modest minicomputer environment which is typical of most CT scanner facilities.

INTRODUCTION

Abnormalities in the structure of internal organs are often the cause for their abnormal function. Hence, in medical diagnosis it is important to distinguish normal structures from abnormal structures. There are several medical imaging modalities which gather data pertaining to internal organs [5], such as x-ray computerized tomography (CT), magnetic resonance CT, and ultrasonic CT. In these modalities the CT scanners compute the value of a physical property of the tissue (such as x-ray attenuation coefficient in x-ray CT) inside each of a number of non-overlapping rectangular volume elements which together

NATO ASI Series, Vol. F19
Pictorial Information Systems in Medicine
Edited by K. H. Höhne
© Springer-Verlag Berlin Heidelberg 1986

represent the region of the body being scanned. The elements are usually referred to as voxels and the measured physical property of the tissue inside each voxel is called its density. The array of voxels together with the density assigned to the voxels is called a scene. The problem addressed in this paper relates to the visualization of structural information present in such scenes.

Several techniques are currently available in the literature for presenting the structural information to a human observer (see [13]). The most popular of these are perhaps surface display methods ([7], [10], [4], [3], [1], [2], [14], [8]). Though the 3D mode provides accurate, reproducible and communicable presentation of 3D structural information, their clinical supremacy over the commonly-used slice-viewing mechanism still remains to be established.

Though the clinical validation lags the development in techniques, we anticipate 3D display techniques to be useful in many biomedical problems. We believe that an effective solution to many biomedical problems depends on the ability to visualize, manipulate and to quantitate structures using 3D scenes. These capabilities are interdependent. For example, often it becomes necessary to manipulate a structure so as to remove obscuring parts and to subsequently visualize the features of interest. Speed of display is an important issue which should be considered in conjunction with the manipulation aspect. The intended use of interactive manipulation is to modify the object by 'trial-and-error'. After each modification the object is visualized, and further modifications are carried out if the resulting display so indicates. Clearly, for the feasibility of interactive modification of objects, it should be possible to carry out both display and manipulation computations quickly, ideally in less than a second. Those software packages which claim to have the convenience of implementability on the minicomputer of CT scanners have not yet come to the state of such speeds of operation. Besides, interactive manipulability has not yet been made available in such packages in a form convenient to use, though some provide some form of slice-by-slice interactive separation of regions ([13]). Some of the special-purpose systems built ([2], [8]) to carry out the display and analysis of 3D structures do provide fast display capabilities and some forms of manipulation and quantitation using 32-bit processors. They are right now in an experimental stage and are quite expensive. Hence, the design of effective data structures and algorithms for the problem of medical display and analysis that are suitable for the minicomputer environment is of both theoretical and practical interest.

In an attempt to devise interactive manipulation techniques that provide satisfactory speeds, we have developed display techniques based on a scheme of representing objects called the directed-contour (DC) representation ([12], [11]). This representation is used both for producing a shaded-surface display and for interactive manipulation of the object. The surface display is derived directly from the DC without going through an intermediate step of fitting a surface. This results in a dramatic improvement in speed, though at the expense of some quality of display. But this does not matter, since, after manipulating the object to the desired level, the final views of the modified object can be created using more refined display algorithms ([6]).

REPRESENTATION

In DC representation, objects are described by their borders in a slice-by-slice fashion. We assume that the given scene is segmented into a binary scene in which each voxel has a value '0' or '1' assigned to it. By a _slice_ we mean the subscene in which all voxels have the same z-coordinate (see Figure 1). In each slice, borders of '1'-regions are represented by a sequence of voxels (essentially pixels in the slice) such that as the border is

Figure 1

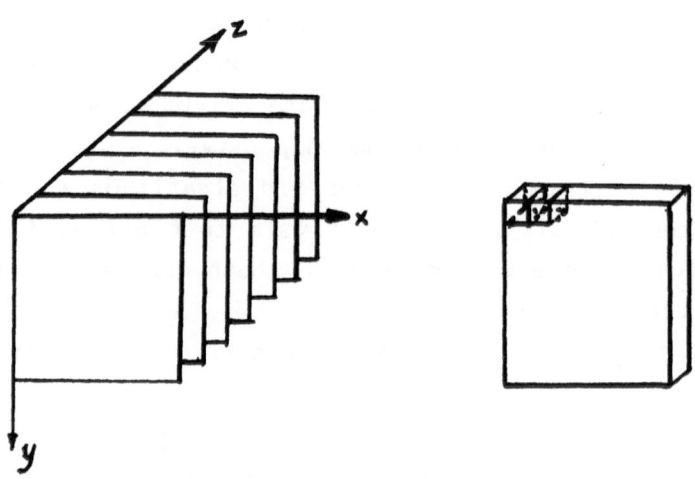

Discrete scene, slice and voxel.

traversed the '1'-region lies to the left. Further, the contours are chain-encoded so that the DC's can be very compactly. With the DC's in each slice we associate a <u>containment tree</u> in which each node represents a contour, and its sons represent contours which are 'immediately inside'. For example, in Figure 2, the border CO of the slice represents the root node, and C1 and C2 are its sons representing contours which are 'immediately inside' CO. Algorithms have been developed to automatically compute the DC representation of a given binary scene [11].

DISPLAY

The main objective of producing a surface display, as far as manipulation is concerned, is (i) to guide the user to indicate interactively the nature of manipulation, and (ii) to give the user feedback by generating quickly a

Figure 2

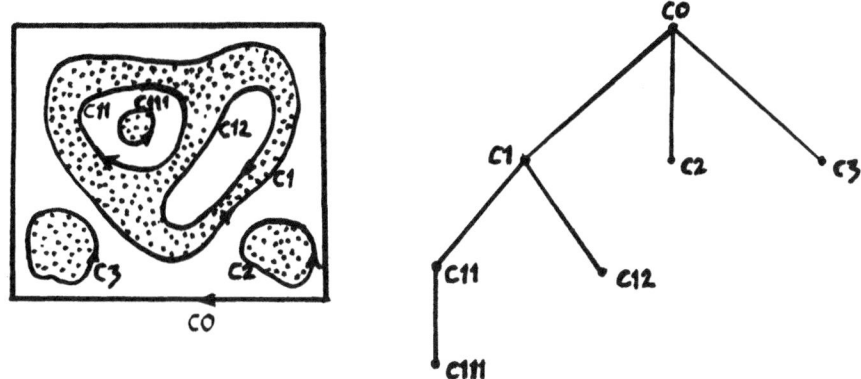

Directed contour representation of binary scene.

display of the modified object so that the user can judge
whether the manipulation carried out was appropriate or not.
The requirement of fast display for interactive manipulation
need not be overstated. In fact, the quality of the display
is secondary if the display can be produced quickly.

    We use the DC representation as the data structure for
both the display and the manipulation algorithms. An advantage of this over using different data structures is that
the time required to convert the data structure used for
manipulation to that used for display is saved. In the
present implementation, viewing is restricted to planes
perpendicular to the slices (see Figure 3). By re-slicing
the data perpendicular to the x- and y-axis, the method is
readily extended to viewing directions perpendicular to the
xy and yz planes.

    A number of properties of the DC representation and the
simplicity of the special viewing mode are made use of order
to minimize the display computations. These are given
below:

1. Only those contours which correspond to the sons of the root of the containment tree need to be considered for display.

2. We assume that the viewing direction is perpendicular to the xz plane (see Figure 3) and that the object is rotated about the z axis. Under these assumptions, generally, only about half of a contour needs to processed (the part lying between the points where the global minimum and the global maximum x coordinates occur for the rotated contour; see Figure 3). Since the 'inside' of the object lies to the left of the directed contour, only those voxels on the contour which are encountered while traversing in the direction of the contour starting from the voxel with the minimum x-coordinate and ending with the voxel with the maximum x-coordinate need to be displayed.

3. The display algorithm determines the pixels on the screen to be painted and the intensity of these pixels for each potentially visible segment of the contour defined by a pair of successive voxels on the contour. Since the contours are chain-coded, a segment corresponds to a chain link. For example, Figure 4 shows two segments v1 v2, v2 v3 with chain codes 1 and 3, respectively. For the assumed viewing direction, segment v1 v2 is not visible, but v2 v3 is visible. Clearly, the criterion for visibility of a segment is that the x-component of the rotated direction vector corresponding to the segment should be negative. This implies that the eight rotated direction vectors may be computed outside the loop on segments and the potential visibility of a segment can be determined by a table look up.

4. The renderings of the part of the object in two different slices project on to two disjoint sets of scan lines. This implies that the object can be processed slice-by-slice, and, at the same time, the image can be generated line-by-line. This nicety almost completely removes the need for the usually time-consuming memory management required to handle input/output.

Figure 3

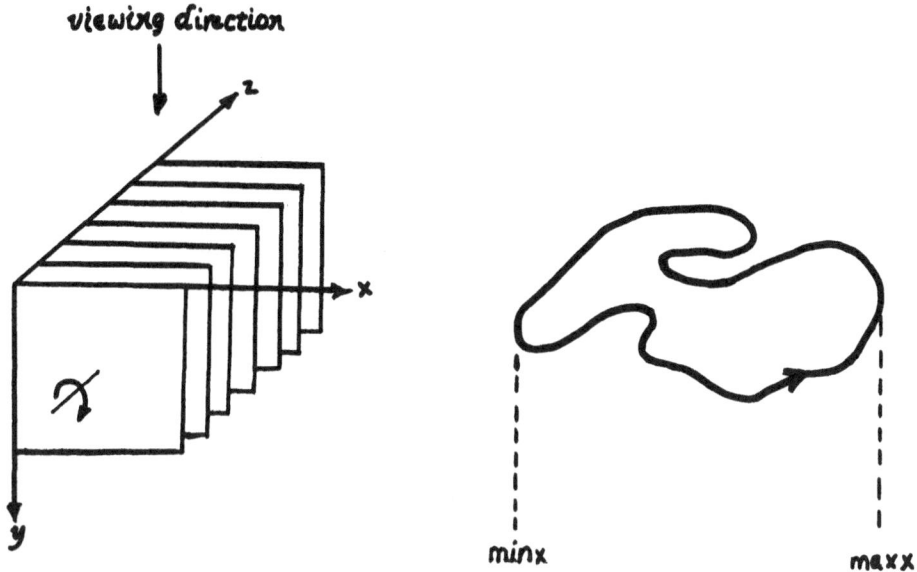

Illustration of the viewing mode for the display
algorithm DC-display.

Figure 4

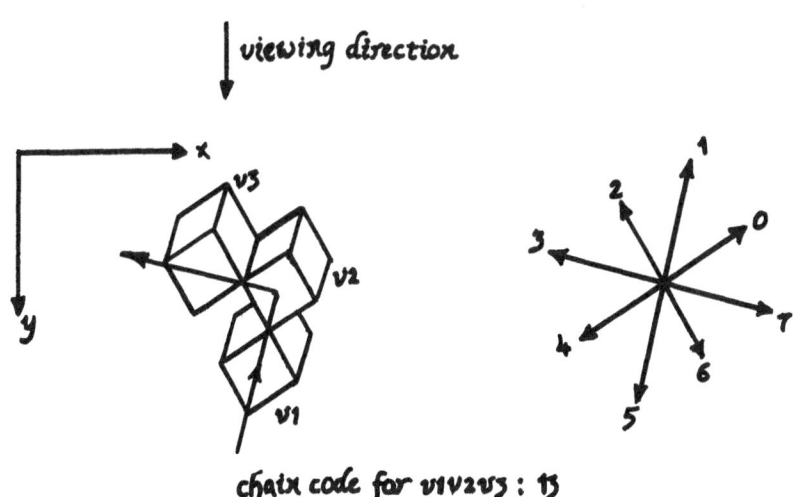

Illustration of how to decide visibility of contour segments.

The actual rendering algorithm uses a z-buffer mechanism for removing hidden parts ([9]). Shading is assigned to the visible segments as an inverse linear function of the distance of the segment from the view point.

Some display examples are presented in Figure 5 using x-ray CT data. For the spine data in Figure 5(a), the original scene was 248x254x55. This was interpolated to yield 248x254x251 scene which is used to generate the display. The original scene takes about 10 seconds for display while the interpolated scene takes 35 seconds. In Figure 5(b), the original 256x256x40 scene is interpolated to yield a 256x256x120 scene. The display is produced using the interpolated scene and takes about 20 seconds per view. The original scene requires 5 seconds per view. The algorithm is implemented in FORTRAN on a Data General Eclipse S200 minicomputer.

MANIPULATION

The ultimate goal of this work is to develop a graphics tool which can provide capabilities for (i) altering objects in the scene (ii) making geometric measurements on objects. These capabilities, we believe, will be useful in surgical planning, in following the changes in shape of organs in response to treatment, and, in general, in studying life processes and evaluating therapeutic procedures. Without restricting to any specific application, the anticipated manipulative capability in its general form should allow its user to 'edit' the objects in a scene in any desired way. In designing such interactive tools a basic issue that needs to be addressed relates to the nature of human interaction provided by the graphics tool. This, in the medical application areas may have to correlate with the basic surgical operative procedures.

We have chosen the shaded-surface display to be the medium of visual communication during the execution of

Figure 5a

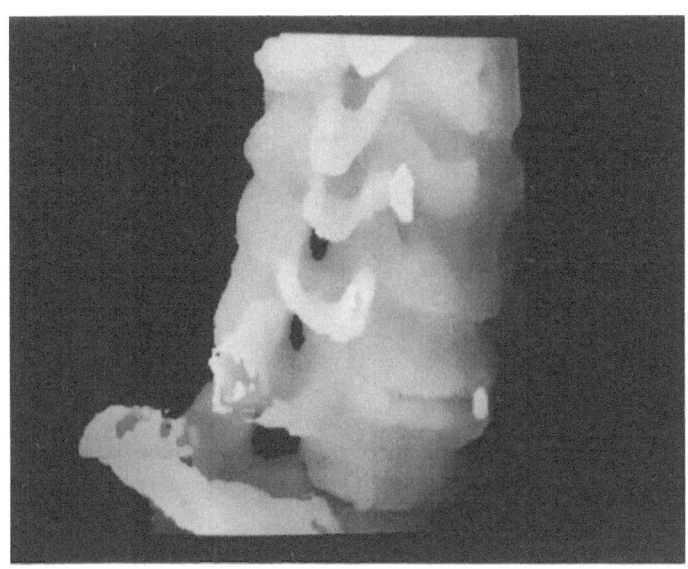

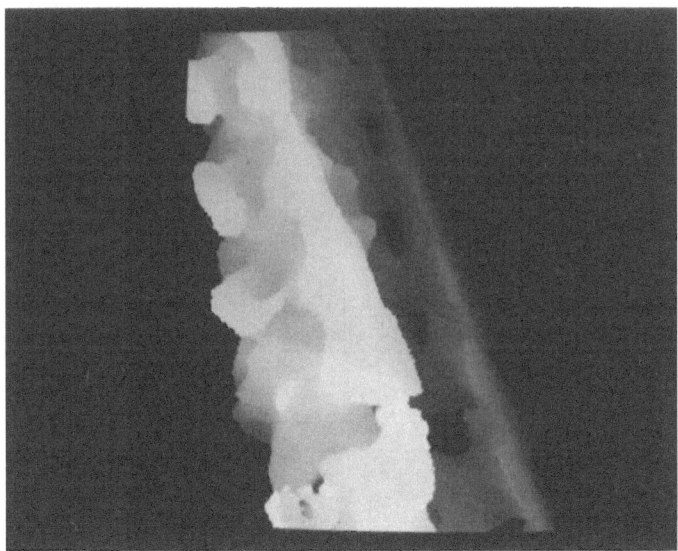

Display and manipulation example using the DC
representation of spine and skull data.

Figure 5b

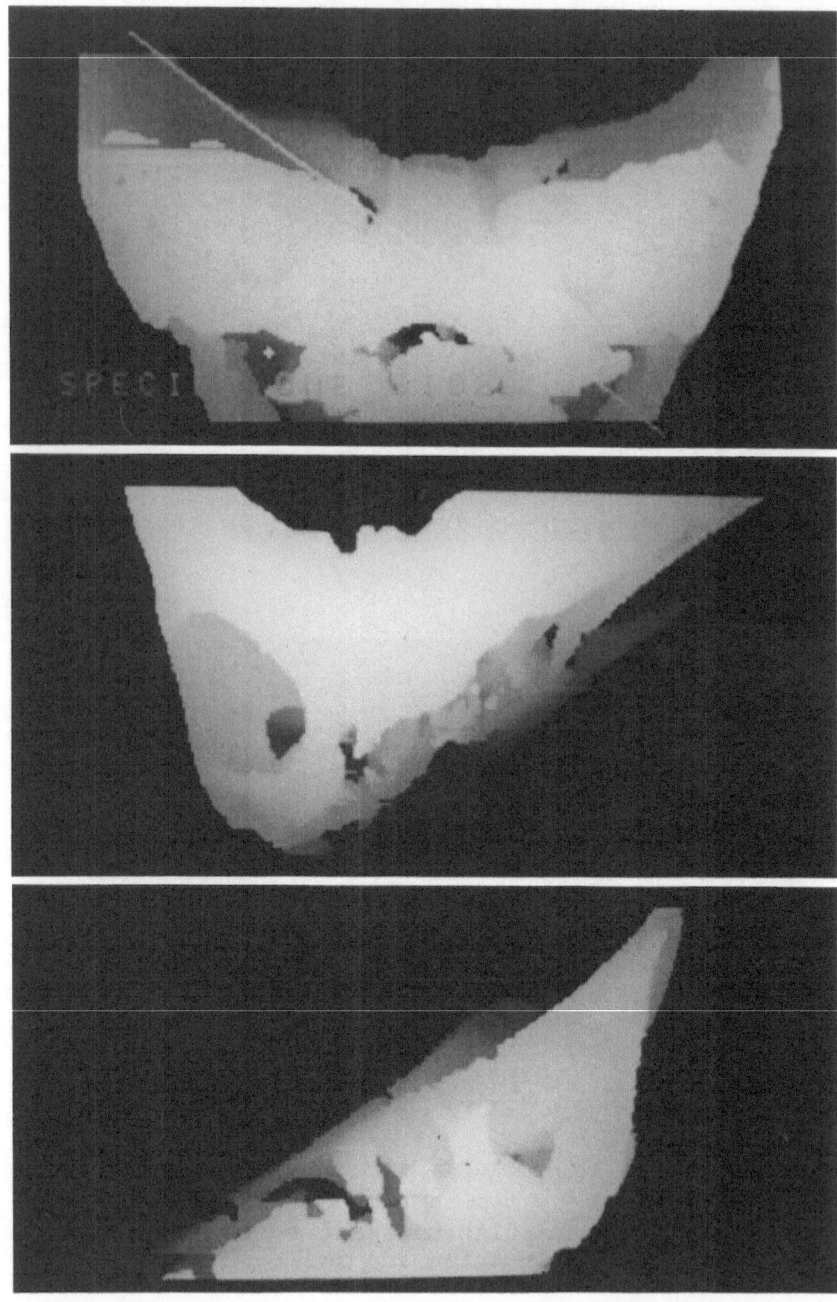

Display and manipulation examples using the DC
representation of spine and skull data.

interactive procedures. The actual manipulation consists of cutting by a plane: the objects in the scene are first displayed at an appropriate orientation. The user then draws a straight line on the display to indicate a plane perpendicular to the screen at the line. The modified objects resulting from the cutting operation may be subseqeuntly displayed.

The inputs to the manipulation algorithms are the DC representation of the binary scene, the cutting plane specified by the user and the information (in the form of a transformation matrix) about the geometric transformations underlying the present depiction of the objects in the display. For each slice the computations are carried out as follows. The line of intersection L of the cutting plane with the plane of the slice is first computed using the known transformation matrix. The points of intersection of the line and the contours in the slice are then computed. Let P and P' be the points of intersection of L with the border of the slice. Starting from P traverse along L (towards P') until a point of intersection is reached. Let this point of intersection be P1. Now traverse along the contour just intersected in the direction of the contour until a point of intersection is reached. Now traverse along L towards P1. Continue this procedure of alternating traversal between L and the contours until P1 is reached. The path traversed starting from P1 and back to P1 along with those contours which are not intersected by L and which are on the side of L in which the modified contour lies together represent the modified DC's for one side of L for the slice under consideration. To compute the DC's for the other side of L, the above procedure is repeated starting from P' and proceeding towards P.

Figure 5 illustrates the 'cutting-by-plane' manipulation on the spine and skull patient data referred to earlier. We believe that such direct 3D manipulability facilities visualization of desired structures, in the examples, the neural foramina of the spine and specific cranial base deformities which are hidden. The computational timings for the manipulation algorithm are similar to that of the display algorithm.

CONCLUSION

We have presented a new scheme of representing discrete scenes by directed contours. We have described algorithms for representing, displaying and manipulating discrete scenes based on the new representation. At present, the generation of each view takes about 5 seconds for typical objects. To our knowledge this timing is much superior to what is reported in the literature for other techniques implemented on comparable computing equipment.

ACKNOWLEDGEMENTS

The research for this paper was supported by NIH grant HL28438. The authors are grateful to M.A. Blue for typing the manuscript.

REFERENCE

[1]   Cook, L.T.; Dwyer, S.J. III; Batnitzky, S.; Lee, K.R.: A three-dimensional display system for diagnostic imaging applications. IEEE Comput. Graphics and Appl. 3, 13-19 (1983).

[2]   Dev, P.; Wood, S.; Duncan, J.P.; White, D.N.: An interactive graphics system for planning recon- structive surgery. Proc. Nat'l. Comput. Graphics Assoc. Conf. 130-135, Chicago, Illinois (June 1983).

[3]   Fuchs, H.; Pizer, S.M.; Tsai, L.C.; Bloomberg, S.H; Heinz, E.R.: Adding a true 3D display to a raster graphics system. IEEE Comput. Graphics and Appl. 2, 73-78 (1982).

[4]   Herman, G.T.; Liu, H.K.: Three-dimensional display
       of human organs from computed tomograms, Comput.
       Graphics Image Process. 9, 1-29 (1979).

[5]   Herman, G.T. (ed.): Proc. IEEE 71, special issue on
       computerized tomography (1983).

[6]   Herman, G.T.: Three-dimensional computer graphics
       display in medicine. Technical Report MIPG88,
       Medical Image Processing Group, Department of
       Radiology, University of Pennsylvania, Philadelphia,
       (July 1984).

[7]   Mazziotta, J.C.; Huang, K.H.: THREAD (three-
       dimensional reconstruction and display) with bio-
       medical applications in neuron ultrastructure and
       computerized tomography. Amer. Fed'l. Info. Process.
       Soc. 45, 241-250 (1976).

[8]   Meagher, D.J.: Interactive solids processing for
       medical analysis and planning. NCGA Conf. Proc. II,
       96-106 (1984).

[9]   Newman, W.M.; Sproull, R.F.: Principles of Inter-
       active Computer Graphics. McGraw-Hill Book Company,
       New York, New York (1979).

[10]   Sunguroff, A.; Greenberg, D.: Computer generated
       images for medical applications, Comput. Graphics 12,
       196-202 (1978).

[11]   Tuy, H.K.; Udupa, J.K.: Representation, display and
       manipulation of 3D discrete scenes. Proc. of 16th
       Hawaii Int'l. Conf. on System Sciences II, 397-406
       (1983).

[12]   Udupa, J.K.: Interactive segmentation and boundary
       surface formation for 3D digital images. Comput.
       Graphics Image Process. 18, 213-235 (1982).

[13]   Udupa, J.K.: Display of 3D information in discrete
       3D scenes produced by computerized tomography. Proc.
       IEEE 71, 420-431 (1983).

[14]   Vannier, M.W.; Marsh, J.L.; Warren, J.O.: Three-
       dimensional computer graphics for craniofacial
       surgical planning and evaluation. Comput. Graphics
       17, 263-274 (1983).

Presentation and Perception of 3-D Images

R. Lenz, P. E. Danielsson, S. Cronström, B. Gudmundsson

Linköping University

S-58183 Linköping

SWEDEN

Abstract

In our paper we describe some experiments in processing and display of 3-D images. The experiments have been carried out on the PICAP II system.

Introduction

In the last few years a number of imaging techniques has been developed that can produce 3-D images. Many of these methodes were specially developed for medical imaging and the future will certainly see an increased application of these techniques in daily clinical use. All these 3-D imaging devices have in common that they produce density-values for each point in a cubic grid. These density distributions constitute a semitransparent world that is totally different from our usual 3-D environment and the world of 3-D computer graphics. Therefore special display techniques are needed to present these volumes to the user. Up to now most systems can only display serial sections of the object and the task of reconstructing the 3-D nature of the object is left to the users. From life-long experience radiologists have a rather good ability to do this reconstruction but for applications like interactive surgical planning and radiotherapeutic dose planning other display methods are needed to make the data also useful for non-radiologists. A pictoral information system that also includes the storage and retrieval of 3-D images should therefore also include a 3-D display station.

At the Department of Electrical Engineering, Linköping University, a project aiming at thedevelopment of efficient algorithms and methods for processing and display of 3-D images from different image sources has been in progress for about one and a half years. In contrast to most previous approaches we do not restrict us to the display of binary volumes but we want to present the density values as well. The basic tool in these efforts is the PICAP II image processing system[1], the architecture of which is shown in Figure 1.

NATO ASI Series, Vol. F19
Pictorial Information Systems in Medicine
Edited by K. H. Höhne
© Springer-Verlag Berlin Heidelberg 1986

PICAP II is centered around a fast, time-shared bus that connects a number of special purpose processors and interfaces to a large RAM for storage of digitized images. PICAP II is a multi-user system operating in MIMD-mode where allocation and scheduling of the processors is controlled by the host computer.

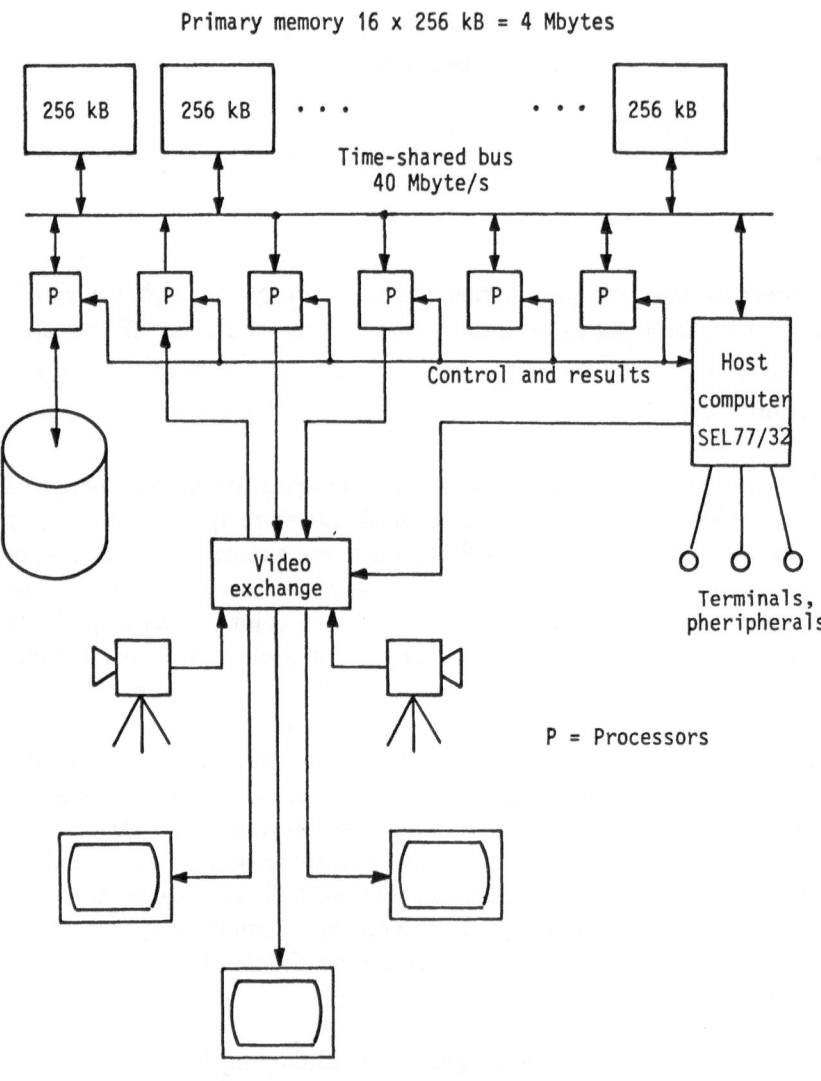

Figure 1

In many foreseeable applications where 3-D image processing is used, user interaction will be an essential requirement[2]. Full interaction with 3-D data in the sense of immediate response to user requests probably calls for heavy investments in new image processing hardware. However, by limiting ourselves to interactively controlled display of precomputed projections, we have been able to use PICAP II in our experiments to great advantage. The fast digital video disk unit connected to the central bus (Figure 1) plays an important role here. This unit has a storage capacity of 300 Mbyte and a transfer rate of 10 Mbytesec., (ten times the rate of conventional disk unit) or full video rate. By precomputing a sequence of projections from a 3-D image and storing it on the disk, the projections can be displayed in rapid sucession under user control creating the illusions of such effects as continous rotation, variable illumination, etc.

So far, we have not investigated any particular application in depth but rather experimented with a number of 3-D images obtained with different imaging techniques. These images are the following:

1. A volume from a pigs head built up from 128 256 x 256 slices obtained from a CT-scanner. Through interpolation 256 slices were produced resulting in a cubic volume.

2. Several volumes from the PHOIBOS[3] laser-microscope of size 256 x 256 x 256 showing neurons.

3. 80 x 80 x 80 transmission electron microscopy tomographic volumes of a part of a cell.

4. Volumes from a gamma-camera with slices of size 64 x 64.

One of the slices from the CT-volume is shown in Image 1 .

Sequence of operations

The following Figure 2 shows the steps involved between the pickup and the display.

As mentioned before, with our present equipment interactive control is limited to the display functions. Also, our work does not include the image pickup; our starting point has been a 3-D volume consisting of parallelepiped volume elements (voxels).

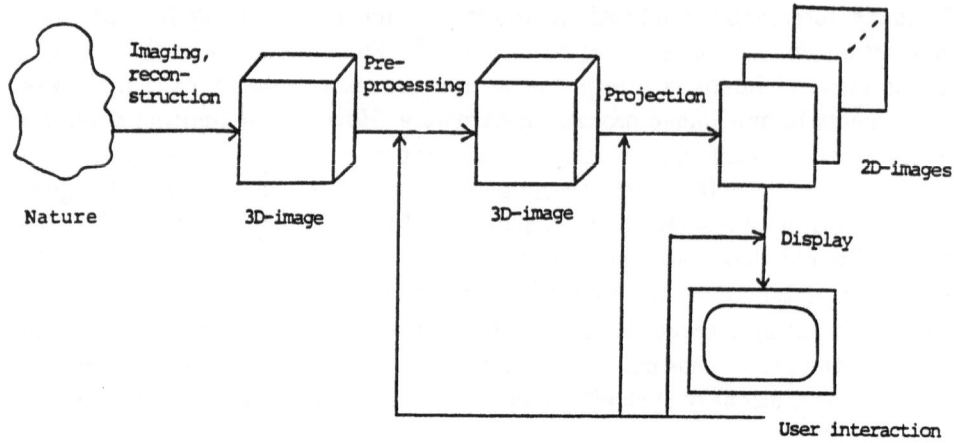

Figure 2

Our experiments with different 3-D volumes have shown that before projection there is often a need to preprocess the volume to enhance the visibility of different structures. After preprocessing one can select from a number of different projections methods. The preprocessing and projection steps are at present implemented in software in the host computer of PICAP II.

Preprocessing

The preprocessing typically involves the application of spatial filters, both linear and nonlinear to compute transformed volumes from the originals. Up to now we have implemented the following operations:

1. gradient filters in a 2 x 2 x 2 neighborhood

2. min,max and median filter in a 2 x 2 x 2 neighborhood

3. grayscale transformations

4. distance map and thinning

5. combination of the original volume and (one or more) preprocessed volumes

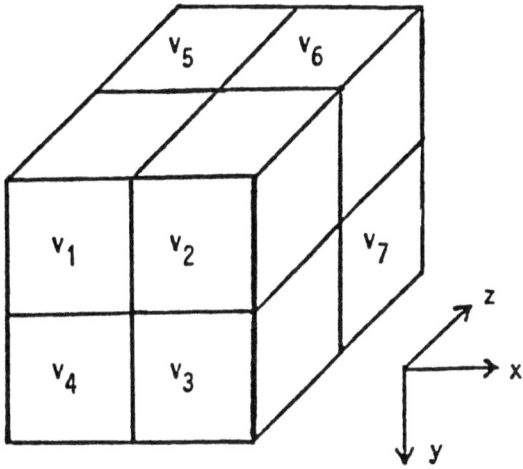

Figure 3

The ability to combine several volumes may be used for instance to weight the density values in the original volume with a function of the distancemap to enhance the central parts of on object.

A *gradient* volume is obtained by first computing the grayvalue differences

$$D_1 = v_7 - v_1$$

$$D_2 = v_8 - v_2$$

$$D_3 = v_5 - v_3$$

$$D_4 = v_6 - v_4$$

along the four main diagonals in a 2 x 2 x 2 neighborhood (see Figure 3) which facilitates the following approximations of the gradients:

$$\frac{\partial}{\partial x} \approx \frac{D_1 + D_2 + D_3 + D_4}{4}$$

$$\frac{\partial}{\partial y} \approx \frac{D_1 - D_2 + D_3 - D_4}{4}$$

$$\frac{\partial}{\partial z} \approx \frac{D_1 + D_2 - D_3 - D_4}{4}$$

In order to make thick objects with a very complex structure more transparent we use a *thinning* procedure to remove the outer layers of the object and to highlight the centerparts. The implemented thinning algorithm consists of three steps:

1. Computation of a distancemap

2. Detection and marking of the local maxima in the distancemap

3. Shrinking with simple masks were local maxima must not be removed

This procedure is a straight forward generalization of the 2-D thinning methodology proposed by Davies and Plummer[4] to 3-D volumes.

Projections

Since we use conventional raster-scan monitors for display, the 3-D image must be projected into 2-D. We compute a whole series of projections with a (numerical) rotation of the volume between two projections. There are, of course, several ways to compute such projections. The projection methods we have worked with are look- through projections, depth coding and shaded surface display.

A *look-through projection* is computed by traversing the volume with 256 x256 projection rays and accumulating the density values along the rays. The computed sums are then mapped into a gray value. By using negative exponential functions to compute the gray values of the resulting projection one can simulate the absorption of x-rays going through the volume and different kinds of films can be simulated. Image 2 shows a look-through projection of the pigs head. When computing this projection all voxels below a certain threshold were ignored and the resulting image corresponds to a x-ray of the bone structure.

A look-through projection of the gradient-filtered neuron volume is shown in Image 3.

Image 4 shows a *depth-coded* image of the pigs head. Here the image is first segmented into object and background. When a projection ray traverses the volume, the depth of the first object voxel along that ray is recorded and the pixel corresponding to that particular ray is assigned a gray value that is inversely proportional to the depth.

The surfaces of the objects can be *shaded* in more elaborate ways as shown in Image 5. A virtual light source illuminates the object and the shading in each point on the surface is calculated from the surface gradient at that point.

If all voxels below a certain depth are neglegted one gets the effect of a *cut* through the volume. Image 6 demonstrates how this technique can be used to look into the brain cavity.

Display session

As mentioned before, we compute a whole series of projections typically with the volume rotated 2 degrees from one projection to the next. These projections are stored on the fast digital disk of the system from which they can be displayed at full video rate. Therefore we can show these projections in a rapid sequence creating the effect of rotation. Or we can display two different projections as a stereo pair. Stereo display can also be combined with rotation and we have found this to be a very powerful display mode .

With the help of a joystick or a tablet the user has full interactive control over the rotation .

In addition to rotating the volume the user can also move around a cursor symbol in the volume. This symbol is overlaid on the current stereo pair and the user can make geometric measurements or delimit subvolumes by putting markers in the volume.

Future work

Our experiments so far have been encouraging and they convinced us that 3-D image processing has a potential in many application areas. We will also investigate novel processor architectures that will enable us to process and project 3-D images in real time.

References

1. P. E. Danielsson, B. Kruse, B. Gudmundsson, "Memory Hierarchies in PICAP II", in *Proc. of Workshop on Picture Data Description and Management*, pp. 275–280, IEEE 1980

2. G. T. Herman and J. K. Udupa, "Display of 3D Digital Images: Computational Foundations and Medical Applications", *IEEE Computer Graphics and Appl.* **Vol. 3, No. 5,**pp.39–46, August 1983

3. N. Åslund, K. Carlsson, A. Liljeborg and L. Majlöf, "PHOIBOS, A Microscope Scanner Designed for Micro-fluorometric Applications Using Laser Induced Fluorescence", in *Proc. of Third Scandinavian Conference on Image Analysis*, Studentliteratur, Lund, Sweden, 1981

4. E. R. Davies and A. P. N. Plummer, "Thinning Algorithms: A Critique and a New Methodology", *Pattern Recognition*, **Vol. 14, No. 1-6,** pp.53–63, 1981

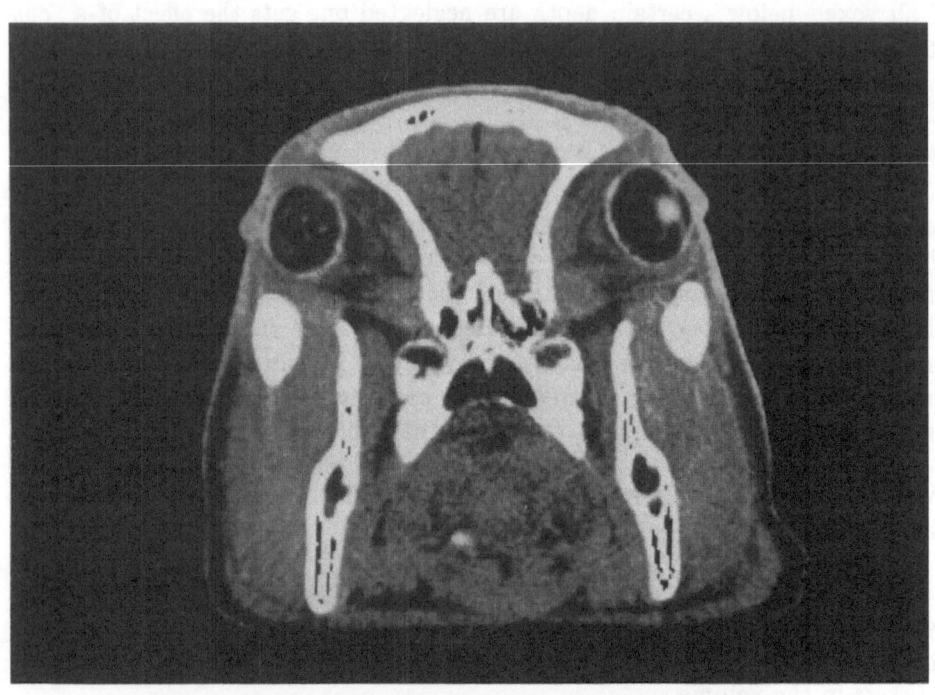

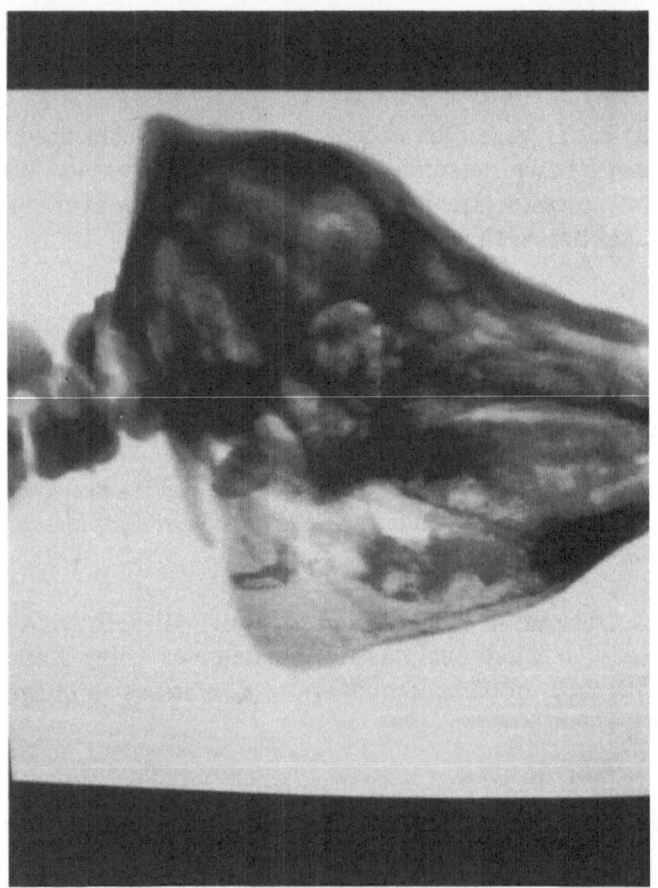

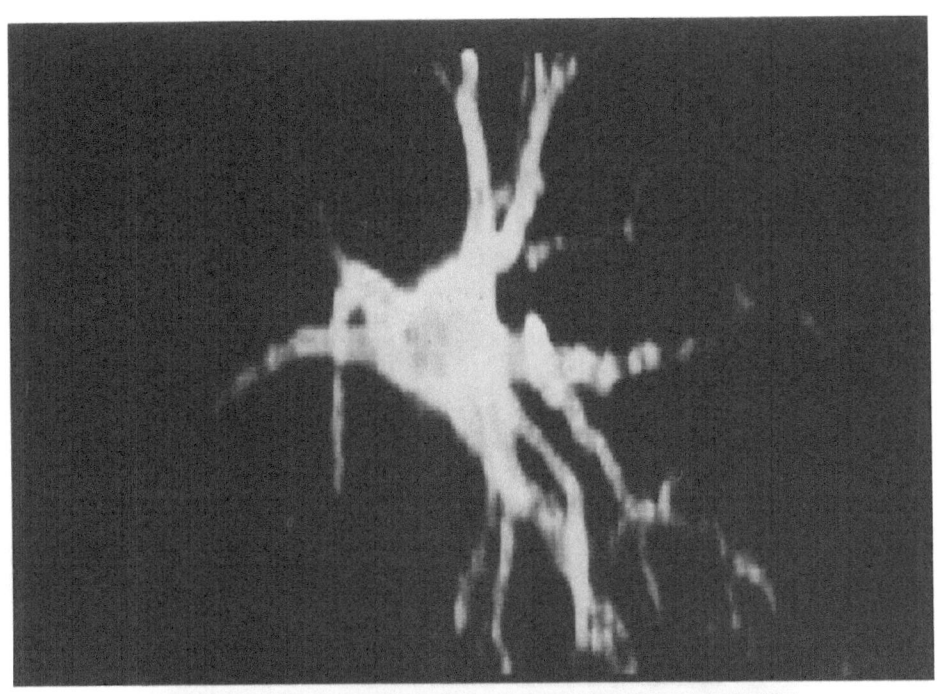

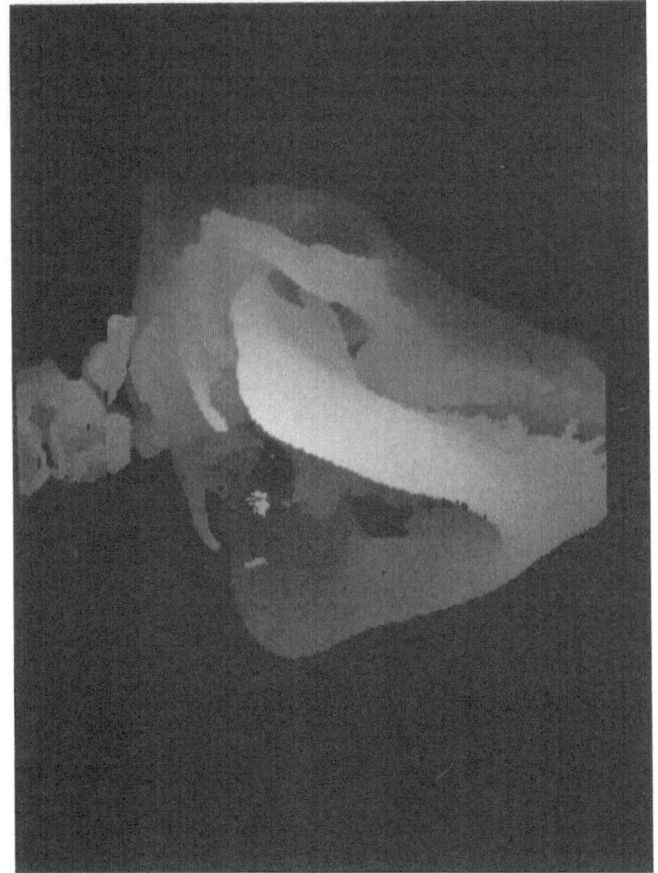

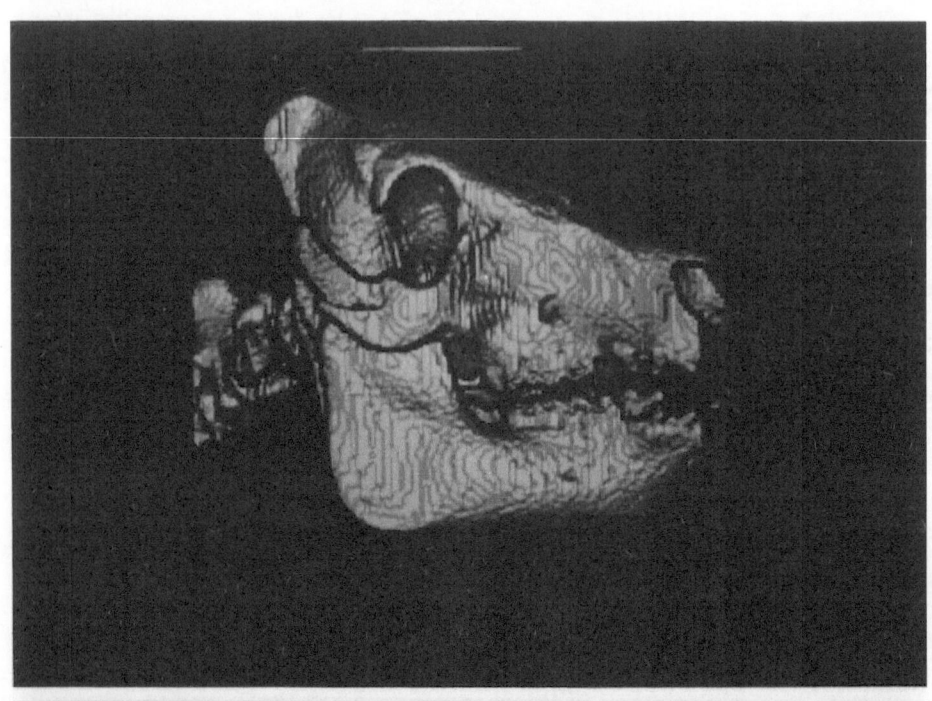

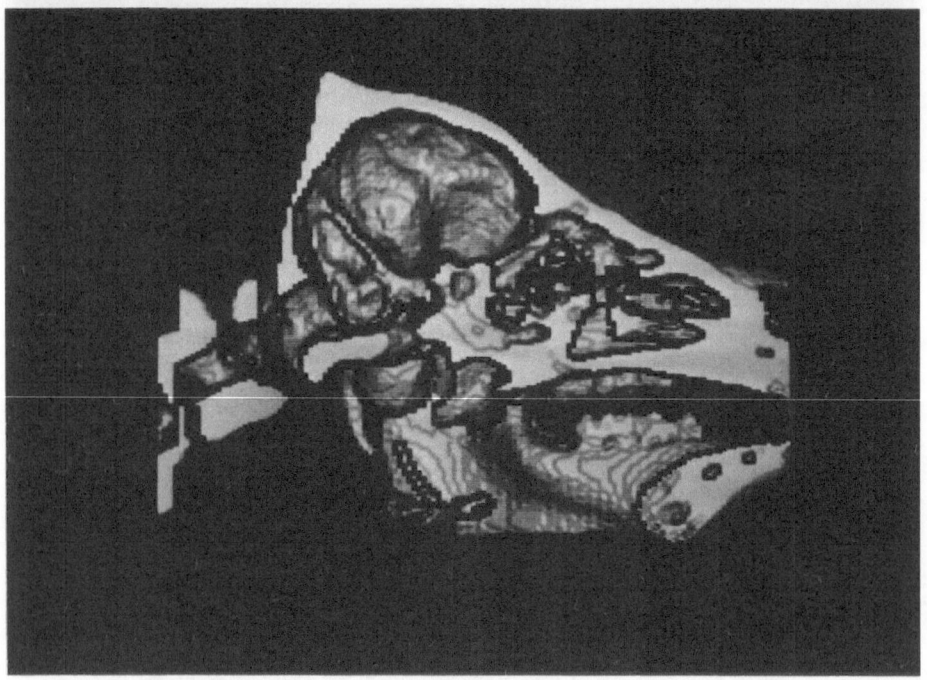

EXPERIMENTS WITH THREE-DIMENSIONAL RECONSTRUCTION OF VENTRICULOGRAMS USING AUTOMATIC CONTOUR DETECTION.

F.R.P. Boecker, G. Witte (*), K.H. Höhne

Institut für Mathematik und Datenverarbeitung in der Medizin und Radiologische Klinik (*)
Universitätskrankenhaus Eppendorf, Martinistr. 52,
2000 Hamburg 20, FRG.

Abstract

In this paper gray level information of the digital subtraction angiography (DSA) image sequence of a human left heart ventricle is used to compute a three-dimensional object of voxels. First the ventricle-contour is determined using the Marr-Hildreth filter. Thickness of the inner ventricle is estimated from the gray levels. For the final volume reconstruction the ventricle is assumed to be mirror-symmetrical along its longitudinal plane. The reconstructed voxel ventricle is displayed with a ray tracing procedure using standard distance shading. The result is a three-dimensional animated scene (50 frames per second) of the inner volume of the left ventricle. From the pattern recognition point of view the Marr-Hildreth operator turned out to be well suited for edge detection in low contrast x-ray-images. From the radiologist's point of view the interpretation of ventricle motion is clarified.

Computing Review Classification: I.3.3 Picture/Image Generation: Display Algorithms, I.3.5 Computational Geometry and Object Modeling: Solid Representation, I.3.7 Three-Dimensional Graphics and Realism: Animation Shading, I.4.6 Segmentation: Edge Detection, J.3 Life and Medical Sciences: Medical Information Systems.
General Terms: Experimentation, Algorithms
Keywords and Phrases: Digital-Subtraction-Angiography, automatic ventricle contour extraction, Three-Dimensional Reconstruction.

1. Introduction

One of the advantages of PACS (Picture Archiving and Communication Systems) is the possiblility of integrating several image modalities in an image data base management system (1) and presenting images in a form most suitable to the

NATO ASI Series, Vol. F19
Pictorial Information Systems in Medicine
Edited by K. H. Höhne
© Springer-Verlag Berlin Heidelberg 1986

current task of the medical user. X-ray images as well as slices of tomographic sequences, functional images, or images synthesized by computer graphics methods can be displayed in one system. Therefore the medical user will increasingly select those representations out of different alternatives which support his diagnostic or therapeutic question most. This is especially true if dynamic processes and three-dimensional structures have to be visualized.

Medical objects are complex. In conventional radiography they can only be displayed indirectly as shadows of x-ray-radiation. The true 3D-information must be logically inferred from the projection images. Physicians have to perform the deduction process conciously, which requires a long education in radiology. The human vision system, however, has a "hardware mechanism" for inferring the actual 3D-structure of objects from the 2D retinal image, working unconciously and apparently without any effort. For that reason it seems to be desirable to extract 3D-information about objects from x-ray images and to produce direct views of the objects which can be interpreted easily by the human visual system.

In the following we present an experiment, in which the 3D-structure of a left ventricle is extracted from a dynamic digital subtraction angiography (DSA) sequence and displayed with computer graphics methods. By means of this the understanding of the dynamic process is facilitated even for unexperienced users. Automatic contour detection is used to isolate the heart region. The 3D-representation uses the voxel approach (See (13) for a survey of 3D representations in medicine). The derivation of the 3D-structure, however, differs radically from the common volume reconstruction method (3). Because of the inherent ambiguities in projected images, it was possible to obtain well defined results only by making simplifying assumptions with respect to morphological parameters and the image generation process.

2. Methods

2.1 Computation of the DSA-image-sequence of a heart cycle

Fluoroscopy of the intravenous ventriculograms was performed after central venous injection of 20 ml 76% Urografin (flow 12 ml/sec) with 10 μR per image. The output sequence of the high resolution image intensifier unit was recorded on video tape. A part of the image containing the left ventricle was digitized afterwards (Fig. 1 left).

Conventional DSA is not applicable because of the heart wall movement. ECG-triggering and the blurred-mask-method are common techniques to overcome those problems. In our case, averaging and subtraction were done phase adjusted, that is only with images in the same phase of the heart cycle (9). For each $i=1,...,n$, let EMPTY(i) be a sequence of images displaying one empty heart cycle. In addition, for each $i=1,...,m$, let CM(i) be an image sequence displaying one contrast-medium filled heart cycle. The computational steps performed are:

1. Pixelwise averaging of several empty heart cycles resulting in an averaged empty cycle L:

$$L = \frac{1}{n} \sum_{i=1}^{n} EMPTY(i)$$

2. Pixelwise averaging of several contrast medium filled heart cycles resulting in an averaged contrast medium filled cycle V:

$$V = \frac{1}{m} \sum_{i=1}^{m} CM(i)$$

3. Pixelwise logarithmical subtraction of the averaged empty cycle from the averaged filled cycle.

$$SUB := \ln (V) - \ln (L)$$

The number of available cycles, however, is too small to improve the signal-to-noise-ratio of the image series substantially. Therefore we finally applied a low-pass (moving average) filter (Fig. 1 right).

While this kind of operator leads to blurring of sharp edges when applied in a spatial domain it works very well in the time domain where no abrupt intensity changes occur (10). The filtering process eliminates primarily artifacts.

Digitization was done with a spatial resolution of 256 by 256 Pixels and 256 different gray levels at a rate of 50 frames per second (real time). Digitization and phase adjusted logarithmic subtraction was done with the system CA-1 (2),(8), having an 8 MB RAM image storage and special purpose image processing hardware and software.

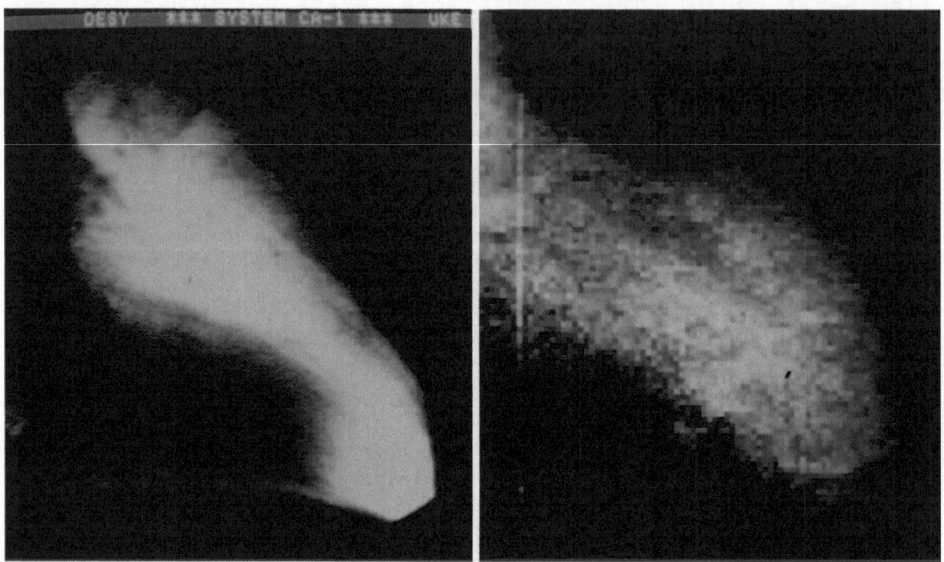

Fig. 1: left: A frame of the original unprocessed sequence.
right: A frame after phase adjusted subtraction.

2.2. Filtering and Contour Detection

In order to derive 3D-information from the DSA-images, it is necessary to drop the non-interesting objects in the images. The ventricle contours have to be extracted. One solution could be the interactive segmentation of each frame in the sequence by a radiologist. This is not practicable for three reasons:

1. High inter-individual differences from physican to physican in contour finding.

2. Low frame-to-frame continuity. This leads to jumping contours in the animated sequence which cannot occur in reality.

3. High work-load. The segmentation cannot be done by non-specialists.

A method was looked for, which

1. finds acceptable ventricle contours (judged by physicians),

2. finds the ventricle contour reproducible,

3. shows continuous transitions from frame to frame.

Heart contours show up as intensity changes in DSA-images. Intensity changes usually occur at different scales. There is no single filter which is able to detect intensity changes at all scales. D. Marr (6) proposes to blur the image, therefore wiping out structures of higher frequency components in order to limit the frequency range, one has to look at. Intensity changes in the image give rise of a peak in its first derivative and to a zero-crossing in its second derivative (Fig. 2). The zero-crossings of the filtered 2nd-derivative correspond to the locations of intensity changes in the image at a certain range of scale.

Intensity change   1st derivative   2nd derivative

Fig. 2: The principle of zero-crossings.

In (6) it is shown that a Gaussian smoothing operator is an optimal compromise between conflicting smoothness constraints. The Gaussian distribution is smooth and localized both in the spatial and frequency domain. Therefore it is unlikely that artifacts are introduced by the filtering process.

Also, conditions are given in (6), where the Laplacian Operator as a second derivative operator is able to detect intensity changes in images. These conditions are fulfilled in most images occurring in practice. It is therefore not necessary to compute computationally expensive directional derivatives in images. For a more detailed analysis from the implementation point of view, see (4). Fig. 3 shows the Marr-Hildreth Operator in pictorial form for Sigma = 7 in pixels.

The Marr-Hildreth-Operator is defined by:

$$O = \nabla^2 G(\sigma) * I$$

where

∇^2: Laplacian Operator,
G: 2D-Gaussian,
σ: standard deviation of G
I: DSA-Image
$*$: convolution-operator

Fig. 3: Example of the Marr-Hildreth operator
for Sigma = 7 pixels

Depending on the choice of Sigma (the standard deviation
of the Gaussian) it operates in different band-limited
frequency ranges. X-ray images - like all shadow images - have
relatively low contrasts. Experiments with different filter
widths showed that a standard deviation around Sigma=7 pixels
was well suited for our image sequence. The contours detected
were acceptable (inspected by physicians) and the transitions
from image to image were smooth. Fig. 4 shows zero-crossings
for different filter widths applied to the same frame of the
image sequence.

The heart contour, represented by selected zero-crossings
of the filter-output (Fig. 5), was traced and filled into a
boolean mask image. A starting point (the same for each image
in the sequence) was specified. The inner part of the contour
in the boolean mask image was filled using a parity check
filling algorithm (11). A boolean operation between masks and
DSA-images allowed to drop the non-heart-region. Finally
(Fig. 6) the sequence was smoothed in order to delimit the
decline of the gray levels of the ventricle border to the
background (zero gray level).

The ventricle extraction from the heart region was done
heuristically by cutting the atrium along an interactively
specified line (Fig. 5). This step can in principle be done
with the Fourier-Transform (5). In phase images atrium and
ventricle show up clearly separated, because of the anticyclic
contraction behavior of both parts of the heart (14).

The result of this processing step after automatic
contouring of the 2D-ventricle region was a DSA-sequence of an
isolated, beating left ventricle.

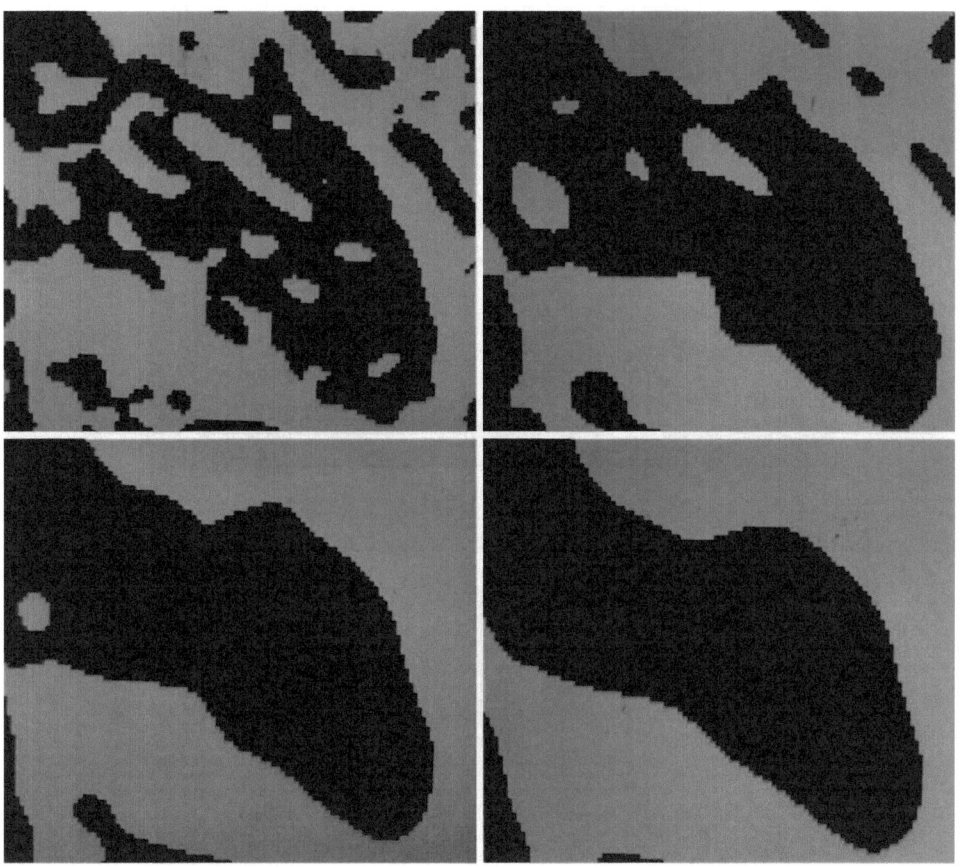

Fig. 4: zero-crossings for different filter widths.
                    top left: Sigma = 3 pixels
                   top right: Sigma = 5 pixels
                bottom left: Sigma = 7 pixels
               bottom right: Sigma =10 pixels
          zero-crossings are represented by the transitions
          from black to white.

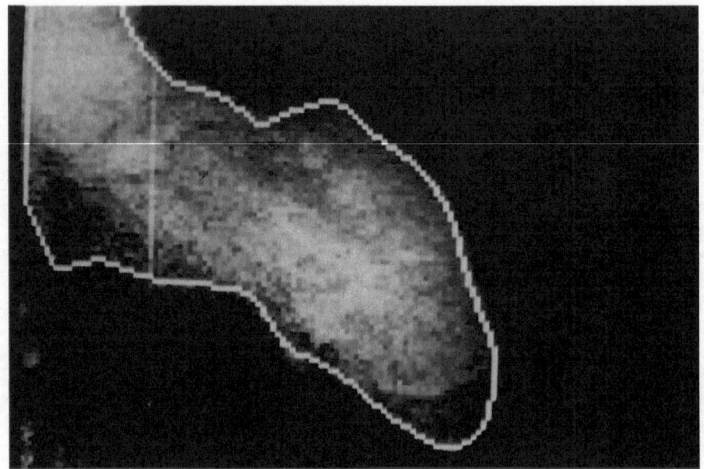

Fig. 5: Zero-crossings of the heart region as contour for Sigma=7 pixels overlaid over the corresponding image.

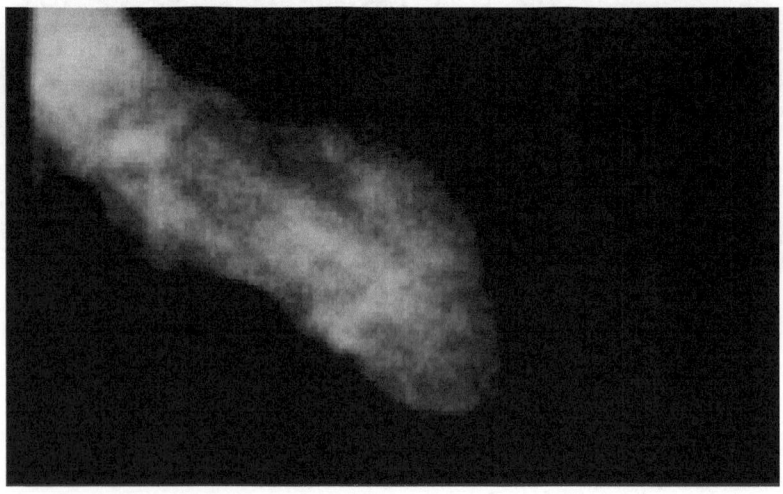

Fig. 6: Basis for the 3D-reconstruction was the smoothed heart region after dropping all non-interesting image parts.

2.3 Volume-Reconstruction

Gray levels in x-ray-images represent absorption properties of the penetrated tissue. Depending on the representation (positive or negative) high (low) gray levels correspond to high (low) absorptivity of the tissue. Assuming that only one kind of tissue is visible in the image, the thickness of the penetrated tissue can be computed from the gray levels up to a constant factor. The DSA-method fulfills this assumption approximately. In the following we assume that only blood filled with contrast medium is visible in our sequence. The following derivation shows how to compute thickness values from gray levels under such circumstances.

The attenuation of x-ray-radiation by matter is:

$$I = I_0 \cdot e^{-\sum_i k_i \cdot c_i \cdot d_i}$$

(1)

I : intensity of radiation behind the object
I0: intensity of radiation in front of the object
ki: absorption coefficient of the material i
ci: density of the material i
di: thickness of a layer of material i

Gray levels depend linearly on I. Assuming a gray level to be caused by one kind of tissue only, it is a measure for the thickness of the penetrated tissue (here: contrast medium filled blood). Ignoring superposition of different kinds of tissue (here: superposition of blood vessels) we have:

$$I = I_0 \cdot e^{-k \cdot c \cdot d}$$

(2)

k,c,d absorption coefficient of the matter, density of the matter and thickness of the tissue layer (here: contrast medium filled blood).

The exponential decline of the intensity values in equation (1) has to be linearized by taking the logarithm. Solving the equation (2) for d leads to:

$$d = -(1/kc) \cdot (\ln I - \ln I_0)$$

(3)

This condition is approximately fulfilled using phase-adjusted subtraction angiography. Intravenous application of the contrast medium guarantees a homogenous distribution of the contrast medium. Logarithmic subtraction linearizes the exponential absorption behavior. Gray levels and intensity ought to depend linearly on each other (a condition which has to be guaranteed by the imaging system). Therefore thickness of layer and gray level are proportional to each other.

A first approximation of the ventricle thickness (up to a constant factor) can therefore be computed with the following function:

$$f: \left\{ g^{Min}, ..., g^{Max} \right\} \longrightarrow \left\{ 0, ..., maximum\,Thickness \right\}$$

$$x \longmapsto \frac{-const}{g^{Min} - g^{Max}} \cdot (x - g^{Min})$$

where

maximum_thickness: heuristically selected value for the maximum
thickness of the ventricle.
gMin: minimum gray level of the image sequence,
gMax: maximum gray level of the image sequence,
const: maximum gray level of the imaging system,
(here 255),

The maximum thickness of the ventricle in the diastolic
state was set to the length of the axis of height of the heart
in that state.

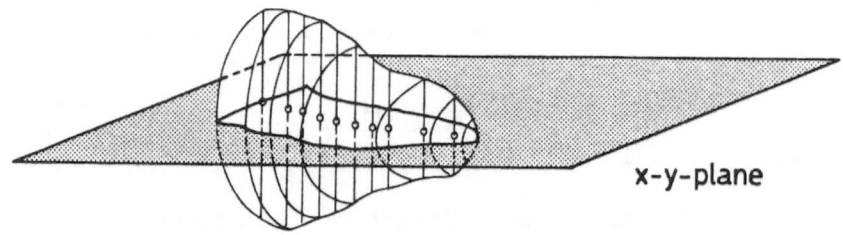

x-y-plane

Fig. 7: The principle of volume reconstruction

The computed thickness of the inner ventricle at a certain
point, as described above, leaves undefined the entrance and
exit point of the path through the ventricle. In order to
resolve this degree of freedom we assume (somewhat arbitrarily)
the position of both points to be symmetrical along the
xy-plane (Fig. 7).

The performed operation corresponds to a scaling of the
gray level mountain with succeeding reflection along the
xy-plane. The whole procedure is based on the following
assumptions:

1. no superpositional structures in the area of the left
ventricle.

2. homogenous distribution of the contrast medium in the
ventricle.

3. The maximum thickness of the heart is equal to the
axis of height of the heart in the diastolic state.

4. symmetrical shape of the heart along its medium longitudinal plane.

5. The minimum thickness is zero.

   While the first 3 assumptions are approximately valid, the last two conditions are of theoretical nature, being necessary in order to determine the position of the ventricle in 3D-space and a scaling factor for the thickness. Information about those two facts are not contained in DSA-images.

   A binary cube was used as data structure for describing the computed ventricle volume (ventricle = logical true).

2.4 3D-Display

   The binary cube, in which the 3D heart volume was marked, defines a 3D-coordinate system, we call the object coordinate system. Observer position, -orientation, and -viewing direction define a second coordinate system, the observer coordinate system.

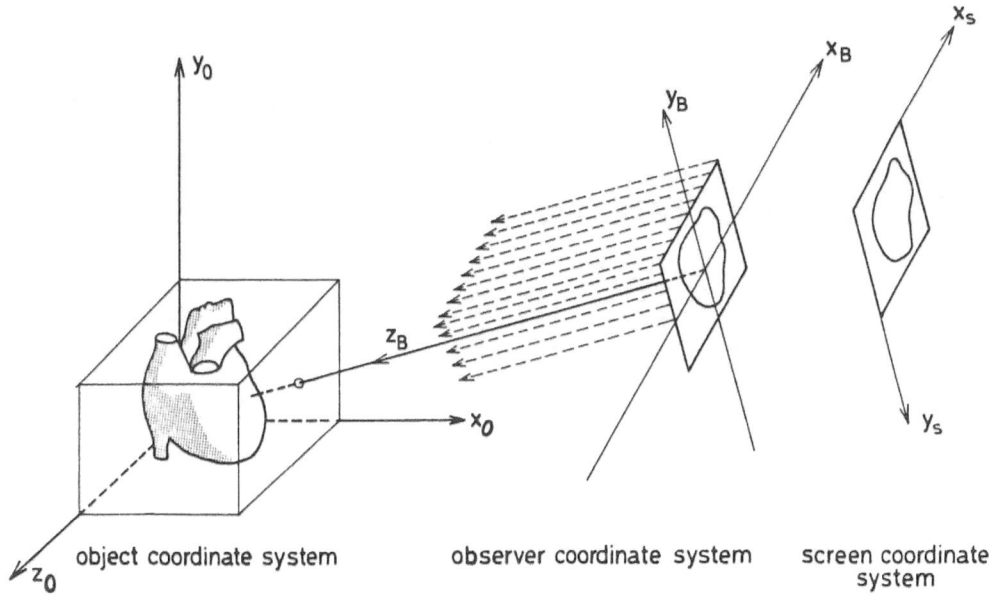

object coordinate system   observer coordinate system   screen coordinate system

Fig. 8: coordinate systems used and the display of the
        reconstructed volume.

According to computer graphics standard we define the object coordinate system to be a right system and the observer coordinate system to be a left system. The xy-plane of the observer coordinate system serves as projection window, which is mapped onto the screen-viewport of the display monitor. We implemented parallel projection.

The 3D-reconstruction was done as follows: for each point in the window of the xy-plane in the observer coordinate system a ray was constructed along the observer z-axis, until it hit a surface point of the heart volume. For any ray which did not intersect the heart volume, the point was set to gray level zero, in the former case a depth shading was done (Fig. 8). The path of the ray was calculated by a 3D line generator (digital differential analyzer).

For each image in the sequence volume-reconstruction and 3D-display was repeated resulting in an output sequence showing the inner ventricle in each position of the heart cycle.

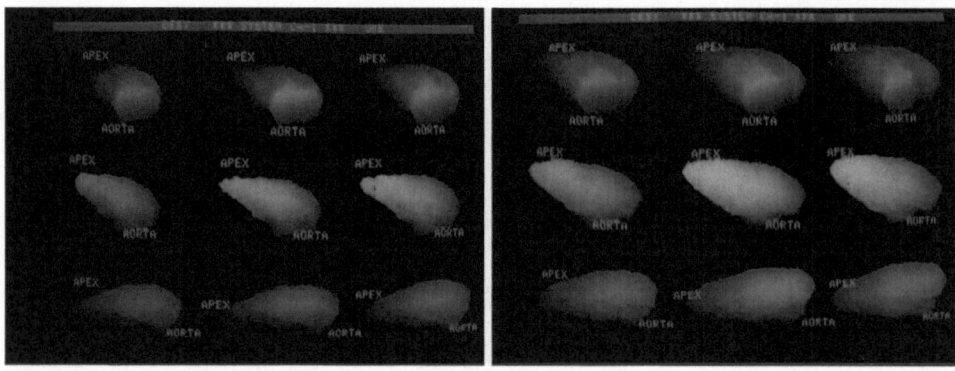

Fig. 9: A subsequence of the ventricle during the diastolic phase from different point of views:
1. line (6 frames): seen from the Aorta to the Apex.
2. line (6 frames): the Apex points diagonally upwards.
3. line (6 frames): side view.

The computed image sequence was recorded on video tape in real time using the CA-1 system. The film shows the movement of the contrast medium filled blood in the left ventricle. Physicians judged the movement to be natural. Contrary to the original x-ray-sequence the movement was easy to interprete even by a non-radiologist.

Fig. 9 visualizes part of the results. Three subsequences consisting of frame 1,4,7,10,13,16 of the ventricle during the diastolic phase are shown from different point of views.

It has to be noted that the shading algorithm is very simple. Watched as a movie, however, the dynamic behavior of the represented volume was fully recognizable (Fig. 9).

3. Discussion

The experiment we have described here allowes us to generate a three-dimensional view of a dynamic process, contrary to the static 3D-views computed from CT-slices. The 3D-information necessary to do this is not completely contained in the DSA-images. This degree of freedom had to be resolved by an (unrealistic) symmetry assumption.

From the radiological point of view, it might be questioned whether the presented animated 3D-sequence gives any new information at all to the viewer. The information contained in the 3D-reconstruction is actually not new at all. In principle the DSA-sequence contains already all the information about depth of the ventricle region the physician might be interested in.

One can doubt, however, whether this information is really used as depth information by the medical user. The time behavior of the contrast medium is difficult to capture, because the contrast medium bolus is flat. The optical impression is an increasing darkening. Problems like non-linearity of absorption, superposition of background structures as well as different flow speeds of the contrast medium have to be taken into account by the viewer.

Therefore we think that the method might be suited as a feedback instrument for the radiologist. He could compare automatic and "mental" reconstruction with each other and correct misinterpretations. In this context the radiologically unexperienced physician could use the reconstruction as a teaching tool.

The DSA-sequence shows only one projection. Different viewpoints can be produced by new x-ray-examinations, increasing the total x-ray exposure to the patient or have to be simulated in the physician's mind by "mental rotation". Mental rotation of complex objects is difficult and cannot be done in real time by humans (12). To the contrary our 3D-reconstruction can be used to compute any projection you like. Once the binary cube is computed, several views can be generated from arbitrary viewpoints. This might be an essential contribution to the better cognition of dynamic 3D medical objects. The method facilitates the image interpretation process, because it releaves the physician from mental rotation. This might even lead to x-ray dose reduction, because fewer examinations might be necessary. In addition the result of the reconstruction is more objective. It is reproducible and can be discussed more precisely.

The method used lends itself to all kinds of ventricle quantification such as volume determination, computation of the ejection fraction etc.

From the pattern recognition point of view we can state that the zero-crossings of the Marr-Hildreth filtered images result in closed contours which can be traced automatically. The directional independence of the Laplacian Operator leads to efficient implementation. Space- and frequency-locality of the smoothing with a Gaussian guarantees few artifacts when looking for intensity differences. The filter takes into account a much bigger range of scale than "classical" edge operators. This is important especially with respect to the smooth contrasts in an x-ray image.

The heart contours found by the operator are more objective than those found interactively. Contours found interactively showed unnatural discontinuities from frame to frame and were not reproducible. The Marr-Hildreth operator is claimed to work in a similar fashion to the early processing of the visual system (4). It gives information about intensity differences without interpreting them. In addition automatic contour tracing is, of course, faster than manual contour tracing and has the advantage of resulting in smooth transitions from frame to frame.

Theoretical examinations (15), (16) show that zero-crossings in a certain sense, represent a "complete" description of images. This could be an important aspect for data compression in the PACS domain. According to our knowledge there are no research activities going in this direction. As proposed by (7) the zero-crossings are suitable as a basis for higher level descriptions in computer vision systems.

In the DSA-sequences we processed it was possible to detect heart contours with one filter channel only. In general this is not true. Experiments with CT- and MR-images showed that one has to combine several frequency channels in order to detect edges contained in the image. Here, there is more work to be done.

4. Conclusion

We have shown that gray level information can be used alone to compute and display a dynamic three-dimensional voxel scene from a DSA-image sequence. Despite an unrealistic symmetry assumption necessary to resolve inherent ambiguities in x-ray-projection images the scene showed the movement of the inner volume of a left ventricle during a complete heart cycle.

Displaying the scene from different point of views greatly enhances the understanding of the movement of the heart even for non-radiologists. The automatically found heart contours were satisfactory from the radiologist's point of view.

The algorithms can be used to serve as a tool in clinical teaching and research.

Acknowledgement

We would like to express our gratitude to K. Aßmann, M. Bomans, C.-J. Peimann, M. Riemer, U. Tiede (Institut für Mathematik und Datenverarbeitung in der Medizin) and Bob Glaser (currently Desy) for constructive comments on earlier versions of this paper.

5. References

1. Aßmann, K.; Venema, R.; Höhne, K.-H.: Software Tools for the Development of Pictorial Information Systems in Medicine - The ISQL Experience. This volume.

2. Böhm, M.; Obermöller, U.; Pfeiffer, G.; Höhne, K.H.: Image Management with the system CA-1. Proceedings PACS I, Newport Beach, California, Jan. 1982.

3. Herman, G.T.: Three-Dimensional Imaging from Tomograms. In: Höhne, K.H.(ed.): Digital Image Processing in Medicine, Proceedings Hamburg, Oct. 1981, Springer Verlag.

4. Hildreth, Ellen: The Detection of Intensity Changes by Computer and Biological Vision Systems. Computer Vision, Graphics and Image Processing, 22 (1983) 1-27.

5. Höhne, K.H.; Obermöller, U.; Riemer, M.; Witte, G., Böhm, M.: Data Compression in Digital Angiography using the Fourier Transform. Medical Physics, 10 (1983) 899-905.

6. Marr, D.; Hildreth, Ellen: Theory of Edge Detection, Proc. Royal Society, London, B 207, (1980) 187-217.

7. Marr, D.: Vision, Freeman Press, San Francisco 1982.

8. Nicolae, G.C.; Höhne, K.H.: Multiprocessorsystem for the real-time processing of video image series. Elektronische Rechenanlagen, 21 (1978) No 4, 171-183.

9. Obermöller, U.; Witte, G.; Höhne, K.H.: Parametrische Bilder aus intravenösen Angiogrammen des linken Herzventricles. 5. DAGM Symposium 1983 in Karlsruhe, DVE Fachberichte, VDE-Verlag, 47-53.

10. Obermöller, U; Witte, G.; Rödiger, W.; Höhne, K.-H.: Parametric Images from digital intravenous Ventriculograms, Internal Paper, IMDM, UKE, Hamburg, 1984.

11. Pavlidis, Theo: Algorithms for Graphics and Image Processing. Computer Science Press, Rockville, MD, 1982.

12. Shepard, R.N.; Metzler, J.: Mental Rotation of three-dimensional Objects. Science 171, 701-703.

13. Udupa, J.K.: Display of 3D Information in Discrete 3D Scenes Produced by Computerized Tomography. Proceedings of the IEEE 3 (1983) 420-431.

14. Witte, G., Obermöller, U., Jacobs, G., Roediger, W., Grabbe, E., Höhne, K.H., Bücheler, E.: Funktionsbilder des Linken Ventrikels mittels venöser Digitaler Angiographie. Fortschritte Röntgenstr. 1 (1984) 49-52.

15. Yuille, A.L.; Poggio, T.: Scaling Theorems for Zero-Crossings, MIT AI Memo No. 722, 1983.

16. Yuille, A.L.; Poggio, T.: Fingerprint Theorems for Zero-Crossings, MIT AI Memo No. 730, 1983.

SYSTEMATIC USE OF COLOUR IN BIOMEDICAL DATA DISPLAY

J.P.J. de Valk*, W.J.M. Epping** and A.Heringa***

*   Department of Diagnostic Radiology, University of
    Nijmegen, current working address:
    BAZIS – AZL, Leiden University Hospital, Bld. 50,
    Rijnsburgerweg 10, 2333 AA Leiden, The Netherlands.
** Department of Medical Physics and Biophysics,
    University of Nijmegen,
*** Department of Cardiology, University of Nijmegen.

1. INTRODUCTION

In recent years the number of methods to process multi-dimensional
graphical biomedical data has been considerably increased (e.g. see [1]).
At the same time the display procedures themselves within pictorial
information systems have become more complex and sophisticated (e.g. see
[2,3]). As the final outcomes of these processes have to be interpreted by
the human observer, the data presentation must be adapted to the human
sensory systems in order to guarantee an adequate data transmission from
output device to man. Often the processed data are visualised using
computer graphics and images, in which some kind of pattern has to be
recognised [4]. It is required that the properties of both the data imaged
and the human visual system are taken into account, in order to process
the data optimally (e.g. see [5]).

In the display of 2-D data values often an intensity mapping procedure is
applied, mostly resulting in an image consisting of grey shaded pixels.
Major problems, met in practice, are:

1. The quantitative interpretation of many grey shades is very difficult
   for the human observer: the value of a particular pixel is very hard to
   identify and memorise by just looking at its greyness. This is not the
   case when different colours are used.

2. Small gradients in greyness over greater distances are not completely
   observed or even not observed at all, contrary to colour gradients.

3. As a result from 1. and 2. grey value subranges cannot well be
   discriminated in the images. The consequence is that the segmentation
   of the image into discernable parts becomes rather difficult.

NATO ASI Series, Vol. F19
Pictorial Information Systems in Medicine
Edited by K. H. Höhne
© Springer-Verlag Berlin Heidelberg 1986

4. When too few grey levels are used in pictures one obtains observable
   transitions from one grey level to another causing boundaries which are
   conspicuous without serving a purpose.

A key to the solution of some of these problems (e.g. see [6]) can be
found in the properties of the human visual system, specifically in its
colour perception. The properties of human colour perception allow
quantitative reference to pixel values because of a better memory for
colours. Moreover it yields the means for image emphasis or segmentation
and discrimination where needed, as well as simultaneous representation of
more than one parameter per pixel, e.g. phase and amplitude of complex
numbers by two independent scales of e.g. redness and blueness.

2. THE COLOUR IMAGE SYSTEM

Usually the data for display is available in some digital format. It is
obvious that the number of bits per pixel determines the display
digitisation effect. Apart from this, image and colour memory
characteristics, such as the number of horizontal and vertical pixels, and
the number of simultaneously available colours (determined by the number
of bit-planes in the video image memory), play an important role in the
image quality.

In our experiments we have used a RAMTEK 9300 colour video display system
with a BARCO CDCT 2/66 display monitor (r, g, b; 625 lines; 50 Hz
interlaced) coupled to a PDP-11/45 minicomputer.

Properties of this system are:

640 x 512 pixels

6 bit-planes (2**6 simultaneous displayable colours)

12 bits/colour address (4 bits per primary colour)

3. COLOUR PERCEPTION AND SPACE

Colour is a perceptive concept, determined by the physiology and
biophysics of the visual system. Its description is based on visual,
psychophysical and electrophysiological experiments and well developed
models determining its phenomenology. Tristimulus (video) tehniques are
relying on the phenomenon that three well-chosen colours (primaries) allow
the reproduction of all possible colour sensations in man by an

appropriate additive mixture of the primaries in a linear colour space [5,7,8,9]. For details on the use of colour in a general context the reader is referred to two recent papers of Murch [10,11].

Since the receptors for red, green and blue stimuli in the human visual system show different sensitivities, perceived intensities of different colours are not the same for equal physical intensities of the colours. Therefore a concept 'Luminance' (Y) has been defined by the Commission Internationale d'Eclairage (CIE) to represent the colour intensity as perceived by the average human observer [7]. A projection of R, G and B values into an (x,y)-plane and a third dimension Y has been worked out by the CIE which has been adopted internationally in 1931. In this (x,y,Y)-space the interpretation of colours becomes much easier with regard to perception. Neither x, y or Y are allowed to adopt negative values.

A transformation of the x- and y-coordinates, such as proposed by MacAdam in 1937 [7,12], has been adopted by the CIE in 1960, which results in a space where steps of equal size correspond as much as possible to equally perceived chromatic changes for constant luminance. This space is known as the 'Uniform Chromaticity Scale' CIE-UCS.

Out of many possibilities we have chosen this particular transformation for three reasons. Firstly, because it is well-documented; secondly, because it is acknowledged internationally and thirdly, because it can be shown to be quite useful in practical situations, as we will demonstrate by three biomedical imaging examples.

Because of the restricted space here, we refer to our recent paper [13] for complete theoretical and numerical details concerning our colour research.

4. THREE PRACTICAL EXAMPLES OF COLOUR SCALE CHOICES

4.1 Pseudocolour as an aid for segmentation of myocard scintigraphs.

In scintigraphs of the heart, pathologies are found in areas with lower intensities compared to a standard (see Fig.1). The segmentation of these areas is an important issue [14]. In practical displays increasing intensities should be shown with such a continuous scale that maximum and minimum values are at the upper and at the lower end respectively, so that full use is made of the dynamic range of the available scale.

A grey-scale possesses little possibilities in the direction suggested. Since colour display systems become more and more commercially available these days (sometimes at nearly the same expenses as black and white display systems before) colour mapping seems worthwhile investigating, especially in concern to its possibilities to meet the requirements pointed at above. Milan and Taylor [15] suggested that an appropriate choice of a 'natural' colour scale should fulfill the following two conditions:

1. Colours representing high intensities should appear of high intensity.

2. Colours in an image area should allow local averaging, resulting in a mean area intensity without too much disrupting effects.

Considering only the colour intensity, known as the luminance, this luminance should be chosen monotonically increasing to obtain a quantitatively correct interpretation. Because of the logarithmical sensitivity for intensity of the human visual system (Weber's law), the luminance of the natural colour sequence chosen should increase exponentially.

Besides this restriction imposed to luminance, three other important conditions are to be fulfilled:

1. The number of perceived colours should be as high as possible.

2. The chromatic steps should be perceptually equal.

3. The colours should be technically feasible.

It is common knowledge (see e.g. [7]) that the human observer can discriminate quite well between colours of a magenta hue. This in combination with the fact that a so-called "heated-object" sequence (from red through orange and yellow to white, e.g. [16] is an acceptable natural sequence for human observers, has led to a scale that seems rather useful. The colours along this combined magenta-heated object path through colour space are selected evenly in the Uniform Chromaticity Scale space. The result is shown in Fig.2.

4.2 Colour representation of the electric field of the heart at the body surface.

The electrical activity of the heart can be measured at the body surface. In clinical practice measurements are usually carried out at no more than 8 sites. More recently in the so-called body surface mapping procedure the potential at the body surface is recorded at a larger number of positions (see e.g. [17]). The resulting data can be interpreted by considering the 64 time signals one at a time, but is seems more adequate to base a diagnosis upon images of the potential distribution at the body surface at subsequent time instants. In that case the distributions can be represented using a grey scale. This kind of representation (Fig.3), however, suffers from some drawbacks. The diagnostic interpretation cannot be optimal because of the small number of perceivable grey shades. Consequently the range of distinguishable potential values is too small. Moreover the zero potential level is not easily recognised over the area if black is reserved for strong negative potentials. Therefore we prefer a colour representation for display of the body surface data.

We selected our colour coding scheme satisfying the following requirements:

1. Easy recognition of the zero level representing non-significant potentials.
2. Clear discrimination of positive and negative potentials.
3. Existence of a direct relationship between the potential value and the luminance.
4. Realisation of a perceptively smooth transition from zero to maximum positive and negative values.

This resulted in the blue-green-red colour scale right of Fig.4. The zero level is represented in green, large positive values in red and large negative values in blue. In the transaction from zero to larger positive values green decreases while red increases. As soon as the green component has become zero, a further increase of the potential is represented by a higher intensity. For the negative potentials the same approach has been chosen with blue instead of red.

4.3 Detection of coherent activity in neural population using colour representation.

An increasing number of laboratories now is recording simultaneous activity of several (2-30) neurons in order to investigate the existence and function of neural assemblies.

Neural assemblies would manifest themselves by the highly correlated activities of the constituent neurons. The investigation of functional properties of neural assemblies is then mainly reduced to a pattern recognition problem: detection of patterns of neural events (an event being the all or none activity of a neuron as measured extracellularly).

Because of several reasons a display technique has been devised which contains all original information by preserving both identity (i.e. neural origin) and temporal relationship of neural events [18]. For this purpose we represent neural identity, which can be considered as a category variable, by colour. The display technique has been called 'Neurochrome' by us. The colour coding scheme has been selected in order to maximise perceptual distance between the colours, with the restriction that these should be of near equal luminance. Accordingly the three principal colours red, green and blue and three complementary ones cyan, magenta and yellow were selected.

As compared to colour representation, the alternative of using black and white symbols has two major drawbacks:

1. Black and white symbols are more difficult to discriminate than coloured dots.

2. Black and white symbols are larger than coloured dots, thereby reducing spatial resolution.

The two representations, coloured dots and black and white symbols, of neural activity are illustrated with an example taken from the auditory midbrain of the grassfrog, containing four neurons [Figs. 5 and 6]. Symbols and colour key are at the right of the figures. An auditory stimulus was applied consisting of tone-pips of duration 48 ms and interval between onset of tone-pips of 1 s. The figures are arranged such that they should be read as a page of text. The n-th line represents the neural response to the n-th tone-pip.

5. CONCLUSION

This paper has illustrated different possibilities of colour coding for biomedical imaging problems. In practice it is required to know the specifications of the colour imaging system and to have prior knowledge about the data to be imaged, in order to match the purpose of the imaging itself. This knowledge will direct the colour choice.

Acceptable results in colour representation can be achieved employing the Uniform Chromaticity Scale transformations in the colour selection procedures. With these instruments pattern recognition problems in biomedical data analysis can be worked out using the discrimination and segmentation powers of colour. The results point at a very wide application area for colour representation, provided that proper colour coding schemes and sequences are selected for the mapping of biomedical data.

ACKNOWLEDGEMENTS

The authors would like to thank several persons in Nijmegen and Leiden for carefully preparing the manuscript and for providing valuable help and comments.

REFERENCES

[1] Huang, H.K.: Biomedical image processing. CRC Critical Reviews in Bio-engineering, 5, (1981) 185-271.

[2] Pratt, W.K.: Digital Image Processing. Wiley, New York (1978).

[3] Hall, E.L.: Computer Image Processing and Recognition. Academic Press, New York (1979).

[4] Graedel, T.E. and McGill, R.: Graphical presentation of results from scientific computer modes. Science, 215, (1982) 1191-1198.

[5] Cornsweet, T.N.: Visual Perception. Academic Press, New York (1970).

[6] Sheppard, J.J.: Pseudo-color as a means of image enhancement. Am. J. Optom. Arch. Am. Acad. Optom, 46, (1969) 735-754.

[7] Wyszeski, G.W. and Stiles, W.S.: Color Science; Concepts and Methods, Quantitative Data and Formulas. Wiley, New York (1967).

[8] Bouma, P.J.: Physical Aspects of Color. Philips Technical Library, Eindhoven (1974).

[9] Foley, J.D. and Van Dam, A.: Fundamentals of Interactive Computer Graphics. Addison Wesley, Reading Mass (1982).

[10] Murch, G.M.: The effective use of color: physiological principles. Tekniques, 7, 4, (1984) 13-16.

[11] Murch, G.M.: The effective use of color: perceptual principles. Tekniques, 8, 1, (1984) 4-9.

[12] MacAdam, D.L.: Projective transformations of ICI color specifications.J. Opt. Soc. Am., 27, (1937) 294.

[13] De Valk, J.P.J., Epping, W.J.M. and Heringa, A.: Colour representationof biomedical data. Med. & Biol. Eng. & Comput., (1985) in press.

[14] Gallagher, J.H., Preston, D.F., Robinson, R.G., Herrin, W.F., Servoss,W. and Fritz, S.: Image processing and pattern recognition in nuclear medicine. IEEE Trans. on Inf. Processing and Pattern Rec., 55, (1977) 55-60.

[15] Milan, J. and Taylor, K.J.W.: The application of the temperature-color scale to ultrasonic imaging. J. of Clin. Ultrasound, 3, (1975) 171-173.

[16] Chan, F.H. and Pizer, S.M.: An ultrasonogram display system using a natural color scale. J. of Clin. Ultrasound, 4, (1976) 335-338.

[17] Heringa, A., Uijen, G.J.H. and Van Dam, R.Th.: A 64-channelsystem forbody surface mapping. In: Electrocardiology '81, Akademiai Kiaido, (1981).

[18] Epping, W.J.M., Van den Boogaard, H.F.P., Aertsen, A.M.H.J., Eggermont, J.J. and Johannesma, P.I.M.: The neurochrome: an identity preserving representation of activity patterns from neural populations. Biological Cybernetics, 50, (1984) 235-240.

FIGURE CAPTIONS

Fig. 1 + 2: Nuclear image (scintigraph) of a normal myocard image. The intensities displayed correspond with recorded gamma-ray acitivities.

Fig. 3 + 4: Potential distribution at the body surface at four different time instances (16, 32, 48 and 64 ms after the onset of the depolarisation of the ventricles). In the precordial lead V2 the moment of the map is indicated by a vertical line.

Fig. 5 + 6: Representation of activity of a neural population, that consists of 4 neurons. The neurons are coded according to the symbols or colours shown to the right of the figures. The duration corresponding with the length of the horizontal axis is 1 s, being the interval between onset of tone-pips, and represents time after a tone-pip, running from left to right. The vertical axis gives the line index.

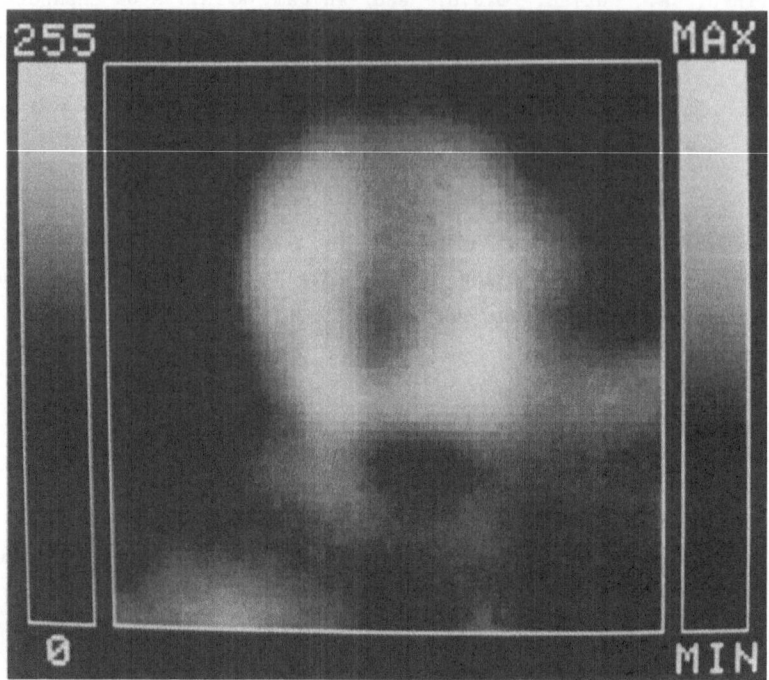

Fig. 1 Nuclear image(scintigraph) of a normal myocardium. The intensities displayed correspond with recorded gamma ray activities.

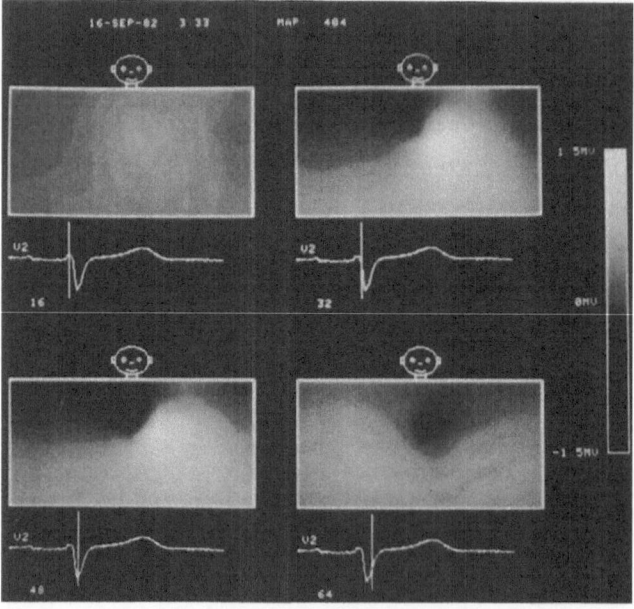

Fig. 3 Potential distribution at the body surface at four different time instances. In the ECG they are indicated as vertical lines.

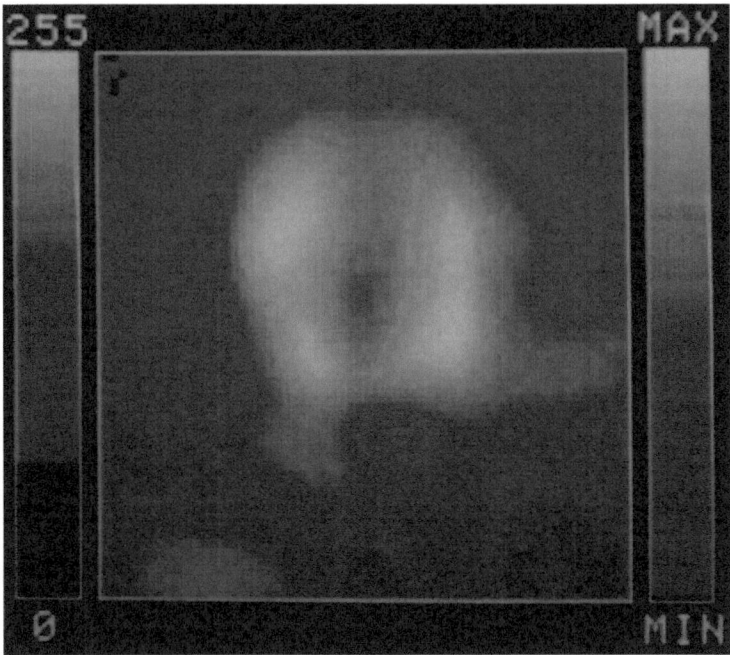

Fig. 2 Nuclear image(scintigraph) of a normal myocardium. The colors displayed correspond with recorded gamma ray activities.

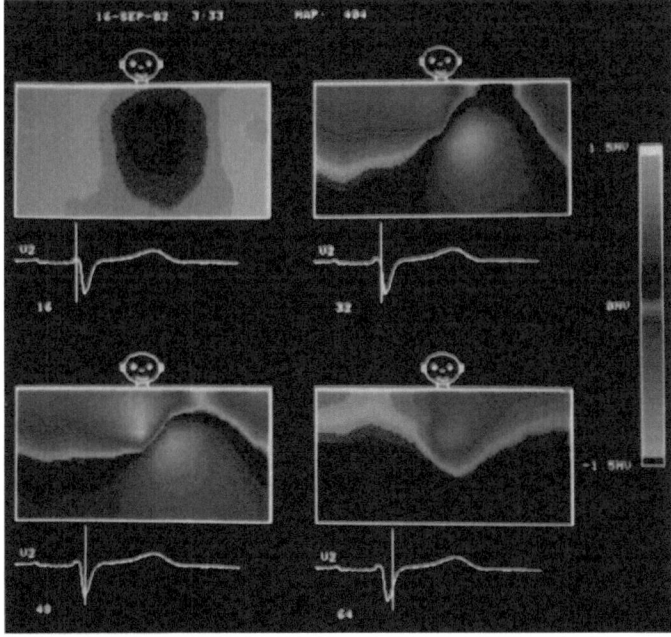

Fig. 4 Potential distribution at the body surface at four different time instances. In the ECG they are indicated as vertical lines.

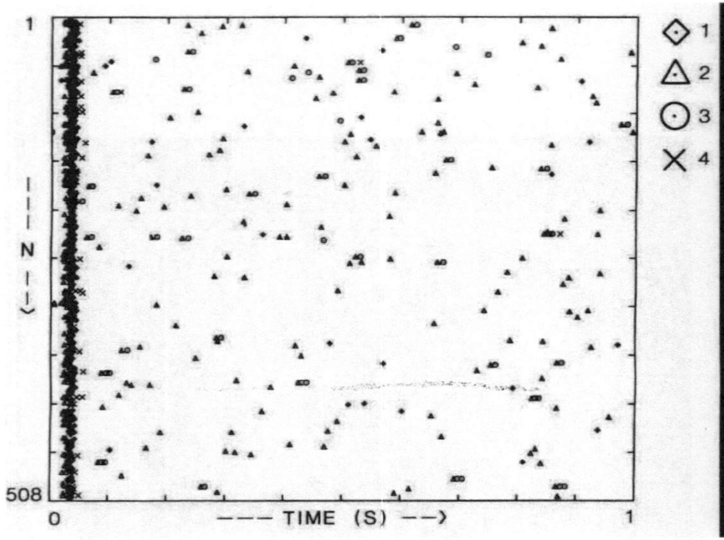

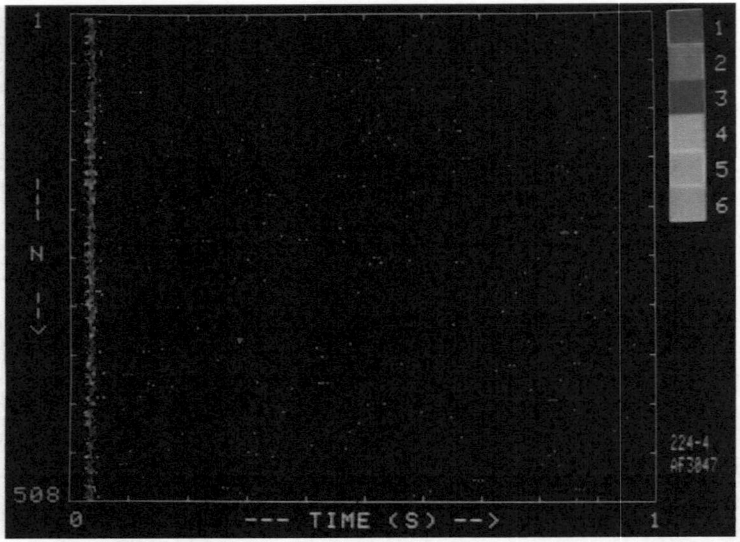

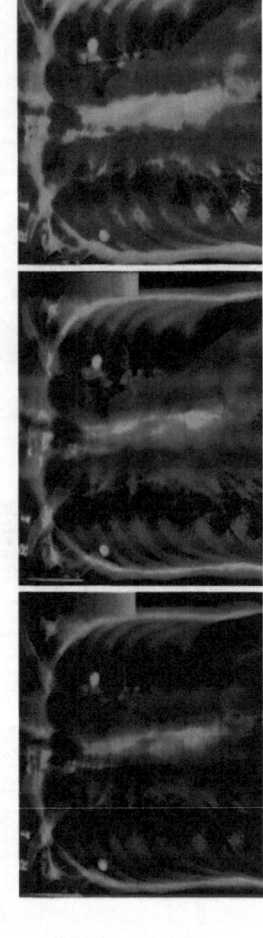

Figs. 5 and 6.   Representation of activity of a neural population, that consists of 4 neurons.

Fig. 8 (S. PIZER: Psychovisual Issues in the Display of Medical Images, p. 211)
An ahe'd chest radiograph in three linearized scales: grey scale, the heated object scale and the rainbow scale. The scale appears next to each image.

3-D MODEL OF VERTEBRA FOR SPINAL SURGERY

Harcharan Singh Ranu

Department of Biomedical Engineering
Louisiana Tech University
Ruston, Louisiana 71272 U.S.A.

INTRODUCTION

The finite element method of stress analysis is a useful tool,
particularly for irregularly-shaped structures such as the vertebra. It is able to identify locations of high stress and to
quantify changes in stress due to surgery. Therefore, a three
dimensional finite element model has been developed which
simulates the in vivo spinal surgery such as (laminectomies).

Harrington rod spinal instrumentation is another technique,
which is commonly used for the correction of the scoliotic
problems of the human spine. Such an in vivo loading situation
is also simulated.

MODEL FORMULATION

A finite element model of a complete vertebra is shown in
Figure 1. Figure 2 shows a side view of the vertebral model. In
both figures 1 and 2, the hidden nodes are identified by numbers
in parentheses.

The vertebral model is made of the following elements:

Eight-noded hexahedral brick elements.
Four-noded general shell elements.
Three-dimensional axial elements.

Three-dimensional eight-noded hexahedral (brick) elements were
used to represent the spongy bone of the centrum, the cortical bone
of the pedicles, the articular facets and the laminae. The vertebral end plates, the cortex of the vertebral body and the transverse and spinous processes were modelled by thin shell elements.
They were quadrilateral in shape and had both membrane and bending
resistance. Three dimensional spring elements were used to simulate
the inferior intervertebral disc which resisted axial or shear
loads. The disc was not simulated on the superior end plate where
concentrated nodal loads were applied. For the centrum, there were
a total of 72 brick elements arranged in three transverse layers,
each containing 24 such elements. The posterior structure was composed of 22 brick elements, four for the two pedicles, 12 for the
two lamina and six for the articular facets. The superior facet

NATO ASI Series, Vol. F19
Pictorial Information Systems in Medicine
Edited by K. H. Höhne
© Springer-Verlag Berlin Heidelberg 1986

was modelled by a single hexahedral while the inferior facet was represented by two such elements. There was a total of 91 thin shell elements, 86 for the cortex and end plates, three for the spinous process and one each for the transverse process. Eighty-four axial spring elements were used across the inferior end plate, of which 32 provided compressive support. Around the outer periphery of the body, 28 axial springs were located which gave shear support. A total of 24 springs were located on the inferior articular process to give compressive and shear support. Material properties of vertebral bone and spinal disc were selected from experimental data obtained by Ranu and King (1), Ranu (2) and Ranu (3) as shown in Table 1.

TABLE 1

MATERIAL CONSTANTS FOR THE STATIC RUN

Element Type	Description	Modulus of Elasticity		Poisson's Ratio	Weight Density	
		kPa	psi		N/cm^3	lb/in^3
Hexahedron	spongy	2.41×10^6	0.35×10^6	.2	.0041	.015
	compact	11.03×10^6	1.6×10^6	.25	.0184	.0679
Thin plate/ Thin shell	all	12.48×10^6	1.81×10^6	.28	.0184	.0679
		STIFFNESS				
		N/cm	lb/in	N cm/rad	lb in/rad	
Boundary	axial element	538.85	302.69	---	---	
	IV disc	14010	8000	---	---	
	shear element	318.41	181.82	---	---	
	IV disc	7005	4000			
	torsional- element	---	---	2054.27	181.82	
	IV disc	---	---	18488	1636	

The NISA finite element code was used to compute vertebral response to a static load and the following cases were studied:

(1) Bilaterally symmetric vertebra subjected to a combined axial and shear load. The values were 2390 and 420 N, respectively.

(2) Asymmetric vertebra with two lamina elements removed on the left side (elements 161 and 162).

(3) Asymmetric vertebra with top element removed on the left side (element 161).

(4) Asymmetric vertebra with bottom lamina removed on the left side (element 162).

(5) Bilaterally symmetric vertebra subjected to an axial load applied by Harrington rod.

MODEL RESULTS

Case 1: Both the geometry and loading were bilaterally symmetric. The displacements and stresses were found to be symmetric.
Figure 3. is a demonstration of bilateral symmetry. All nodal
displacements were exaggerated 40 times relative to the vertebral
model scale. A left lateral view of the displacements is shown
in Figure 4. Figure 5 shows the stress contour pattern. High
stress distribution was found in the region of the laminae and at
the junction of the pedicle with the lamina.

Case 2: Figure 6 shows the tensile stress patterns when the
entire left lamina is removed (elements 161 and 162). Although the
displacement patterns did not change significantly, there were
increases in tensile stresses in the region of the pars. Figure 7
compares the tensile stress patterns in the lamina region of cases
1 and 2.

Case 3: When a superior portion of the left lamina (element
161) was removed, the tensile stress at nodes 225 and 223 increased
by a large amount. Figure 8 compares the tensile stress patterns
in the lamina region of cases 1 and 3.

Case 4: When an inferior portion of left lamina (element 162)
was removed, the tensile stresses at most of the nodes increased.
It can be seen from Figure 9 that the stress contours in case 4 run
all along the superior portion of the lamina whereas for case 1,
the stress contours run only along the inferior region of the
lamina.

Case 5: Concentrated nodal loads totalling 365 N were applied
to the left lamina. The tensile stress pattern is given in Figure
10. High tensile stresses were found at the inferior junction of
the pedicle and lamina and at the inferior tip of the lamina
(element 162).

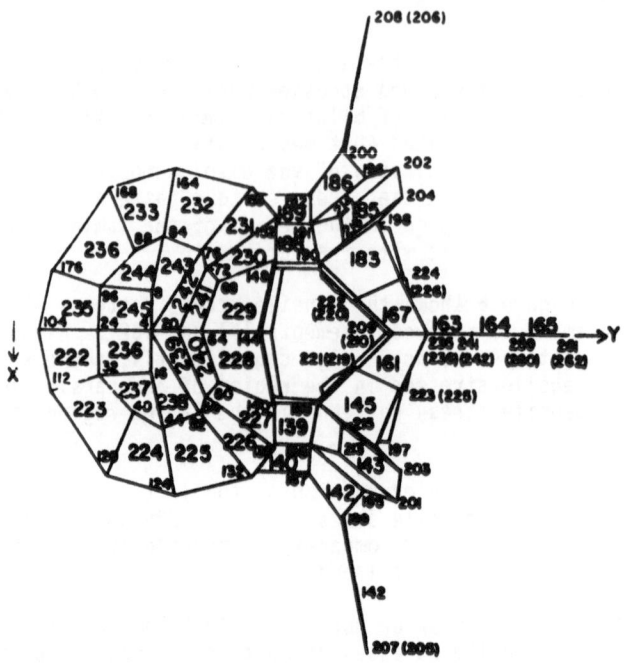

Figure 1. Top view of a complete vertebral model.

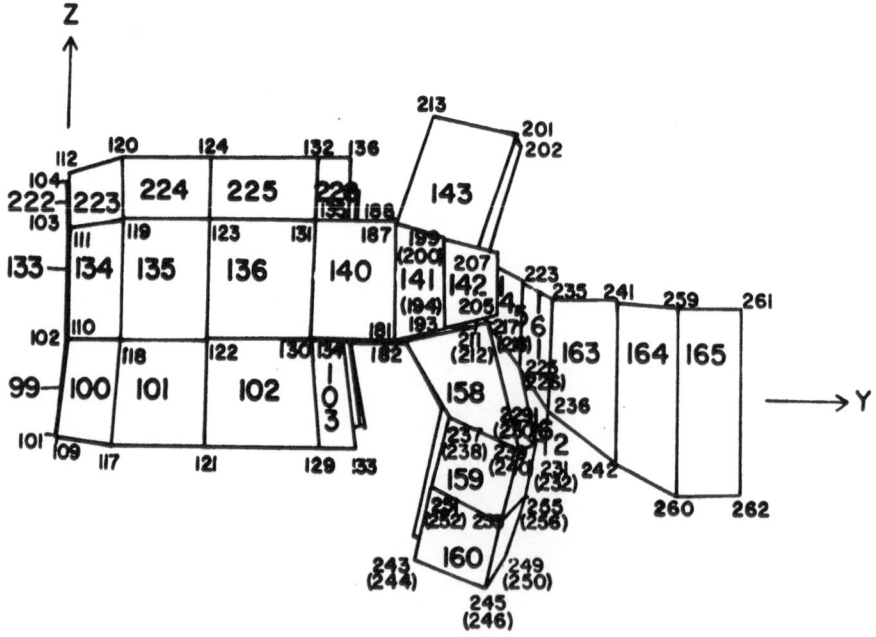

Figure 2. Side view of a complete vertebral model.

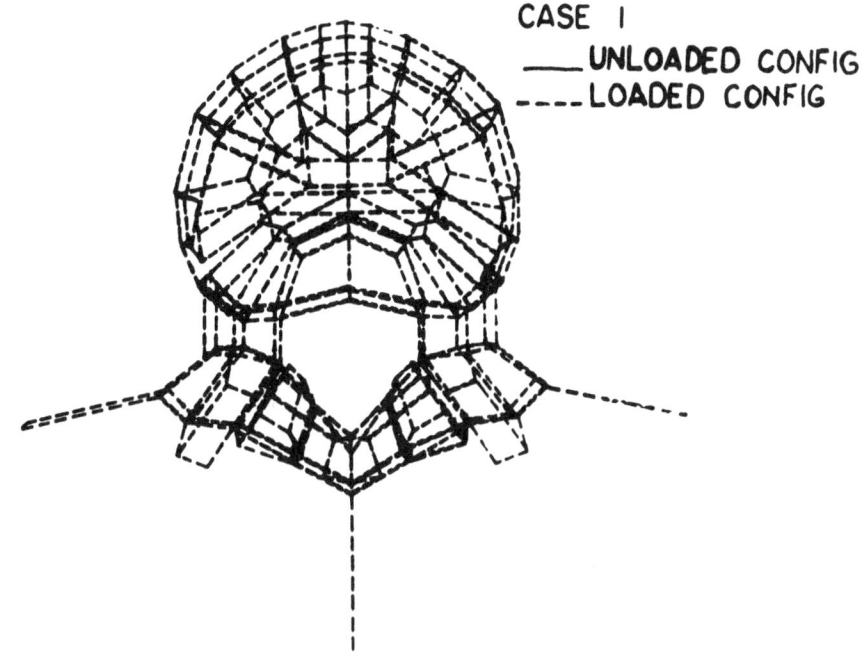

CASE I
___ UNLOADED CONFIG
---- LOADED CONFIG

Figure 3. Bilaterally symmetric displacements for the model
   (case 1)-top view.

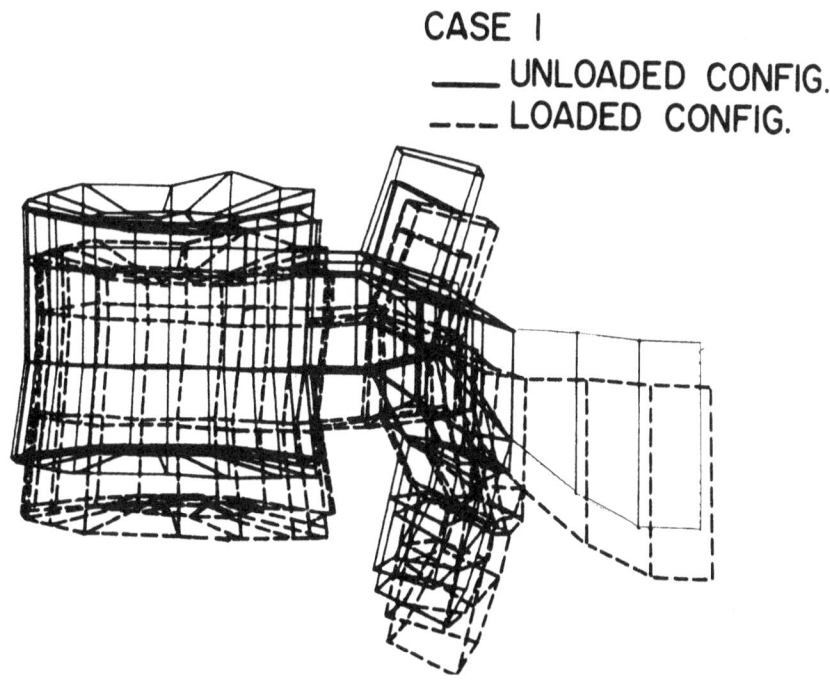

CASE I
___ UNLOADED CONFIG.
--- LOADED CONFIG.

Figure 4. Side view of displacements for case 1.

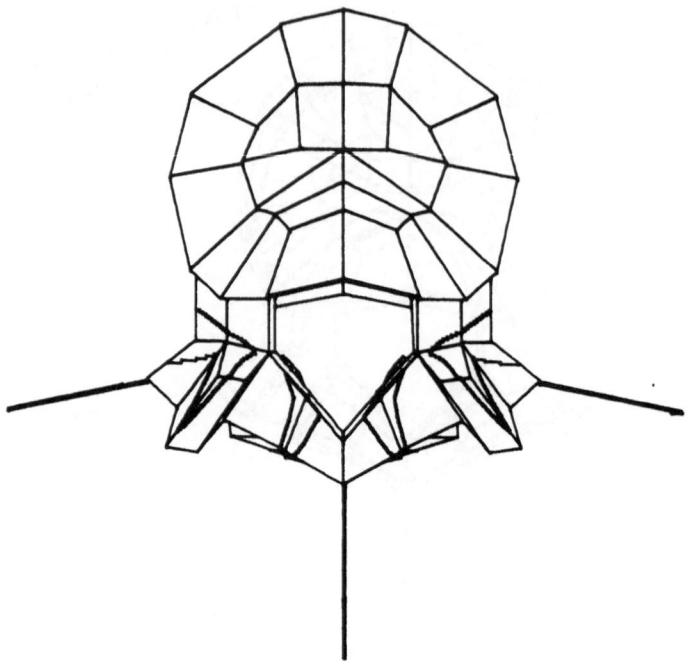

Figure 5. Case 1 predicted tensile stress contour patterns. Top view.

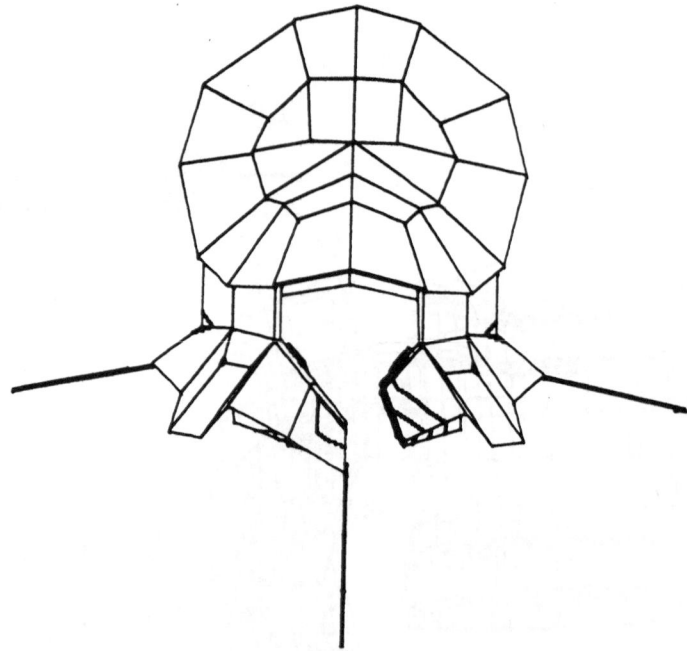

Figure 6. Case 2 predicted tensile stress contour patterns. Top
view.

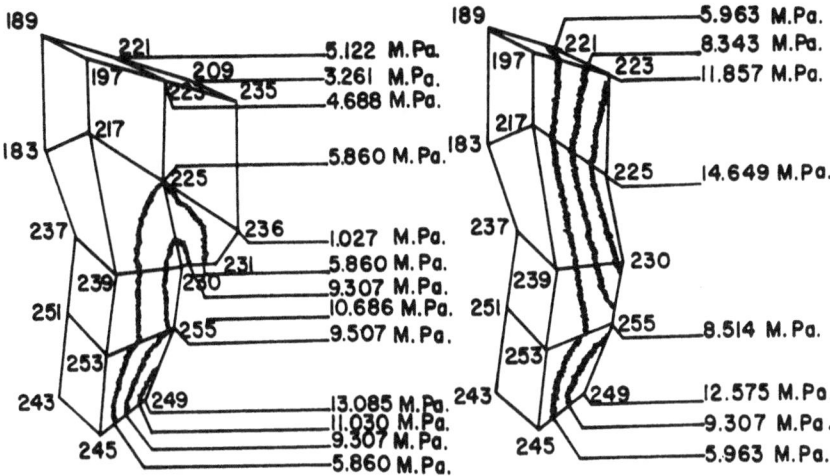

Figure 7. Comparison of tensile stress (MPa) contour patterns in the laminar region of case 1 and case 2. Side view.

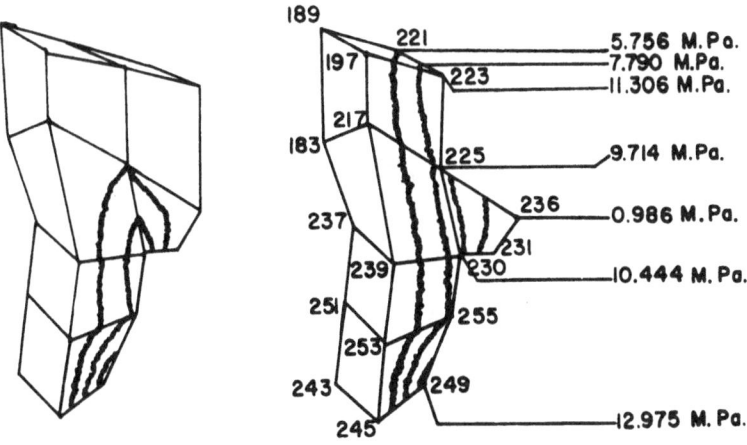

Figure 8. Comparison of tensile stress (MPa) contour patterns in the laminar region of case 1 and case 3. Side view.

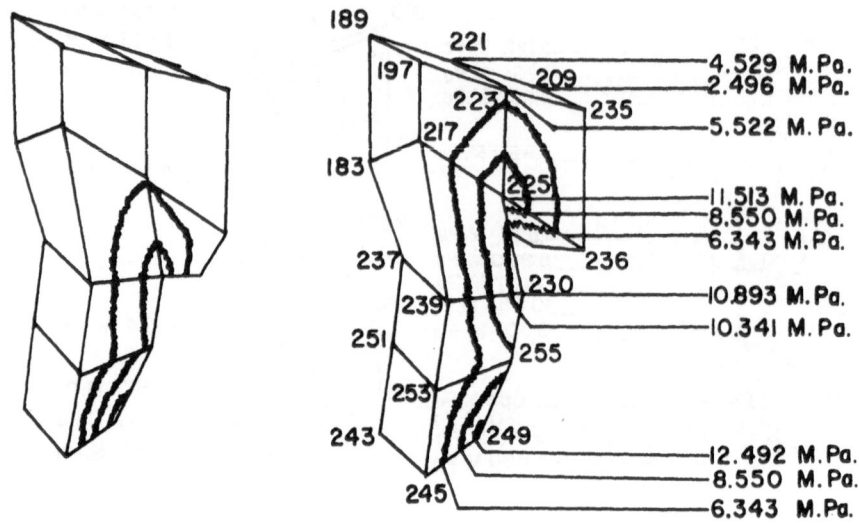

Figure 9. Comparison of tensile stress (MPa) contour patterns in the laminar region of case 1 and case 4. Side view.

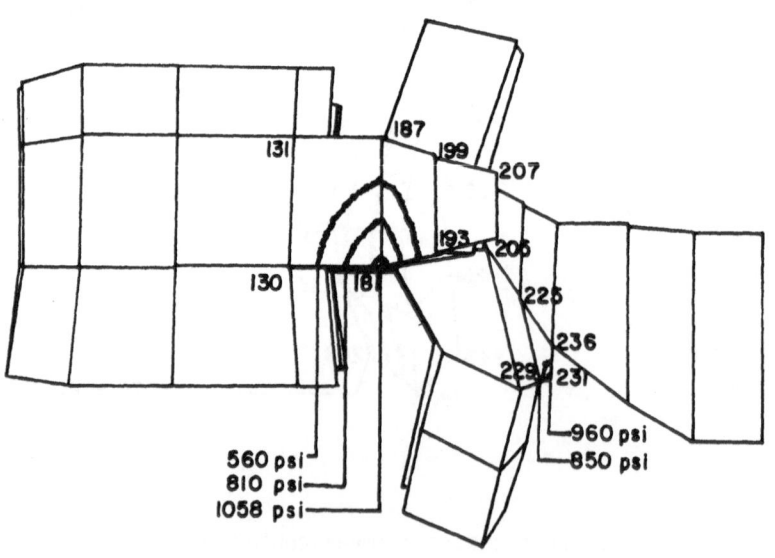

Figure 10. Case 5 predicted tensile stress contour patterns. Side view.

CONCLUSIONS

The conclusions drawn from this study are:

(i) A three dimensional model of vertebra has been developed. It yielded results which compared favorably with strains measured by Ranu and King (1), and Ranu (2).

(ii) A laminectomy was performed to indicate changes in stress distribution and locations of high tensile stress. A complete laminectomy on one side or removal of the inferior or superior lamina increased the stress concentration in the laminar region.

(iii) During laminectomy high tensile stresses occured on the affected side in the region of the pars interarticularis.

(iv) Simulation of Harrington rod concentrated loads on the lamina produced high tensile stresses near the pars interarticularis and at the posterior junction of lamina and pedicles.

ACKNOWLEDGEMENT

The assistance of Dr. Harpal Kaur Ranu is fully acknowledged.

I would like to thank Martha Klipping and Sharon Washington for typing this paper.

REFERENCES

1. Ranu, H.S. and King, A.I. Correlation of Intradiscal Pressure with Vertebral Endplate Pressure, in Engineering Aspects of the Spine Conference Proceedings. I. Mech. E. Publication, 1980-2, pp. 37-42.
2. Ranu, H.S. Viscoeleastic Behaviour of the Human Spinal Disc and the Vertebra, In Proceedings of the 1st Southern Biomedical Engineering Conference. (Ed. S. Saha), Pergamon Press, New York, 1982, pp. 85-88.
3. Ranu, H.S. Time Dependent Response of the Human Intervertebral Disc to Loading. Engineering in Medicine. 1984 (In Press).

Correlation Between CT, NMR and PT Findings in the Brain

C. Nahmias, E.S. Garnett
Nuclear Medicine, Section of Radiology
McMaster University
Hamilton, Ontario, Canada

Introduction:

The last ten years have seen the advent of three new technologies, each of which can probe a different aspect of the intact brain. Computerized axial tomography (CT) defines the intracerebral anatomy on the basis of differences in mass attenuation coefficients. It differentiates some tumours and the consequences of cerebrovascular accidents from normal brain tissue[1]. Nuclear magnetic resonance (NMR) also defines intracerebral anatomy, but in greater detail than CT. It has the ability to discriminate well between grey and white matter and to distinguish very small intracerebral lesions such as plaques of multiple sclerosis[2]. Unlike CT, NMR has the potential to measure some parameters of brain metabolism. These relate, so far, to measurements of the phosphorous containing molecules associated with energy metabolism[3]. The third technique, Positron tomography (PT) has, at least theoretically, no limit to the molecules of biological interest that can be studied. Regional blood volume, blood flow and energy requirements have been measured using ^{11}CO, $H_2^{15}O$ and ^{18}F-2-FDG or $^{15}O_2$ respectively. More recently, the distribution of neurotransmitters and of their receptors in the brain has been obtained[4,5].

The first two technologies, CT and NMR, show a significant degree of overlap in their applications and often lead to the same clinical information: specific tissue characterization. There is little overlap between these two technologies and PT that depicts functional anatomy of the brain. However, the three techniques are complementary and information from any one can be used to augment and clarify the interpretation of the others.

NATO ASI Series, Vol. F19
Pictorial Information Systems in Medicine
Edited by K.H. Höhne
© Springer-Verlag Berlin Heidelberg 1986

Any useful pictorial information system should provide facilities for easy and convenient storage and retrieval of digitally encoded images. In order to achieve high quality presentation, it should allow for the appropriate manipulation of the images, such as thresholding, enhancement or choice of monochrome or colour displays. It should also include the ability to display images obtained by different modalities, side by side or superimposed, to compare and contrast them and to facilitate the identification of corresponding morphological and functional structures. Although some of these goals have not yet been satisfactorily resolved, the purpose of this paper is to illustrate the usefulness of the correlation between the three imaging modalities in providing a better understanding of the working brain in health and disease.

The normal brain:

We are using CT, NMR and PT to study a group of normal volunteers. The morphological information obtained from a CT or an NMR study is comparable (fig. 1a,1b), even though the detail obtained by NMR, such as the differentiation between white and grey matter, is superior to that obtained by CT. The physiological information obtained from a PT study depends on the agent used; for example, ^{18}F-2-FDG will map the energy requirements of the brain whereas ^{18}F-FDOPA will map the distribution of dopaminergic neurones in the brain (fig. 1c, 1d). However, the ability to compare the distribution of these agents in a slice with NMR pictures obtained at the same level is very useful in delineating the anatomy of the brain. Of more importance, the correlation between PT and NMR images provide an explanation for certain inhomogeneities of ^{18}F-2-FDG accumulation which are shown to be constrained to follow the general pattern of sulci and gyri as depicted by NMR.

509

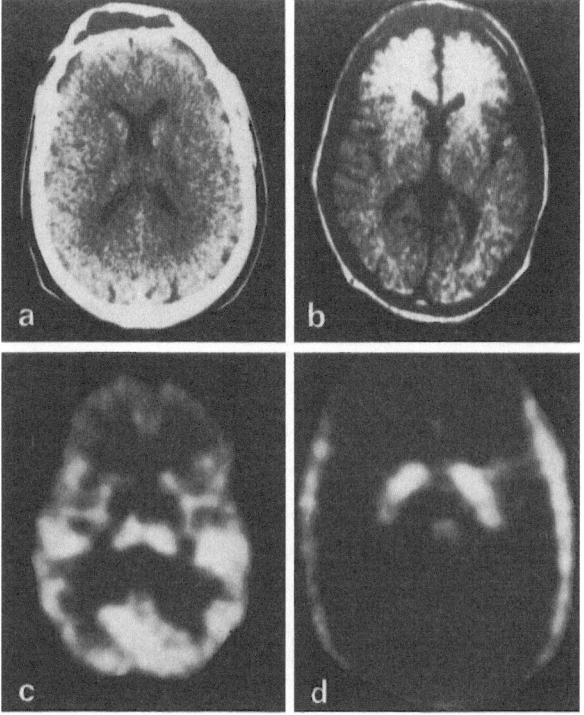

Fig. 1. The normal brain: horizontal section obtained at the
level of the basal ganglia.
a) CT scan
b) NMR scan
c) PT scan (^{18}F-2-FDG)
d) PT scan (^{18}F-FDOPA).

The diseased brain:

In general, the morphological information obtained either
from CT or NMR studies is invaluable in the correct
interpretation of the findings obtained by PT. On the other
hand, complementary metabolic information obtained using
different positron emitting agents and PT is invaluable for
the correct understanding of the diseased state. Two examples
will be discussed, the first resulting from the examination
of the nigro-striatal pathway and the other from the
examination of the anoxic brain.

The caudate nucleus and putamen (the striatum) are large
subcortical masses innervated in part by dopamine containing

nerve endings that originate in the substantia nigra. In the
striatum the dopaminergic input system and the intrinsic
cholinergic neurones are functionally interdependent. Normal
functioning of the striatum as a whole is determined by the
proper balance between these two systems; elimination of the
nigral dopamine input to the striatum effectively disrupts
normal function. The function of the caudate nucleus may be
associated with behaviour, while the putamen is predominently
associated with movement.

Huntington's disease is an inherited, degenerative,
progressive disorder characterized by behavioural changes and
choreiform movements. At autopsy, there is a marked loss of
neurones in the caudate nucleus accompanied by some loss of
glia. This loss of tissue appears on CT as an enlargement of
the lateral ventricles and is evident in the later stages of

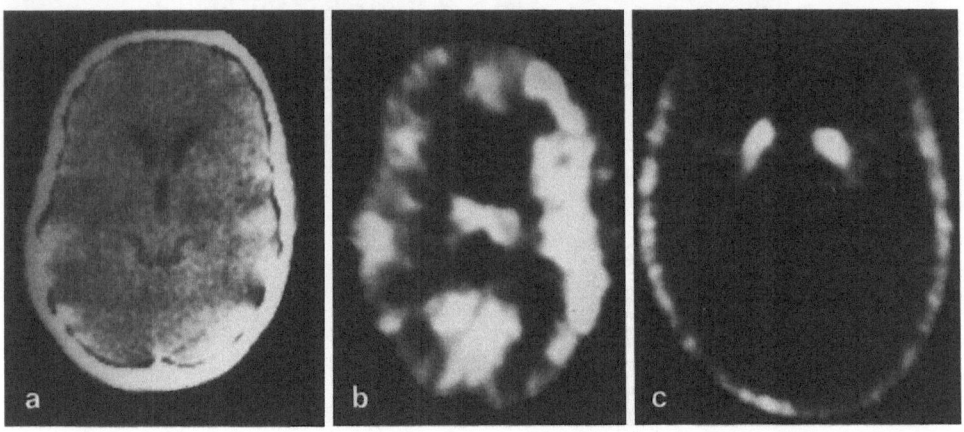

Fig. 2. CT scan (a), ^{18}F-2-FDG scan (b) and ^{18}F-FDOPA scan (c)
of a young patient suffering from Huntington's
disease.

the disease. In the early stages of the disease, the CT scan
is essentially normal and therefore not helpful in making a
diagnosis (fig. 2a). However, the PT study done with ^{18}F-2-FDG
shows that striatal glucose consumption is markedly reduced
both in the caudate and putamen (fig. 2b). In contrast, the
PT study done with ^{18}F-FDOPA shows an obvious accumulation
of the tracer in the region of the striatum, the caudate and
putamen accumulating the tracer to the same extent (fig. 2c).
These findings indicate that a profound reduction in striatal

energy consumption is an early feature of Huntington's disease that antedates any definitive changes in CT^6. Our findings also suggest that the dopaminergic pathway continues to function even though the neurones with which this pathway synapses no longer consume glucose.

Parkinson's disease is also a degenerative, progressive disease characterized by an interference with all aspects of voluntary motor performance. The CT scan of a patient suffering from Parkinson's disease is essentially normal (fig. 3a). The PT study done with ^{18}F-2-FDG shows that glucose metabolism in the striatum is normal (fig. 3b), at least in the early stages

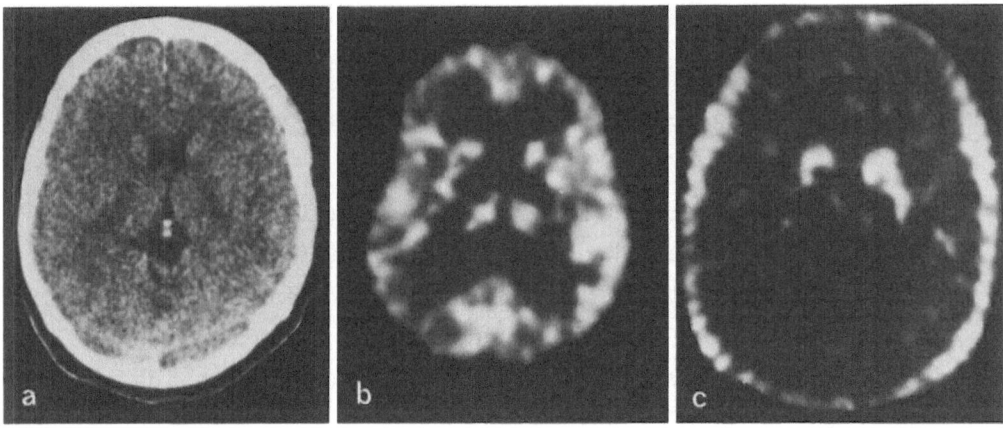

Fig. 3. CT scan (a), ^{18}F-2-FDG scan (b) and ^{18}F-FDOPA scan (c) of a patient suffering from unilateral Parkinson's disease.

of the disease. However, the PT study done with ^{18}F-FDOPA shows that dopamine accumulation is markedly reduced in the putamen contralateral to the side of the body exhibiting the signs of the disease[7] (fig. 3c). In this case, the neurones of the striatum are normal with respect to glucose consumption; however, they lack the input of the essential neurotransmitter, dopamine.

Dystonia is a disease marked by involuntary, irregular contorsions of the muscles of the trunk and of the extremities. Again, the CT scan is usually not useful in establishing a diagnosis (fig. 4a). However, in secondary dystonia, both the accumulation of ^{18}F-2-FDG and of ^{18}F-FDOPA is markedly reduced

in the region of the striatum, predominantly in the putamen
(fig. 4b,4c).

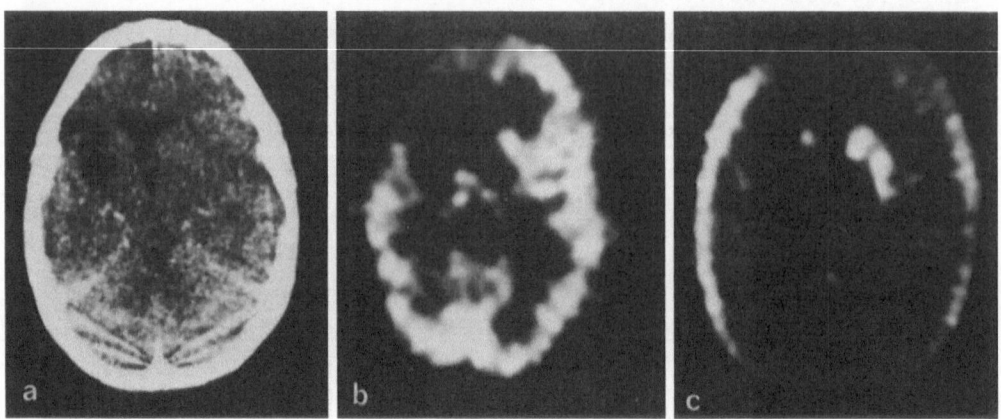

Fig. 4. CT scan (a), ^{18}F-2-FDG scan (b) and ^{18}F-FDOPA scan (c)
of a young patient suffering from secondary dystonia.

Hypoxia of the brain may result from a lack of oxygen in
the blood delivered to the brain, or from a disruption in the
circulation that supplies the brain with oxygen. The
consequences of this lack of oxygen can be an interference
with the normal activity of the brain, or a complete loss of
function in a region of the brain.

We have used CT and PT performed using ^{18}F-2-FDG to study
a group of infants who have suffered some degree of asphyxia
at birth[8]. In these newborn infants, there is a lack of
information about external landmarks for the structures in the
brain. A CT scan performed before the PT examination has
allowed us to position the baby more properly. Our results
show that, in general, the region of the brain that is
abnormal on CT also shows abnormal glucose consumption
compared to the same region on the opposite side of the brain
(fig. 5a,5b). It is notable that the area of metabolic
impairement demonstrated on PT extends considerably beyond the
anatomical area of the abnormality seen on the CT scan.

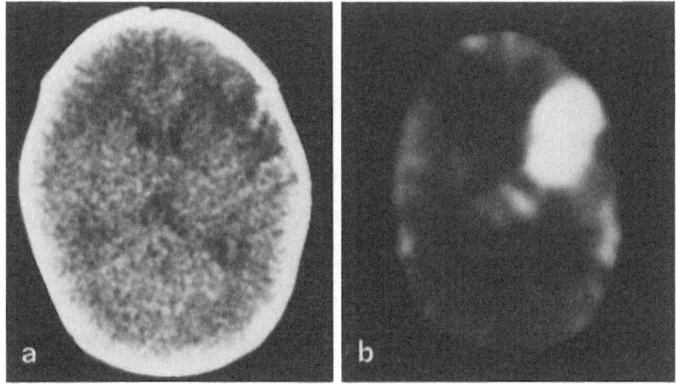

Fig. 5. CT scan (a) and ^{18}F-2-FDG scan (b) of a six day old
infant.

We have used NMR and PT performed using ^{18}F-2-FDG to study
a group of patients with cerebral infarction. In this group
of patients, the region of metabolically inactive tissue as
demonstrated by PT is more extensive than the anatomical
lesion as depicted by NMR (fig. 6a,6b).

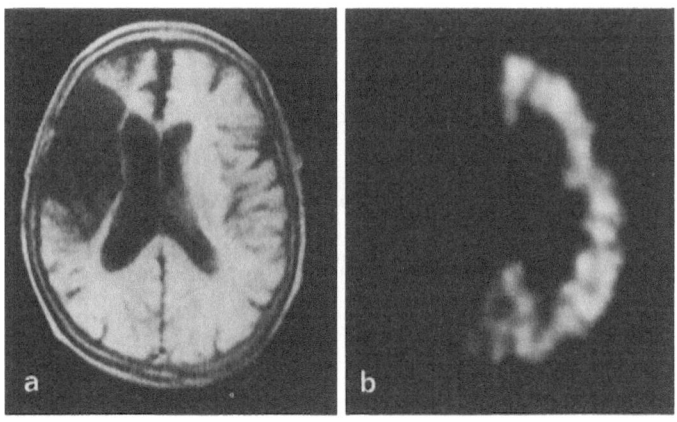

Fig. 6. NMR scan (a) and ^{18}F-2-FDG scan (b) of a patient who
had suffered a large right frontoparietal infarct
five years prior to the studies.

Conclusion:

In general, we have found the ability to correlate the morphological information available from a CT or an NMR study and the metabolic information available from a PT study to be invaluable in the investigation of normal and disease processes in the living brain. To be able to correlate the findings obtained by CT, NMR and PT the position at which the head is imaged must be reproduced using each technique. Similarly, the level at which each slice is obtained must be the same. Ideally, the proportions of the image of the brain obtained by each technique should be matched to allow easy superposition of the various images.

Although some of these problems have not yet been satisfactorily resolved, the correlation between the three imaging modalities is proving to be very useful in providing a better understanding of the working brain in health and disease. Advances in pictorial information systems will make these comparisons much more easy to obtain.

References:

1- M. Gado, J. Hanaway and R. Frank. J. Comput. Assist. Tomogr. 3 (1979) 1.

2- G.L. Brownell, T.F. Budinger, P.C. Lauterbur and P.L. McGeer Science 215 (1982) 619.

3- E.B. Cady, A.M. de L. Costello, M.J. Dawson et al. Lancet 1 (1983) 1059.

4- E.S. Garnett, G. Firnau and C. Nahmias. Nature 305 (1983) 137.

5- K.L. Leenders, S. Herold, D.J. Brooks et al. Lancet 2 (1984) 110.

6- E.S. Garnett, G. Firnau, C. Nahmias et al. J. Neurol. Sci. 65 (1964) 231.

7- E.S. Garnett, C. Nahmias and G. Firnau. Can. J. Neurol. Sci. 11 (1964) 174.

8- L.W. Doyle, C. Nahmias, G. Firnau et al. Develop. Med. Child Neurol. 25 (1983) 143.

DIGITAL IMAGING FOR PLANNING OF RADIATION THERAPY

- PRACTICAL CONSIDERATIONS

Hans Dahlin*, Per Ekström* and Inger-Lena Lamm**

*Medical Computer Physics, Uppsala Datacenter,
    Box 2103, S-750 02 Uppsala, Sweden

**Dept. for Radiation Physics, Lunds lasarett,
        S-221 85 Lund, Sweden

PRESENTATION OF THE CART PROGRAM

The last decade has proved the usefulness of computer technology
in radiotherapy procedures. It is today possible to utilize
modern diagnostic and therapy devices for a more accurate and
safe treatment of malignant cancer. The statistical evaluation
of treatment results during the next years will probably also
prove that statement.

In spite of the powerful progress in computer technology
there are still thresholds to overcome before full integration
of computer technology in radiotherapy can be presented. One of
the main problems that commercial and scientific developers face
today is the definition and communication of data. Most of the
existing systems are mainly built for solving one part of the
total problem without taking necessary input/output parameters
from neighbouring disciplines into account. Each manufacturer
too often invents his own nomenclature and data storage
structure of data handling. The increasing number of different
solutions on the market has therefore necessitated inventions of
general concepts and standards to guarantee future progress in
this field.

A Nordic collaboration, CART (Computer Aided Radio
Therapy), between industry, research institutes and hospital
departments has been initiated and gives one possible way of
overcoming the problem.

The scope of CART is to define a logical network model in
which the different steps in the radiotherapy procedure can be
seen as different nodes (applications). The function of each
node, as well as the required input/output, format
specifications, are described in the CART pre-study report /1/.

NATO ASI Series, Vol. F19
Pictorial Information Systems in Medicine
Edited by K. H. Höhne
© Springer-Verlag Berlin Heidelberg 1986

The CART basic logical model holds the following applications (Fig. 1):

1. Image processing
2. Treatment modelling
3. Beam data acquisition
4. Treatment control
5. Clinical registry
6. Oncological database

The continuation of the CART program is coordinated through 5 experimental systems, "demonstrators", located to Tampere in Finland, Uppsala and Malmö in Sweden, Oslo in Norway and Reykjavik in Iceland. Each system is designed from different main themes, namely: Treatment modelling, Image processing, Treatment control and Data base technology. At the different demonstrators scientific and industrial projects are coordinated and developed within a time period of 2 years, starting from January 1, 1985. A project for "CART Standards", holding members from the different demonstrators, will coordinate the information transfer between the different demonstrators, international research centers and standardization committees.

The program is financed by the Nordic Industrial Fund, Nordic high-technology industry and national research funds.

Image processing for three-dimensional therapeutic dose planning is located to the Uppsala system and will be concentrating on development of functional prototypes for:

- Image handling, presentation, database structures and transfer of 3D volume information
- Interactive image work stations (human computer interface)

In the following some practical considerations are pointed out for the image handling problem in radiotherapy.

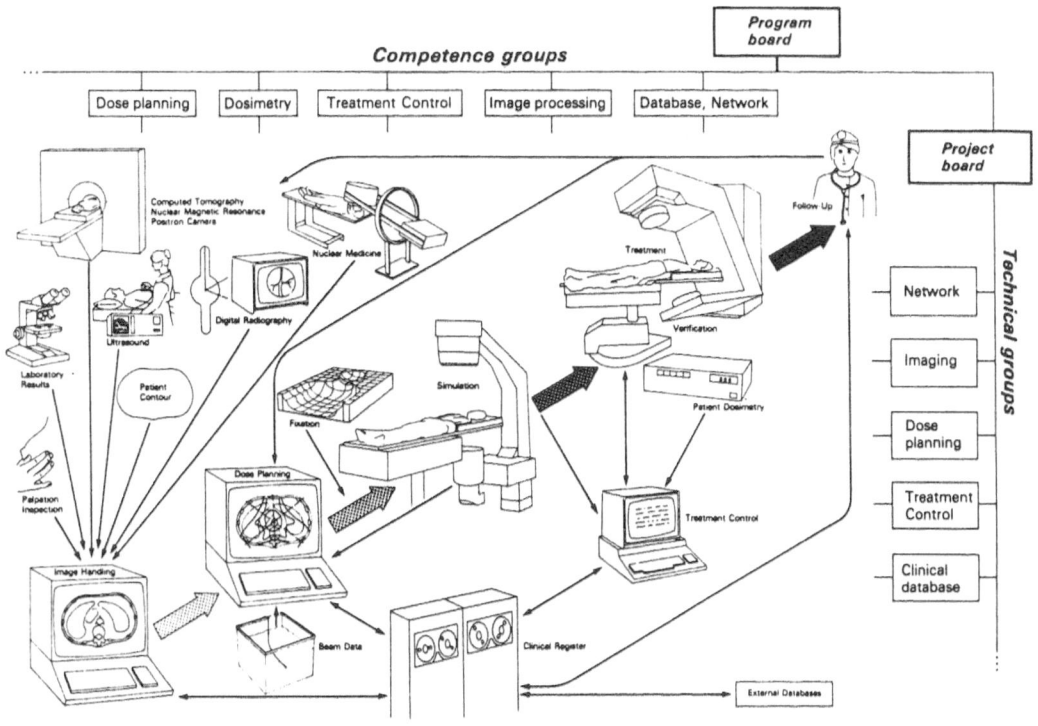

Fig. 1.

THE RADIATION TREATMENT PLANNING PROCEDURE

The treatment planning, understood in a broad sense, can be divided into three parts characterized as diagnostic, therapeutic and physical, summarily described below as a basis for the subsequent discussion.

Diagnostic part

The first step in treatment planning is to determine not only that the patient has (had) a tumour but also where this tumour is (was) situated, its extent and relation to other organs in the body. It is important to recognize that image information must be combined with information from other diagnostic procedures, such as clinical examination, pathology, etc. in this evaluation. However, this point will not be elaborated on further here.

Therapeutic part

The second step is to define the target volume (volumes) and organs at risk, together with prescribed dose levels and fractionation scheme (prescribed biological effects). In general terms, the target volume for a curative radiation treatment includes demonstrated tumour if present, presumed occult tumour and appropriate regional lymph nodes /2/. In the definition of target volume, the effects of patient movement during each treatment session (e.g. breathing), the uncertainty in treatment set-up between treatment sessions and also variations during the whole treatment in patient configuration (e.g. loss of weight) and target volume must be taken into account.

Physical part

The third step is to find the optimum way of delivering the prescribed 4-dimensional (space-time) dose distribution to the patient, using the treatment modalities available. At present this problem is generally restricted to finding the optimum spatial dose distribution, while the aim for the future is to optimize the biological effect in 3D space. Together with some appropriate model of the interaction between radiation field (treatment beam) and matter (patient), beam data and patient data are needed to give the dose distribution. Models relating biological effect and dose-time patterns have to be established as well.

The problems concerning beam data and models for interaction and biological effect will not be dealt with further here. For the dose distribution calculations, the 3D patient data used for tumour and target volume definition have to be translated into relevant physical parameters.

The prescribed dose distribution can in this context be regarded as subsidiary conditions to be satisfied in the optimization procedure.

As mentioned, the patient is not a rigid body, but changes both during each treatment session (short time aspect) and also on the average during the whole treatment period (long time aspect). These variations have to be taken into account also in the dose calculations. Further problems to consider are the use of contrast media, necessary for optimal diagnostic procedures but disturbing the translation to physical parameters, the random occurrence of bowel gas etc.

DEFINITION OF COORDINATE SYSTEMS

The determination of tumour extention, target volume and optimal radiation treatment for a patient has been described in general terms above. In order to relate the patient, in radiotherapy position, to the geometry of the treatment units and of the various information collecting units used, we have to define a set of coordinate systems, thus allowing transformations to be made from one system to another.

For therapy application purposes we define the following basic Cartesian coordinate systems: A fixed system CF, an information collecting unit system CI, a patient system CP and a treatment unit system CG.

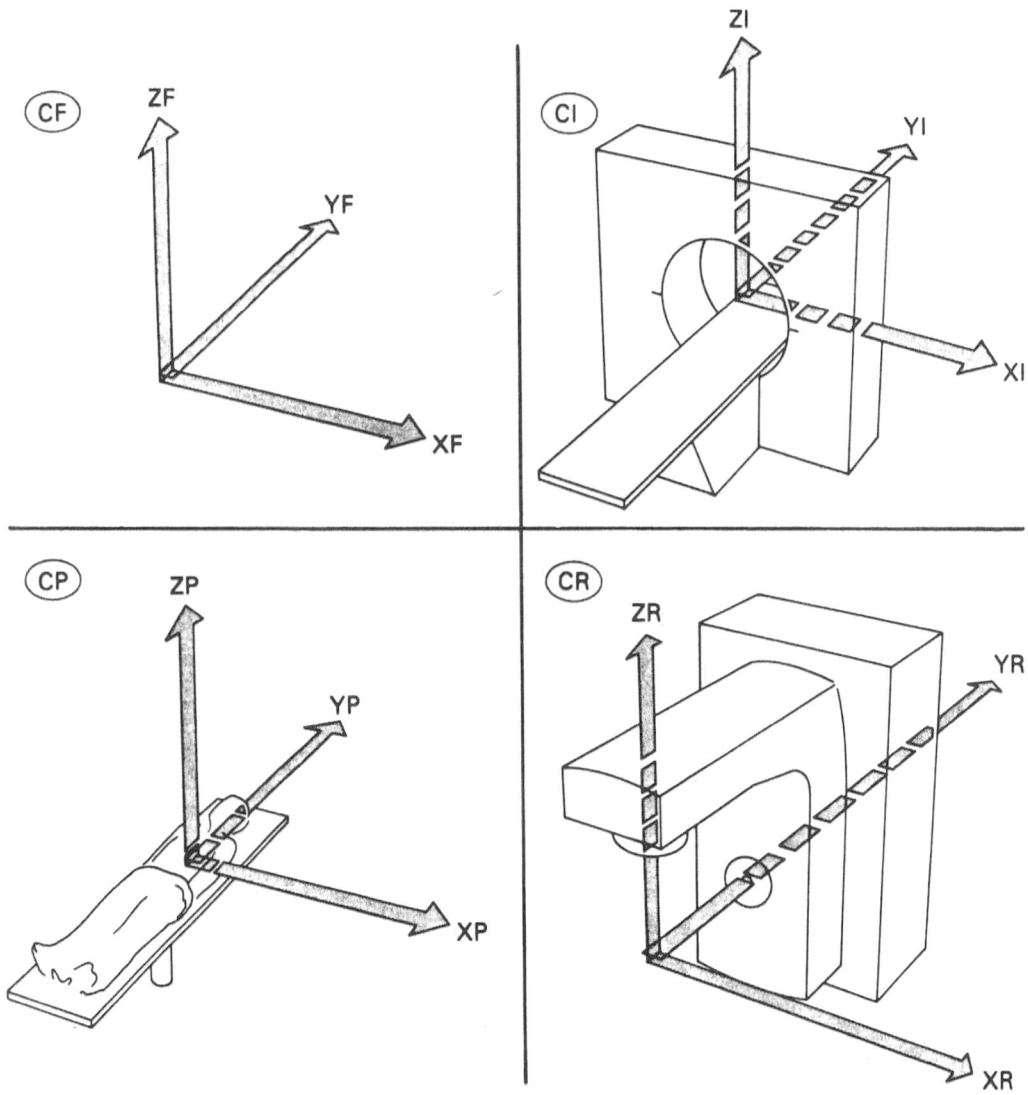

Assume, that these coordinate systems coincide (or only differ by simple translations) for the most common patient position, i.e. a supine patient with the head towards the gantry of the treatment unit (or head first on a CT-scanner), and with all the units in "zero position". These systems are then defined so that the Z-axes point vertically upwards, ZP pointing ventrally and ZG towards the radiation source parallel to the central ray. For the Y-axes, the YP-axis points cranially and the YG-axis towards the gantry. The YG- and YI-axes are parallel to the axes of rotation for the gantry and CT-scanner respectively. The X-axes are then also defined, the XP-axis pointing to the left side of the patient. (These definitions are in accordance with Siddon /3and Möller et al. /4/) It is now easy to describe for instance a prone patient, head first: Rotate CP 180 degrees about the YF-axis, or a supine patient, feet first: Rotate CP 180 degrees about the ZF-axis.

The fixed system is used as a reference system. As an example, consider a CT-scanner used to collect volume patient data, not just "slices". Between each scan, the couch with the patient is moved relative to the scanner. If the couch is rotated, then the patient system CP is first rotated about the ZF-axis followed by successive translations along the new YP-axis. A change of couch height between scans corresponds to a translation of CP along the ZF-axis. An angulated CT-gantry is defined by a rotation of the CI-system about the XF-axis. Thus each tomagram, primarily defined in the CI-system and parallel to the XI-ZI plane, is defined in the patient system CP via the fixed system CF.

To be noted is the necessity of defining the exact position of the patient during treatment in the CP-system (i.e. arms along the sides, chin up etc.). As mentioned, a human being is not a rigid body, but is on the contrary continually moving, both on a psychological and voluntary basis. It has therefore been customary to use fixation devices in radiotherapy to minimize movements and to maximize the reproducibility in patient positioning.) The most rigid element of the human body is the bony structure. Consequently, the origin of the CP-system should be defined relative to the skeleton.

One way of describing the patient geometry is to give the transformation from the CF- to the CP-system together with the CP-coordinates of a set of anatomical reference-points. The definition of the reference points will not be discussed here.

The only transformations considered so far have been translations and rotations, based on the treatment position of the patient during therapy (regarded as a rigid body). We have to recognize, however, that patient information is collected with the patient in other positions as well, for instance to achieve optimal diagnostic conditions. Based on the set of anatomical reference points, it is theoretically possible to make a non-linear transformation from an arbitrary patient position to the established treatment position. It remains to be seen if this approach will be useful. Otherwise, we will have to continue to rely on the human computer to make this transformation in the evaluation procedure.

DIAGNOSTIC-THERAPEUTIC INFORMATION

Information collecting units

The diagnostic imaging systems can be divided into two main categories: 3D subspace systems (CT, NMR, PET, SPECT, US, etc.) and 2D projection systems (digital and conventional radiography, gamma cameras etc.).

For radiotherapy applications it is an advantage if the patient can be placed in treatment position on all the information collecting units, reducing the problems of non-linear patient position transformations. However, when collecting information for the medical and physical treatment planning, the patient position must as closely as possible resemble the therapy conditions. This raises the question if it is possible to use the same set of patient information for both diagnostic, therapeutic and physical procedures, thus saving information collection time. The requirement above on treatment position of the patient for the medical and physical planning will probably lead to the necessity of double information collection, both therapeutic and diagnostic, when the optimal diagnostic position differs from the treatment position. Diagnostic evaluation generally requires "sharp" images, i.e. motionless patients. The ideal situation would be to collect information from the whole volume of interest instantaneously. Some 2D projection systems fall in this category, but for existing 3D subspace collecting systems, the situation is different. A CT-scanner with a scan-time of a few seconds can give sharp tomograms, if the patient holds his breath during scanning. But to cover the whole volume of interest unambiguously, the patient must be kept from moving between scans and also be scanned each time at the same point of the breathing cycle. Diagnostic CT-examinations are in most cases made with the patient holding his breath in maximum inspiration. During radiation therapy, on the other hand, the patient is breathing normally.

NMR-units can collect information simultaneously from a whole volume of interest. The corresponding collection times are in the order of minutes, the patient has to breathe and the result is thus a "breathing average" proportional to the average of the instantaneous sets of information during the breathing cycle.

For therapy purposes, the movements during breathing of the rigidly determined tumour and target volumes must be known. It is also of interest to establish the variation of the physical parameters inside the irradiated volume (defined in the fixed coordinate system CF). Theoretically, a true breathing averaged dose distribution should be calculated for "each instantaneous frame in the 3D movie of the patient", and then averaged.

In practice only the extreme positions are used to determine the movements of tumour and target, from diagnostic information. Compared to the diagnostic demands the demand on spatial resolution is reduced for dose calculations, and a resolution of 2 mm is commonly regarded as sufficient for megavoltage photon beam therapy. We have discussed the theoretical possibility to make a non-linear breathing-transformation from diagnostic to therapeutic 3D images based both on the individual diagnostic information and on a general breathing transformation established for an average patient. Besides the already mentioned resolution reduction and breathing transformation the transformation of a diagnostic image to a therapy-image to be used for dose calculations, requires a contrast- and a parameter-transformation. The influence on the diagnostic image parameters of contrast media, necessary for optimal diagnosis, are to be eliminated via the contrast transformation and also the influence of for instance randomly occurring bowel gas. The parameter transformation translates from image parameters to relevant physical parameters, i.e. parameters necessary for the dose distribution calculations. A practical implementation of this transform has been made for CT-information, where the possibility to translate from ordinary CT-information to relative electron density for megavoltage photon therapy is well proven. With the advent of NMR the question naturally arises whether NMR-information as well can be translated in a consistent way. The anatomical information can of course be used to identify structures such as lungs, bone etc., which as in CT-applications can be given global values of relative electron densities. But, it is yet too early to answer the question above.

Information work station

The image work station must be able to handle all kinds of diagnostic information. It must be able not only to replace the old light boxes, where X-ray films were compared side by side, but to do even better, i.e. to be of help in extracting only the currently relevant part from the vast amount of information present in the digital volume information.

The requirements on interactivity for the work station are great. Several images are to be displayed simultaneously, both 2D sections through the 3D volume and 2D projected images. This requires spatial database structures, large memories, fast image processors, powerful interactive software etc. /4/. It is necessary to have the possibility to "draw" interactively in the images while defining tumour, target volumes and other structures of interest. The interactive "drawing" should be combined with some semi-automatic delineation procedure useful for organ definition, elimination of contrast media etc. In order to compare information from volume collecting systems and projecting systems, projections calculated from the volume information are necessary and conversely back projection of projections into the corresponding generating volume (the latter when patient positions are equivalent). These issues have been extensively discussed by M. Goitein et al. /6,7/. Their image work station has been tested in clinical operation for several years and found very useful for planning radiotherapy. Tumour and target volumes are defined in 3D and displayed in various sections. Radiation beams are positioned interactively in the beam's eye-view and checked via back projection. Port films are calculated.

It is to be noted that for daily routine work speed is principally more important than "fancy representations". The latter might be useful for demonstrations and developments, where speed is of secondary significance. The interaction between viewer and work station must also be taken into account especially for a routine work station, not only from the program facility aspect but also ergonomically.

The calculated dose distribution can be regarded as another set of volume information to be handled in the image work station.

It is thus possible to simulate in 3D both patient and radiotherapy on the computer, "digital treatment simulation". The traditional image work station presents 3D information as 2D section or 2D projections through the volume, in grey scale only or by using colour. The use of colour in itself does not enhance the information content, as displayed for instance in a CT-section. However, if two (or more) parameters are to be superimposed, a presentation using colour to distinguish between the parameters is useful. The development of digital holography might give the possibility of 3D presentations of 3D parameters.

CONCLUDING REMARKS

An advanced image work station on one hand and repeated relevant examinations during and after treatment on the other hand, give us a powerful tool in the search for treatment response and dose-time relationships. With accurate dose distribution calculations (which might include yet another non-linear transformation of the patient to correct for loss of weight during treatment) and accurate treatments, we can truly follow the reactions of both tumour and other organs in relation to given dose levels. Together with the oncological database, where other treatment modalities than radiotherapy are recorded as well, evaluation of combined treatment modalities will be possible. This is of great importance judged from present development trends.

524

REFERENCES

/1/ CART - Computer Aided Radio Therapy - Presentation of an integrated information system in radiotherapy . Report of the pre-study phase 1984-03-15. NORDFORSKs publikationsserie 1984:1. ISBN 91-7596-011-7.

/2/ ICRU Report 29, "Dose specification for reporting external beam therapy with photons and electrons", 1978.

/3/ R.L. Siddon, "Solution to treatment planning problems using coordinate transformations", Med. Phys. 8(6), pp. 766-774, 1981.

/4/ T. Möller, U-B Nordberg, T. Gustafsson, J-E Johnsson, T.G. Landberg and G. Svahn-Tapper, "Planning control and documentation of external beam therapy, Acta Radiol., Suppl. 353, Stockholm 1976.

/5/ Digital image archiving in medicine, Computer Magazine, vol. 16, no. 8, IEEE August 1983.

/6/ M. Goitein and M. Abrams, "Multi-dimensional treatment planning: I. Delineation of anatomy", Int. J. Radiation Oncology Biol. Phys., vol. 9, pp. 777-787, 1983a.

/7/ --, D. Rowell, H. Pollari, J. Wiles, "Multi-dimensional treatment planning: II. Beam's eye-view, back projection, and projection through CT sections", Int. J. Radiation Oncology Biol. Phys., Vol. 9, pp. 789-797, 1983b.

Author Index

NATO ASI Series F